AF686367

Strategic Planning for Urban Sustainability

Strategic Planning for Urban Sustainability

Guest Editors

Carlos José Lopes Balsas
Awais Piracha

Basel • Beijing • Wuhan • Barcelona • Belgrade • Novi Sad • Cluj • Manchester

Guest Editors
Carlos José Lopes Balsas
Belfast School of Architecture
and the Built Environment
Ulster University
Belfast
UK

Awais Piracha
School of Social Sciences
Western Sydney University
Sydney
Australia

Editorial Office
MDPI AG
Grosspeteranlage 5
4052 Basel, Switzerland

This is a reprint of the Special Issue, published open access by the journal *Land* (ISSN 2073-445X), freely accessible at: https://www.mdpi.com/journal/land/special_issues/LKQF826CSA.

For citation purposes, cite each article independently as indicated on the article page online and as indicated below:

Lastname, A.A.; Lastname, B.B. Article Title. *Journal Name* **Year**, *Volume Number*, Page Range.

ISBN 978-3-7258-4963-5 (Hbk)
ISBN 978-3-7258-4964-2 (PDF)
https://doi.org/10.3390/books978-3-7258-4964-2

© 2025 by the authors. Articles in this book are Open Access and distributed under the Creative Commons Attribution (CC BY) license. The book as a whole is distributed by MDPI under the terms and conditions of the Creative Commons Attribution-NonCommercial-NoDerivs (CC BY-NC-ND) license (https://creativecommons.org/licenses/by-nc-nd/4.0/).

Contents

About the Editors

Carlos José Lopes Balsas

Carlos José Lopes Balsas, Ph.D., AICP, is a Lecturer in Planning at the Belfast School of Architecture and the Built Environment at Ulster University. He is an urban and regional planner trained in Europe and the U.S. His main research interests are Urban Revitalization and Resilience, Urban Governance, Non-Motorized Transportation Planning, Mega-Event Development, City and Culture, and International Planning. Carlos has worked in multiple professional capacities in Portugal, Massachusetts, California, Arizona, and New York. He has published books on commercial urbanism, sustainable urbanism, walkability, and city center revitalization. He held prior teaching and research appointments at the University at Albany, SUNY, and Arizona State University, Tempe (USA). Several of his advanced planning studios and report-directed courses have won professional planning awards in Arizona and New York.

Awais Piracha

Awais Piracha, Ph.D., is a Professor at Western Sydney University, Australia. He is a renowned international expert in the areas of sustainable urban development and use of spatial analysis in land use and transport planning and other areas. His research interests are the Geography of Religion, Livability of Cities, Spatial Analysis of Cities and Regions, and Land Use and Transport Planning. Awais Piracha was trained as a civil/environmental engineer as well as a town planner. He is a Planning Institute of Australia Registered Practicing Planner and a NSW Local Planning Panel Expert. He also serves on the Western Sydney Transport Infrastructure Panel appointed by the Prime Minister of Australia. He previously worked as a researcher with the United Nations University (UNU) in Tokyo, Asian Institute of Technology (AIT), Bangkok, and University of Dortmund, Germany. In his professional career, spanning over two decades, he has led or participated in numerous research and consultancy projects as an urban and transport planning expert. Awais Piracha has conducted contract research for the World Bank, Nanjing City, Landcom NSW, Transport for NSW, Penrith City Council, the Australian Housing and Urban Research Institute, and other entities.

land

Editorial

Strategic Planning for Urban Sustainability

Carlos J. L. Balsas [1,*] and Awais Piracha [2]

1 Belfast School of Architecture and the Built Environment, Ulster University, York Street, Belfast BT15 1ED, UK
2 School of Social Sciences, University of Western Sydney, Penrith, NSW 2751, Australia; a.piracha@westernsydney.edu.au
* Correspondence: c.balsas@ulster.ac.uk

1. Introduction

The inception of strategic planning in the Western world can be traced to the post-World War II period [1]. Strategic planning in public administration and territorial management has been borrowed mostly from the private sector, where reducing a large number of variables to only the most plausible and prioritising short- and medium-term actions according to the most consensual criteria are the norm [2]. Strategic planning is a variation of rational comprehensive planning [3] without the need to collect and analyse large amounts of data, arriving at the most important priorities and only then formulating desirable courses of action [4].

As a direct consequence of private sector protagonism in the US and UK during the 1980s, strategic planning in the public sector experienced a resurgence during the 1990s. For instance, the Portuguese Program for the Consolidation of the National Urban System and Support for the Execution of Municipal Master Plans (PROSIURB) was aimed at strengthening medium-sized cities' roles in the structuring of the national territory. As such, medium-sized cities were eligible to receive public development funds pending the making and adoption of both master and strategic plans. The strategic plans were aimed at bolstering not only a network of medium-sized cities, especially in the interior part of the country near the Spanish border, but also strengthening the sustainability of territorial development elsewhere [5].

Also in the early 1990s, Lisbon was perceived to be a pioneer of strategic planning in Europe when it prepared its master plan in tandem with the city's strategic development plan. The master plan was spatial and comprehensive in nature, covering the whole municipality, while the strategic plan identified a limited but implementable number of strategic projects and programmes, which would anchor most of the urban regeneration of the country's capital, including the redevelopment of the EXPO'98 site, on the western side of the Tagus River [6,7].

Across the Atlantic Ocean in the US, strategic planning was equally perceived as capable of delivering more focused territorial development results with fewer resources than those needed to produce extremely detailed and lengthy general plans [8]. Said master plans in the US have traditionally been the outcomes of extensive land use, transportation, infrastructure, economic development, housing, and environmental studies and analyses. Anchor territorial projects, coupled with the identification of key stakeholders and developmental scenarios, became fundamental components of most strategic plans. For instance, Porter's Regional Development Study for the Pioneer Valley of Western Massachusetts [9] comprised the identification of sources of regional competitiveness, the characterisation of their economic performance within a business environment, and the establishment of priorities for the government in hopes of strengthening regional growth clusters.

Over the years, strategic planning gained its place within established planning theories taught in universities' Urban and Regional Planning programmes and implemented in the world of planning practice at all territorial levels, from the neighbourhood to the city, metropolis, region, state, and country [10].

John Bryson at the University of Minnesota in the US and Louis Albrechts at KU Leuven in Belgium have been some of the most prolific key thinkers on the value of strategic planning in urban and regional planning, public administration, and spatial development [2,11]. Strategic planning initiatives in the city-region of Los Angeles, California [12], and Antwerp in the Belgian region of Flanders [13] provide additional food for thought on the advantages and shortcomings of this particular style of planning.

On a more global scale, the edited volume titled '*Situated Practices of Strategic Planning*', co-edited by Albrechts, Balducci, and Hillier [4], gathers worldwide perspectives on some of the latest advances in strategic planning in such places, regions, and countries as China's booming Pearl River Delta, East Africa, the Oresund region of Denmark and southern Sweden, France, the USA, Australia, South Africa, Egypt, Italy, Wales, Greater Vancouver, and Rio de Janeiro, among others.

2. Rationale for the Special Issue

It is well known that land for urbanisation, analogously to water [14], exists in high abundance in certain locations and is extremely scarce in others. The water cycle comprises evaporation and condensation. However, contrary to water, land in general does not change stages by itself. Usually, land for urbanisation is readily and affordably available in locations where one is not supposed to build and quite scarce and extremely expensive where it barely exists. Excessive urbanisation in suburban locations can lead to the potential decline of more centrally located neighbourhoods.

Land in the developed world is covered by a wide array of regulatory instruments. Depending on its location, land may be utilised intensively or left alone as open space. The utility and production components of land ought to benefit not only their owners but also the administrative jurisdictions with management and oversight responsibilities over their use. The administration of land uses can at times be contentious. The same can be said about the timing of land markets and the community gains that such land transformation processes give rise to. Anticipatory and integrated strategic land use and transportation planning mechanisms are needed to obviate opportunistic and speculative land markets.

In many cities of the Global North, the iatrogenic nature [15] (p. 50) of uncoordinated policy–market imbalances has resulted in sprawl, leap-frog development, spoiled countrysides, vehicle dependence, long commutes, unfeasible mass transit, unhealthy environments for vulnerable populations, environmental externalities, and overall poor urban quality of life for everybody. Major societal transformations, motivated by lifestyle preferences, technological possibilities, the COVID-19 pandemic, and post-growth scenarios, have transformed the ways we live, work, and play [16]. While the level of social interaction has resumed, collective consciousness and environmental pressures now require more strategic approaches to the built and natural environments where we spend most of our time [17].

3. Overview of the Special Issue

The Special Issue of *Land* on Strategic Planning and Urban Sustainability sought to fill a gap in how land and the transformation of the built environment have utilised strategic planning to enhance urban sustainability. Though initially we were seeking novel approaches to land, land use planning, urbanisation, land retrofit, regulatory systems, urban redevelopment, urban regeneration, and ecological approaches to improving and conserving land and natural resources, given the journal's worldwide scope, we ended up publishing ten research articles and one review article on the following seven themes: (i) transportation planning and policy; (ii) climate change adaptation and nature-based solutions; (iii) urban amenities, business vitality, and cooperative innovation networks; (iv) neighbourhood gentrification; (v) sustainable urban planning; (vi) low-impact development; and (vii) stormwater and water-sensitive urban design.

Zhang et al. (contribution 1) is a study of the relationship between bus travel and land use around bus stops and along bus routes. The research paper explores a novel method to estimate passenger flow volumes from and to bus stops based on land use elements. Using a square in the Chinese city of Dalian, the authors built and evaluated a gravitational logic estimation method for passenger flow volumes. Regression fitting and fuzzy estimation were used to analyse bus travel patterns based on land use elements, which, according to the authors, serve to facilitate strategic alignment in transit-oriented development (TOD) planning decisions.

Olgun et al. (contribution 2) is an examination of nature-based solutions scenario planning for climate change adaptation in cities with arid and semi-arid climates. Using the rapidly urbanising city of Phoenix, Arizona, as a case study, the research paper examines how the strategic benefits of ecosystem services provided by blue–green infrastructure (BGI) can be maximised in desert cities. The methods comprised GIS-based multicriteria decision-making techniques with special emphasis on a Green Infrastructure Spatial Planning model integrated with the city's existing water structures at the US census scale. The model's mapping of BGI hotspots and an expert stakeholder-driven weighting system revealed that social vulnerability and environmental risks have a positive correlation in Phoenix, which, according to the authors, suggests that stormwater management and urban heat island effects are the criteria to be considered first in BGI planning.

Giedych et al. (contribution 3) assess whether eco-spatial indices can be effective tools for climate change adaptation in residential neighbourhoods. The research paper analyses the effectiveness of the Ratio of Biologically Vital Areas (RBVA) mechanism in a residential district in the southern part of Warsaw, Poland, compared to similar deployments in two Norwegian cities. The study employed a serious simulation game developed under a Norwegian-funded project comprising these five environmental parameters: air temperature, oxygen production, CO_2 sequestration, rainwater harvesting, and biodiversity. The authors claim that the method by which each indicator is calculated is fundamentally a strategic and political choice, which also influences the type of nature-based solutions (NbSs) put forward. The study concludes that higher levels of adaptation to climate change depend on the comprehensiveness of environmental indices and measurement metrics.

Ji et al. (contribution 4) is a study of how urban amenities impact business circle vitality in 32 Chinese metropolises. It is known that urban business circles influence economic and social activities in city-regions. The study's authors argue that even though the promotion of sustainable economic development is high on most city managers' agendas, there is a paucity of research on the specific factors influencing business circle vitality. They also claim that government officials need integrated strategic land use planning to obviate market-oriented speculative land markets. The results suggest that even though the effect of urban amenities on business circle vitality varied considerably across different cities and business districts, the number of residential, transportation, educational, cultural, and recreational amenities and the diversity of catering and retail amenities had significant positive impacts on business circle vitality. The research paper concludes with ways to improve the vitality of business circles by bolstering urban amenities at the street and block levels.

Astore and Tricarico (contribution 5) analyse recent urban transformations in Rome's San Lorenzo neighbourhood. The research paper is an exploratory investigation of how gentrification tendencies are related to the effectiveness of the urban planning tools deployed by the municipality. The authors relied on a theoretical framework to trace the evolution of urban transformations and associated planning interventions. They also conducted interviews with key informants (representatives of local institutions, residents, other researchers, etc.). The findings suggest broader implications for urban governance, namely that the failure of the 2008 New General Regulatory Plan (NGRP) to provide clear and strategic directions has led to inconsistent development outcomes in San Lorenzo.

Gavanas et al. (contribution 6) is a multicriteria analysis of transport policy pathways for autonomous road vehicles (AVs) in the European Union. The research paper evaluates different policy pathways of AV deployment for sustainable urban development in the next EU programming period (2028–2034). It begins with a review of the main strategic priorities of the EU transport policy for the implementation of AVs in urban areas. And it concludes with a study combining the Analytic Hierarchy Process (AHP), *VIseKriterijumska Optimizacija I Kompromisno Resenje* (VIKOR), the participation of experts across Europe, and the weighting of evaluation criteria. The authors suggest that the proposed methodology provides a very important tool for public debate and decision making on how AVs can help to promote sustainable urban development in the European Union.

Muñoz et al. (contribution 7) is a review article of a conceptual model of floodable parks as instruments for urban stormwater management using nature-based solutions. Cities are increasingly utilising floodable parks as effective strategies to mitigate water-related hazards. The systematic review was fourfold: (i) to analyse the conditions needed for the implementation of floodable parks, (ii) to assess their design criteria, (iii) to review the connection between design and ecosystem services, and (iv) to perform an examination of the main processes occurring within a floodable park. The authors conclude that floodable parks can help to create more sustainable, climate-resilient, and circular urban areas in cities.

Balsas (contribution 8) advocates the need to make hidden sustainable urban planning and landscape knowledge visual and multisensorial. The research paper analyses how four educational and research programmes implemented at the University at Albany, New York (USA), built upon visual and multisensorial research methods to (i) augment educational outcomes, (ii) bolster scholarly significance, and (iii) provide up-to-date knowledge to empower concerned stakeholders. According to the author, the deployment of visual and multisensorial knowledge can help to change (a) unsustainable urbanisation practices, (b) social inequities and injustices in cities, and (c) car dependence and sedentary lifestyles, as well as (d) protect natural resources and livelihoods while (e) reducing environmental and social harm. The research methods comprised literature reviews, self-reflections on the implemented programmes, and analyses of participant evaluations of various outreach activities. The author concludes with remarks for other researchers interested in making their pedagogical and scholarly activities more effective and rewarding for themselves and the intended beneficiaries.

Rostamzadeh et al. (contribution 9) examine the effects of low-impact development best management practices (LID-BMP) on reducing stormwater caused by urban land use transformations in Tehran City, Iran. The research methods comprised a combination of geographic information system-based multicriteria decision analysis (GIS-MCDA) and spatial criteria from 2000 to 2022. Rainfall–runoff modelling was also applied to calculate changes in the area using the soil conservation service curve number (SCS-CN) method. The authors suggest that LID-BMP serves as a multifaceted tool within strategic urban planning frameworks, providing environmental, social, and economic benefits that are essential for sustainable urban development.

Tasnia and Growe (contribution 10) conducted a systematic literature review of water-sensitive urban design and flood risk management in contexts of strategic urban sustainability planning. The research paper consists of a PRISMA literature search and bibliographic analysis of 44 articles relevant to water-sensitive urban design (WSUD) and flood risk management (FRM) worldwide from 2013 to 2023. The authors concluded that WSUD and FRM share sustainability goals but differ in their focus and applicability depending on the settlement type and indicators.

Finally, Zhou et al. (contribution 11) investigate the spatial formation and evolution of cooperative innovation networks in strategic emerging industries in northeast China. Using invention patent data and spatial econometric models, the research paper constructs and analyses inter- and intra-city cooperative innovation networks from 2009 to 2023. The authors suggest that the findings offer policymakers empirical evidence to strategically optimise cross-regional innovation ecosystems in resource-dependent regions while supporting urban sustainability and fostering innovation-driven development.

4. Conclusions

As collaborative and communicative planning and just city theories [18] have evolved to occupy centre stage in research and scholarly circles [19], strategic planning can now be clearly found in the private and public sectors, as well as anywhere in between. It is our conviction that this Special Issue provides avenues to help revive strategic planning in the Global North [20] and to test it within distinct geographic realities and processes, most, if not all, being conducive to a greater understanding of urban sustainability, climate change, urban expansion and shrinkage, community resilience, social cohesion, and territorial innovation, as well as post-conflict, post-truth, and degrowth scenarios [21,22].

Author Contributions: Conceptualisation, C.J.L.B. and A.P.; writing—original draft preparation, C.J.L.B.; writing—review and editing, C.J.L.B. and A.P. All authors have read and agreed to the published version of the manuscript.

Funding: This research received no external funding.

Conflicts of Interest: The authors declare no conflicts of interest.

List of Contributions:

1. Zhang, J.; Cai, J.; Wang, M.; Zhang, W. An Estimation Method for Passenger Flow Volumes from and to Bus Stops Based on Land Use Elements: An Experimental Study. *Land* **2024**, *13*, 971. https://doi.org/10.3390/land13070971.
2. Olgun, R.; Cheng, C.; Coseo, P. Nature-Based Solutions Scenario Planning for Climate Change Adaptation in Arid and Semi-Arid Regions. *Land* **2024**, *13*, 1464. https://doi.org/10.3390/land13091464.
3. Giedych, R.; Maksymiuk, G.; Cieszewska, A. Eco-Spatial Indices as an Effective Tool for Climate Change Adaptation in Residential Neighbourhoods—Comparative Study. *Land* **2024**, *13*, 1492. https://doi.org/10.3390/land13091492.
4. Ji, Y.; Wang, Z.; Zhu, D. Exploring the Impact of Urban Amenities on Business Circle Vitality Using Multi-Source Big Data. *Land* **2024**, *13*, 1616. https://doi.org/10.3390/land13101616.

5. Astore, A.; Tricarico, L. Understanding the Transformations of San Lorenzo, Rome: An Attempt at Conceptual Order between Gentrification and Urban Policy. *Land* **2024**, *13*, 1632. https://doi.org/10.3390/land13101632.
6. Gavanas, N.; Anastasiadou, K.; Nathanail, E.; Basbas, S. Transport Policy Pathways for Autonomous Road Vehicles to Promote Sustainable Urban Development in the European Union: A Multicriteria Analysis. *Land* **2024**, *13*, 1807. https://doi.org/10.3390/land13111807.
7. Muñoz, S. M.; Elliott, S.; Schoelynck, J.; Staes, J. Urban Stormwater Management Using Nature-Based Solutions: A Review and Conceptual Model of Floodable Parks. *Land* **2024**, *13*, 1858. https://doi.org/10.3390/land13111858.
8. Balsas, C.J.L. Making Hidden Sustainable Urban Planning and Landscape Knowledge Visual and Multisensorial. *Land* **2025**, *14*, 1. https://doi.org/10.3390/land14010001.
9. Rostamzadeh, S.; Malekmohammadi, B.; Mashhadimohammadzadehvazifeh, F.; Arsanjani, J.J. The Effects of Low-Impact Development Best Management Practices on Reducing Stormwater Caused by Land Use Changes in Urban Areas: A Case Study of Tehran City, Iran. *Land* **2025**, *14*, 28. https://doi.org/10.3390/land14010028.
10. Tasnia, T.; Growe, A. A Systematic Literature Review of Water-Sensitive Urban Design and Flood Risk Management in Contexts of Strategic Urban Sustainability Planning. *Land* **2025**, *14*, 224. https://doi.org/10.3390/land14020224.
11. Zhou, X., Liu, T., Zhang, P.; Zhang, X.; Chu, N. Evolution and Mechanism of Cooperative Innovation Networks Based on Strategic Emerging Industries: Evidence from Northeast China. *Land* **2025**, *14*, 691. https://doi.org/10.3390/land14040691.

References

1. Mintzberg, H. *The Rise and Fall of Strategic Planning*; The Free Press: New York, NY, USA, 1994.
2. Bryson, J.M. *Strategic Planning for Public and Nonprofit Organizations*; Jossey-Bass: San Francisco, CA, USA, 1995.
3. Veal, A.J.; Piracha, A. Meeting the need for a rational basis for open space and recreation planning in new high density residential areas: The recreational activity benchmark model. *Aust. Plan.* **2022**, *58*, 63–68. [CrossRef]
4. Albrechts, L.; Balducci, A.; Hillier, J. (Eds.) *Situated Practices of Strategic Planning—An International Perspective*; Routledge: Oxon, UK, 2017.
5. Cabral, J.; Sá Marques, T. Do planeamento estratégico ao desenvolvimento sustentável: Experiência em Portugal. *Inforgeo* **1996**, *11*, 107–116.
6. Alden, J.; da Rosa Pires, A. Lisbon: Strategic planning for a capital city. *Cities* **1996**, *13*, 25–36. [CrossRef]
7. Balsas, C. Expo'98: From idea to legacy. *Plan Can.* **2005**, *45*, 23–25.
8. Kaufman, J.L.; Jacobs, H.M. A public planning perspective on strategic planning. *J. Am. Plan. Assoc.* **1987**, *53*, 21–31. [CrossRef]
9. Porter, M.E. Regional Competitiveness in the Pioneer Valley. In Proceedings of the Talk to the Pioneer Valley Regional Competitiveness Council, Chicopee, MA, USA, 10 October 2003.
10. Healey, P. *Collaborative Planning: Shaping Places in Fragmented Societies*; Palgrave MacMillan: New York, NY, USA, 2006.
11. Albrechts, L. Ingredients for a more radical strategic spatial planning. *Environ. Plann. B Plann. Des.* **2015**, *42*, 510–525. [CrossRef]
12. McCutcheon, J.M. Utopian Visions in United States Urbanism. *Moreana* **2005**, *42*, 157–186. [CrossRef]
13. Broeck, J. Spatial planning in Flanders and Antwerp 1940–2012: Movements, clashing values and expertise. In *Situated Practices of Strategic Planning—An International Perspective*; Albrechts, L., Balducci, A., Hillier, J., Eds.; Routledge: Oxon, UK, 2017; pp. 233–254.
14. Harvey, D. *Marx, Capital and the Madness of Economic Reason*; Oxford University Press: Oxford, UK, 2018.
15. Badami, M.G. Urban transport policy as if people and the environment mattered: Pedestrian accessibility the first step. *Econ. Political Wkly.* **2009**, *xliv*, 43–51.
16. Sharifi, A.; Khavarian-Garmsir, A.R. The COVID-19 pandemic: Impacts on cities and major lessons for urban planning, design, and management. *Sci. Total Environ.* **2020**, *749*, 142391. [CrossRef] [PubMed]
17. Bleicher, J.; Kumar, P. A novel framework integrating resilient and sustainable urbanism for designing climate resilient sustainable urban neighbourhoods. *Sustain. Cities Soc.* **2025**, 106577. [CrossRef]
18. Fainstein, S.S. New directions in planning theory. *Urban Aff. Rev.* **2000**, *35*, 451–478. [CrossRef]
19. Gunder, M.; Madanipour, A.; Watson, V. (Eds.) *The Routledge Handbook of Planning Theory*; Routledge: London, UK, 2018.
20. Olesen, K. Reviving strategic spatial planning for the challenges ahead. *Eur. Plan. Stud.* **2023**, *31*, 2318–2326. [CrossRef]

21. Bryson, J.M.; Edwards, L.H.; Slyke, D.M. Getting strategic about strategic planning research. *Public Manag. Rev.* **2018**, *20*, 317–339. [CrossRef]
22. Savini, F. Strategic planning for degrowth: What, who, how. *Plan. Theory* **2025**, *24*, 141–162. [CrossRef] [PubMed]

Disclaimer/Publisher's Note: The statements, opinions and data contained in all publications are solely those of the individual author(s) and contributor(s) and not of MDPI and/or the editor(s). MDPI and/or the editor(s) disclaim responsibility for any injury to people or property resulting from any ideas, methods, instructions or products referred to in the content.

land

MDPI

Article

An Estimation Method for Passenger Flow Volumes from and to Bus Stops Based on Land Use Elements: An Experimental Study

Jianming Zhang, Jun Cai *, Mengjia Wang and Wansong Zhang

School of Architecture and Fine Art, Dalian University of Technology, Dalian 116024, China; zhangjming@mail.dlut.edu.cn (J.Z.); wangmengjia@mail.dlut.edu.cn (M.W.); 13703396973@163.com (W.Z.)
* Correspondence: caimans@dlut.edu.cn

Abstract: To unravel the general relationship between bus travel and land use around bus stops and along bus routes and to promote their coordinated development, this paper explores a method to estimate passenger flow volumes from and to bus stops based on land use types, intensities, and spatial distributions around bus stops and along bus routes. Firstly, following the principle of the gravity model, which considers traffic volumes analogous to gravity based on trip generation and distance impedance between traffic analysis zones (TAZs), a gravitational logic estimation method for passenger flow volumes from and to bus stops was constructed with land use elements between bus stop TAZs and the upstream and downstream collections of bus stop TAZs. Building upon this, the passenger flow volumes from and to 38 bus stops in the Xueyuan Square area of Dalian during weekday morning peak hours were taken as the experimental objects. The basic estimation models of two gravity sets corresponding to passenger flow volumes from and to bus stops were constructed using the bus travel generation based on the aggregation of area-based origin unit method and the bus travel distance impedance based on the probability density method. Finally, the reliability of the estimation method of passenger flow volumes from and to bus stops was verified by regression fitting between the surveyed values of passenger flow volume and the estimated values of the basic models. The results indicate that the fuzzy estimation and transformation of bus travel based on land use elements, which serves as a crucial lever for facilitating strategic alignment in transit-oriented development (TOD), can be effectively achieved by using the area-based origin unit method to aggregate bus travel generation and the probability density method to evaluate the bus travel distance impedance.

Keywords: passenger flow volume estimation method; passenger flow volume from and to bus stop; land use; trip generation and travel impedance; transit-oriented development; gravity model

1. Introduction

TOD is an urban planning concept aimed at achieving intensive, efficient, and sustainable urban mobility by aligning public transportation with land use around the stations [1,2]. It can be divided into station area TOD and citywide TOD based on the research scope [3]. Station area TOD focuses on developing high-density, mixed-use, and pedestrian-friendly environments around specific transportation nodes, such as railway stations, bus rapid transit (BRT) stops, and bus stations [4,5]. It prioritizes factors like land use configuration, the built environment, and station accessibility to enhance ridership and foster sustainable growth [6]. Citywide TOD seeks to optimize the synergy between public transport, land use, and resource allocation across urban areas [7,8]. It emphasizes adapting land use patterns, transit systems, and transit routes (stations) to meet the effective organization and strategic development of travel demand [9,10]. Due to the characteristics listed above, TOD has become an important direction of urban renewal efforts [11,12].

As an integral component of public transportation systems and a precursor to high-capacity transit systems, the role of bus systems in TOD is undeniable [13,14]. Passenger flow volumes destined to and originating from service areas of bus stops, referred to as passenger flow volumes from and to bus stops, serving as a link between the bus system and its service area, play a vital role in urban bus network planning and optimization, particularly in TOD contexts where land use is designed to support efficient transit systems [15]. However, constrained by the unclear relationship between land use along bus routes and bus travel [16,17], urban renewal planning efforts for station area TOD, which involve altering land use structures and optimizing pathways from and to bus stops, as well as those for citywide TOD, which entail adjusting urban functional structures and upgrading bus routes and stops, face challenges in assessing current issues of passenger flow volumes and predicting the benefits of proposed updates. Consequently, this situation often leads to corresponding work either being shelved or confined to the vague growth in passenger flow volumes, which exacerbates issues like job–housing separation [18], pendulum commuting, and operational losses [19]. The challenge of regulating the relationship between land use along bus routes and bus travel significantly impedes the high-quality coordinated development of the two.

The aforementioned unclear relationship corresponds to research inadequacies in analyzing bus travel. According to relevant studies, existing analyses of bus travel can be categorized into three levels, micro, meso, and macro, each offering different analytical perspectives. At the micro level, short-term passenger flow prediction models are crucial components of intelligent transportation systems for analyzing dynamic variations of passenger flow. These models are typically divided into parametric and non-parametric models. Parametric models rely on theoretical assumptions and specific conditional parameters to analyze linear models of historical passenger flow data, providing greater interpretability. Examples include linear time series models [20,21], historical average models, Kalman filter models [22], and multi-source data regression models [23]. Non-parametric models, on the other hand, utilize machine learning methods to analyze nonlinear models of historical passenger flow data, offering better objectivity. Examples include nonlinear time series models [24], support vector machine models [25], and neural network models [26–30]. At the meso level, passenger flow forecasting models are used to support the adjustment and optimization of bus networks by analyzing bus travel relationships. These models are typically divided into aggregation and decomposition models. Aggregation models iteratively calculate recent travel distributions based on statistical regularities of historical data and the relatively stable relationship of bus travel, featuring a relatively simple structure. Examples include the constant growth factor method and the average growth factor method. Decomposition models, on the other hand, are based on detailed traffic distribution surveys and behavioral analysis to conduct systematic analysis of traffic flow distribution corresponding to comprehensive models [31], with stronger regularity. Examples include the gravity model method [32], intervention opportunity model method, and maximum entropy model method [33]. At the macro level, bus travel demand forecasting models are based on the analysis of land use relationships between bus stops to support integrated development of land use and bus travel. However, perhaps influenced by micro-level development orientation, compared with macro-strategic forecasting based on the classical scientific reductionist approach, which aims to uncover the general principles and underlying laws behind the complex travel phenomena to support strategic planning initiatives, existing research advocates micro-intelligent predictions based on the contemporary scientific paradigm of complexity theory, which focuses on unraveling the dynamics and mechanisms inherent in complex travel phenomena to aid precise control in design solutions. Moreover, passenger flow forecasting models, which are associated with relationships between land use along bus routes and bus travel, typically rely on the four-stage method [34,35], which involves dividing TAZs according to the service areas of bus stops [36], utilizing survey results of the bus travel OD matrices between TAZs as a simulation basis, and constructing complex bus network models based on trip generation and distance impedance of bus travel [37–40]. Perhaps it is the unclear logical relationship in distance decay [41–43] that leads to misuse of passenger flow forecasting theories and methods. Alternatively, the complexity of bus network models superimposes various differences in land use attributes, and stringent standards of bus travel OD matrix volume simulation magnify the differences in land use attributes [44,45]. This results in the segmentation, alienation, and mingling of land use elements in the analysis of bus travel demand. Consequently, the predominant role of land use generation relationships and bus travel distance impedance is weakened, while the subjective influence of non-land use factors such as household and individual attributes is strengthened. Land use idiosyncrasies gradually replace common patterns, resulting in a state of overfitting. In addition to increasing the difficulty of data acquisition and interpretation, as well as issues regarding model portability, this also leads to a disconnect between land use and bus travel demand. Consequently, it becomes challenging to advance bus travel demand management at the land use level.

The systematic analysis of the time-and-distance decay law [46], which reflects the impedance of bus travel distance in passenger flow forecasting models, contained within residents' travel time and distance distributions, lays the groundwork for understanding the general relationship between land use along bus routes and bus travel. In order to unravel the general relationship above and to foster coordinated development, the rest of this paper is organized as follows.

Based on the principle of the gravity model, Section 2 deconstructs the point–line units of bus travel, taking bus stop service areas as bus stop TAZs and passenger flow volumes from and to bus stops as estimation objects. The gravitational logic estimation method corresponding to passenger flow volumes from and to bus stops is constructed by land use types, intensities, and spatial distributions between bus stop TAZs and the respective upstream and downstream collections of bus stop TAZs.

Subsequently, Section 3 takes the passenger flow volume from and to 38 bus stops in the Xueyuan Square area of Dalian during weekday morning peak hours as the experimental object, and the basic estimation models of two gravity sets corresponding to passenger flow volumes from and to bus stops are constructed using the bus travel generation aggregation of area-based origin unit method and the bus travel distance impedance of the probability density method.

With passenger flow volumes from and to bus stops surveyed and the walking and bus travel distance impedance obtained, Section 4 verifies the reliability of the estimation method of passenger flow volumes from and to bus stops through regression fitting between the surveyed values of passenger flow volume and the estimated values of the basic models.

2. Theoretical Foundation

2.1. The Estimation Principle of Passenger Flow Volumes from and to Bus Stops

The gravity model method, rooted in trip generation and distance impedance between TAZs, provides a theoretical framework for estimating passenger flow volumes from and to bus stops based on land use types, intensities, and spatial distributions around bus stops and along bus routes. The improved gravity model formula can be expressed as Equation (1):

$$Q_{\alpha\beta} = k \cdot P_{\alpha} \cdot A_{\beta} \cdot f(l_{\alpha\beta}), \quad (1)$$

In Equation (1), $Q_{\alpha\beta}$ represents traffic volume (corresponding the gravity) from TAZ α to TAZ β; k is the normalization factor; P_{α} denotes traffic production volume at TAZ α; A_{β} signifies traffic attraction volume at TAZ β; and $f(l_{\alpha\beta})$ denotes the traffic distance impedance function from TAZ α to TAZ β.

According to the gravity model principle mentioned above, the service area of a bus stop (serves multiple routes and nearby stops) can be regarded as a TAZ, specifically termed the bus stop TAZ. Passenger flow volumes from and to the bus stop correspond to the total bus travel volumes between the bus stop TAZ and its upstream and downstream collections of bus stop TAZs, as illustrated in Figure 1a. By translating the above corresponding relationship into the gravitational logic of bus travel, passenger flow volumes from and to the bus stop can be translated into bus travel generation and distance impedance based on land use elements between the experimental bus stop TAZ and its upstream and downstream collections of bus stop TAZs, as depicted in Figure 1.

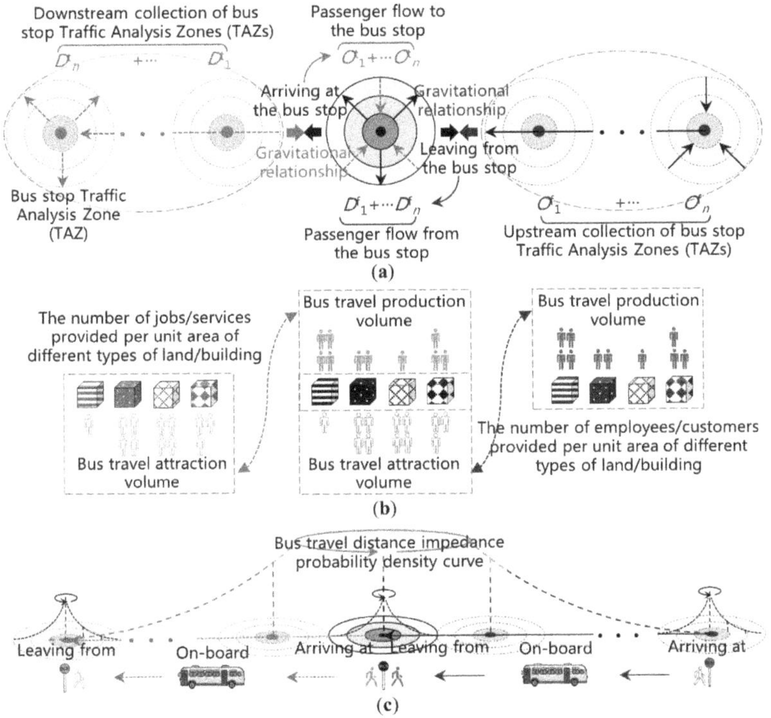

Figure 1. Schematic of the estimation method for passenger flow volumes from and to bus stops: (**a**) illustration of the gravitational relationship of passenger flow volumes from and to the bus stop; (**b**) bus travel generation of area-based origin unit method; (**c**) bus travel distance impedance of probability density method.

As commonly understood, the primary bus travel demand of residents within a specific area and time remains relatively stable, alongside a similar stability in the correlation between land use (building space) catering to residents' activity requirements and the number of activities. Therefore, bus travel generation is closely correlated with land use types and their area scales (intensities), which can be effectively aggregated using the area-based origin unit method. This method treats areas of the same land use type as homogeneous under certain conditions, while considering the average production and attraction volume of bus travel per unit area of land use or building, as illustrated in Figure 1b. Similarly, the correlation between residents' bus travel spatial distributions and the proportionate distribution of residents' bus travel distance remains relatively stable. Therefore, bus travel distance impedance is closely correlated with land use spatial distributions. It can be effectively aggregated using the probability density method, which deciphers the decay probability of different bus travel distances under certain conditions [47]. Due to the complete process of bus travel includes the three stages of "to (from)–onboard–from (to)", its distance impedance can be decomposed into "from and to" distance impedance related to land use around the bus stop and "on-board" distance impedance related to land use around the surrounding areas of the bus stops upstream and downstream, as illustrated in Figure 1c.

2.2. Travel Time and Distance Decay Law

The analysis [46] of the time and distance decay law contained within the proportionate distributions of residents' travel time and distance (by a certain travel mode and for a certain purpose) that ensures relatively accurate estimation of bus travel distance impedance is a crucial theoretical basis for the estimation method proposed in this paper. The research process is depicted in Figure 2.

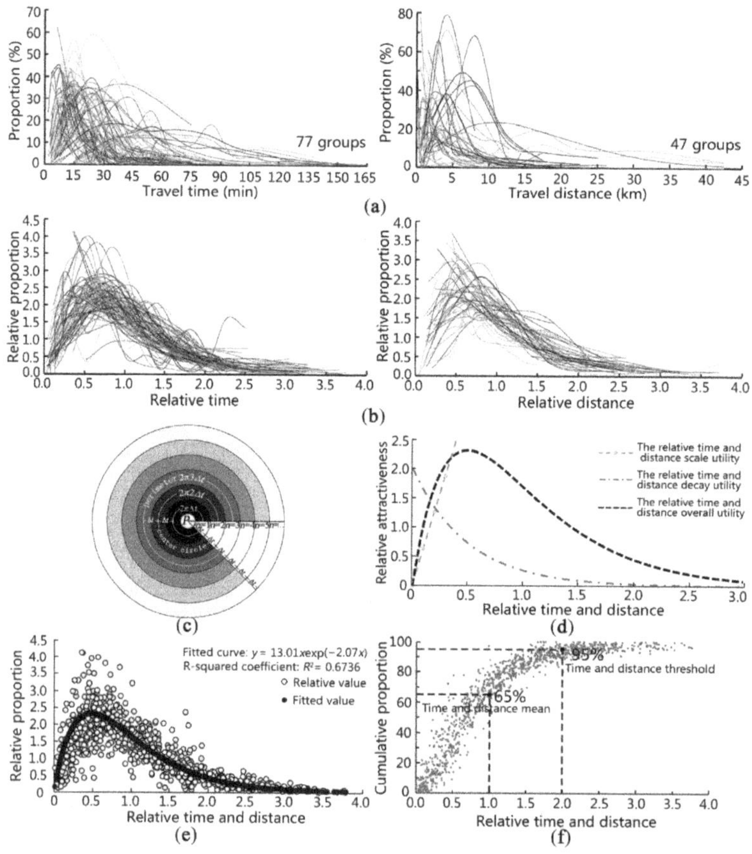

Figure 2. The research process of the residents' travel time and distance distribution law: (**a**) 124 groups of residents' travel time and distance distributions; (**b**) relative time and distance distribution; (**c**) schematic of the relative time and distance scale utility; (**d**) schematic of the utility relationship of relative time and distance; (**e**) regression fitting curve of relative time and distance distribution; (**f**) cumulative proportional distribution of relative time and distance.

Firstly, the 124 groups of residents' travel time and distance distributions with significant differences depicted in Figure 2a were standardized relative to their own mean travel time and distance, resulting in the stable and regular law of residents' travel time and distance distributions relative to their mean values, as shown in Figure 2b. From the holistic spatiotemporal perspective of "supply–travel time and distance–demand", the relative time and distance distribution abstractly represents an attraction relationship between standard supply and demand within the context of relative time and distance, while twice the mean of travel time and distance abstractly represents the scale of the supply–demand domain.

Secondly, the explanatory mechanism for the relative time and distance distribution shown in Figure 2d was constructed based on the utility of relative time and distance scale (illustrated in Figure 2c) and the utility of relative time and distance decay. For a facility point P in Figure 2c, the demand from individuals from different layers of relative time and distance (denoted as x) should exhibit a scale growth proportional to $2\pi x$. However, due to the influence of relative time and distance decay following a pattern of $\lambda\exp(-\lambda x)$, the distribution stabilizes as depicted in Figure 2d.

Finally, the relative time and distance distribution data shown in Figure 2b were fitted with the equation $y = 2\pi x \cdot \lambda\exp(-\lambda x)$, validating the interpretability of the relative time and distance distribution mechanism (as seen in Figure 2e), and yielding a relative time and distance decay parameter λ of 2.07 (after adjustment, the parameter value is 2.08). Additionally, from the cumulative proportion distribution of relative time and distance shown in Figure 2f, it was deduced that the threshold of resident travel time and distance is twice the mean travel time and distance.

3. Experimental Methods and Materials

3.1. The Estimation Method of Passenger Flow Volumes from and to Bus Stops

Building on the aforementioned estimation principle, this paper selects 19 pairs of conventional bus stops (totaling 38 bus stops as shown in Figure 3) in the Xueyuan Square area of Dalian, Liaoning Province, China, as experimental subjects, takes the passenger flow volumes from and to these bus stops during weekday morning peak hours (primarily for commuting) as estimation objects, and aims to examine the feasibility and effectiveness of the estimation principle and method by constructing an estimation model of passenger flow volumes based on land use between the experimental bus stop TAZs and upstream and downstream collections of bus stop TAZs.

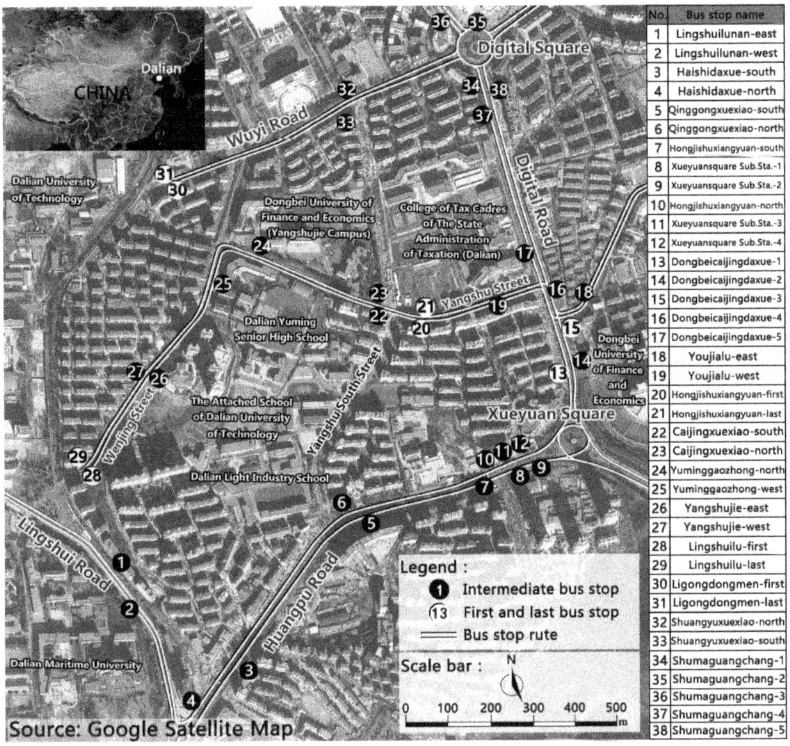

Figure 3. Distribution of 38 experimental bus stops.

By treating passenger flow volumes from and to bus stops as analogous to gravitational forces between the experimental bus stop TAZs and their respective upstream and downstream collections of bus stop TAZs, the estimation model can be broken down into two gravity sets: the walking from-and-to models, which account for walking distance impedance within the experimental bus stop TAZs (since walking constitutes the absolute majority of travel from and to the bus stops in the survey area); and the bus travel from-and-to models, which incorporate bus travel distance impedance within the upstream and downstream bus stop TAZs. In this context, the corresponding relationship is illustrated in Figure 4, and the model formulas are expressed as Equations (2) and (3):

$$Q^{\mathrm{f}}{}_{s} = A^{\mathrm{f}}{}_{s} \cdot P^{\mathrm{f}}{}_{s} \tag{2}$$

$$Q^{\mathrm{t}}{}_{s} = P^{\mathrm{t}}{}_{s} \cdot A^{\mathrm{t}}{}_{s} \tag{3}$$

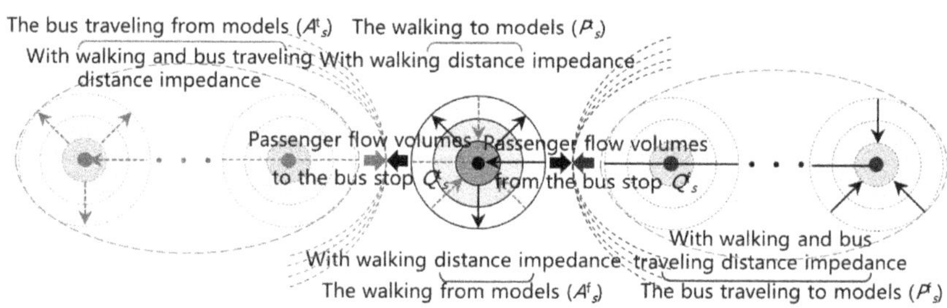

Figure 4. Schematic of the estimation method of passenger flow volumes from and to bus stops.

In Equations (2) and (3), $Q^{\mathrm{f}}{}_{s}$ represents the passenger flow volumes from the bus stops, where "*s*" is the number of experimental bus stops and "f" stands for "from", measured in passengers per hour (p·h^{-1}); $A^{\mathrm{f}}{}_{s}$ represents the "walking from" models; $P^{\mathrm{f}}{}_{s}$ represents the "bus travel to" models; $Q^{\mathrm{t}}{}_{s}$ represents the passenger flow volumes to the bus stops, where "t" stands for "to", measured in passengers per hour (p·h^{-1}); $P^{\mathrm{t}}{}_{s}$ represents the "walking to" models; and $A^{\mathrm{t}}{}_{s}$ represents the "bus travel from" models.

3.2. Key Data for the Estimation Method

3.2.1. Passenger Flow Volumes from and to Bus Stops

The passenger flow volumes from and to experimental bus stops during weekday morning peak hours, which are the objects of fitting and validating the estimation model (corresponding to two basic estimation models), can be obtained by deducting the transfer passenger flow volumes from the drop-off and pick-up passenger flow volumes of experimental bus stops. Following the experimental design, the process for obtaining these volumes is outlined below, with Figure 5 illustrating the relevant content.

Firstly, from 10 to 11 October 2018, video surveillance was conducted at 38 experimental bus stops and 1 transfer subway station (Xueyuan Square) to capture passenger flow information. The video equipment is shown in Figure 5a, the equipment setup is shown in Figure 5b, and the video footage is shown in Figure 5c.

Secondly, images and spatiotemporal information of drop-off and pick-up of passengers at each bus stop and the transfer subway station were captured and annotated during the morning peak hours (from 07:00 to 09:00) over the two days, as shown in Figure 5d.

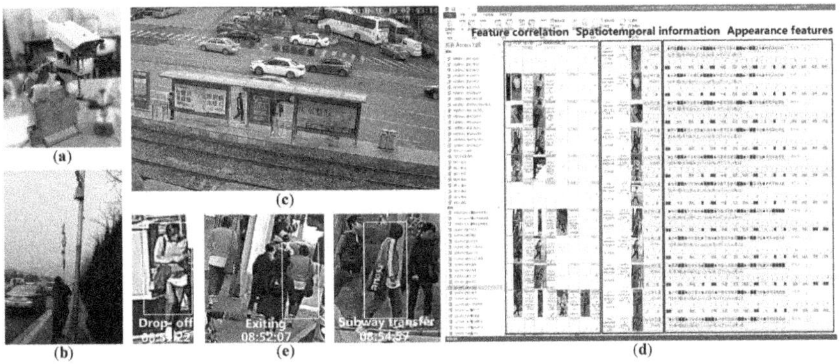

Figure 5. Description of relevant content of passenger flow volume survey: (**a**) video equipment; (**b**) equipment setup; (**c**) video footage; (**d**) the Access database containing appearance features; (**e**) schematic of a transferring passenger.

Finally, based on the technique of pedestrian appearance feature re-identification, an Access database containing appearance features such as age, gender, attire, headwear, hairstyle, and carried items was constructed. Through feature correlation queries traversing image data, it identified transferring passengers, including both intra-stop transfers and inter-stop transfers, among drop-off and pick-up passengers of bus stops (as shown in Figure 5e). The final passenger flow volumes were calculated using the following Equations (4) and (5), with the corresponding survey data shown in Table A2 of Appendix A.

$$Q^{\mathrm{f}}{}_{s} = G^{\mathrm{f}}{}_{s} - H^{\mathrm{f}}{}_{s}, \tag{4}$$

$$Q^{\mathrm{t}}{}_{s} = G^{\mathrm{t}}{}_{s} - H^{\mathrm{t}}{}_{s}, \tag{5}$$

In Equations (4) and (5), $G^{\mathrm{f}}{}_{s}$ and $G^{\mathrm{t}}{}_{s}$ represent the drop-off and pick-up passenger flow volumes, respectively, during the morning peak hours, measured in passengers per hour ($\mathrm{p{\cdot}h^{-1}}$). $H^{\mathrm{f}}{}_{s}$ and $H^{\mathrm{t}}{}_{s}$ represent the drop-off and pick-up transfer passenger flow volumes, respectively, during the morning peak hours, measured in passengers per hour ($\mathrm{p{\cdot}h^{-1}}$).

3.2.2. Walking and Bus Travel Distance Impedance

Based on the distribution law of residents' travel time and distance mentioned above, the distance thresholds and the distance decay probabilities of walking and bus travel from and to the experimental bus stops, which correspond to the impact range of land use and distance impedance in the method proposed in this paper, can be obtained through a questionnaire survey on residents' walking and bus travel distance from and to bus stops during the morning peak hours in the research area, conducted simultaneously with video surveillance. We collected 158 valid questionnaires, and the corresponding distribution results, along with their distance thresholds and impedances, are shown in Figure 6.

According to the survey results shown in Figure 6a,c, the mean distances of walking and bus travel from-and-to distance distributions are 0.4 km and 5.0 km, respectively. Consequently, the distance threshold for walking from and to experimental bus stops is set at 0.8 km, with the probability distance impedance function represented as $\exp(-2.08x/0.4)$, while the distance threshold for bus travel from and to experimental bus stops is set at 10.0 km, with the probability distance impedance function represented as $\exp(-2.08x/5)$.

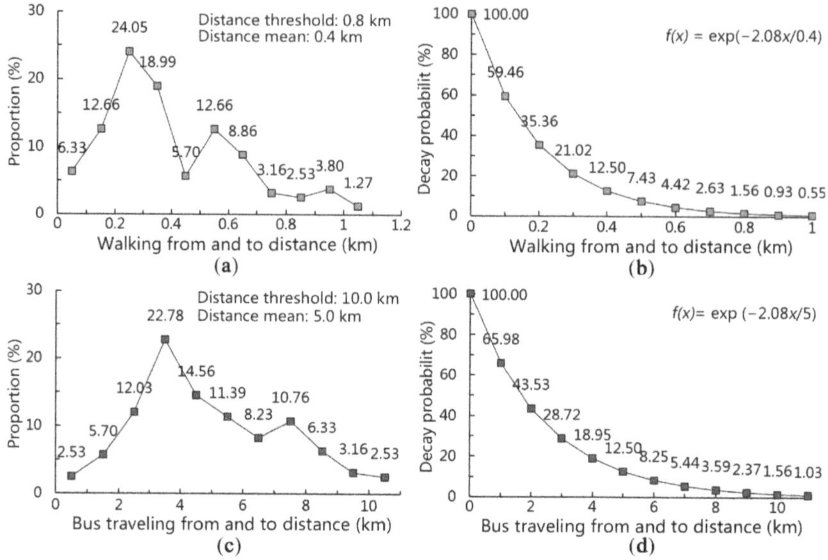

Figure 6. Survey results along with distance thresholds and impedance: (**a**) walking from-and-to distance distribution; (**b**) walking from-and-to distance impedance; (**c**) bus travel from-and-to distance distribution; (**d**) bus travel from-and-to distance impedance.

3.3. Construction of the Basic Estimation Models

3.3.1. The Walking from-and-to Models

The walking from-and-to models correspond to the gravity set associated with the generation and distance impedance of land use within the 38 experimental bus stop TAZs. Taking into account considerations such as the commuting and schooling service characteristics of bus travel during the morning peak hours, as well as the moderate simplification of the aggregation models, the generation estimation in the area-based origin unit method categorizes land use within experimental TAZs into four types: educational, residential, office, and commercial land uses. Their building areas are then aggregated to calculate the production volume and attraction volumes of bus travel in TAZs.

In the distance impedance estimation using the probability density method, where the probability decay function and distance threshold for walking distance from and to the experimental bus stops are known, the walking from and to distances are uniformly segmented, and the median distance decay probability of each segment is used to estimate the spatial distance impedance within the walking from-and-to distance segment. Based on the actual conditions of land use, walking path, and the competitive relationship of the experimental bus stop services, this paper constructs the walking from ($A^f{}_s$) and to ($P^t{}_s$) models. The calculation formula and model example (Figure 7) are as follows:

$$A^f{}_s = \Sigma^n{}_i[S_i\cdot(M^e{}_iT^e + M^r{}_iT^r + M^o{}_iT^o + M^c{}_iT^c)], \tag{6}$$

$$P^t{}_s = \Sigma^n{}_i[S_i\cdot(M^e{}_iO^e + M^r{}_iO^r + M^o{}_iO^o + M^c{}_iO^c)], \tag{7}$$

In Equations (6) and (7), S_i represents the distance decay probability for walking from-and-to distance segments, where i ranges from 1 to 4; $M^e{}_i$, $M^r{}_i$, $M^o{}_i$, and $M^c{}_i$ represent the areas of educational, residential, office, and commercial buildings within the respective distance segments of bus stop TAZs, measured in hectares (ha); and O^e, O^r, O^o, and O^c represent the relative bus travel production parameters, while T^e, T^r, T^o, and T^c represent the relative bus travel attraction parameters, all measured in passengers per hectare per hour ($P{\cdot}h^{-1}{\cdot}ha^{-1}$), with "e", "r", "o", and "c" denoting educational, residential, office, and commercial buildings, respectively.

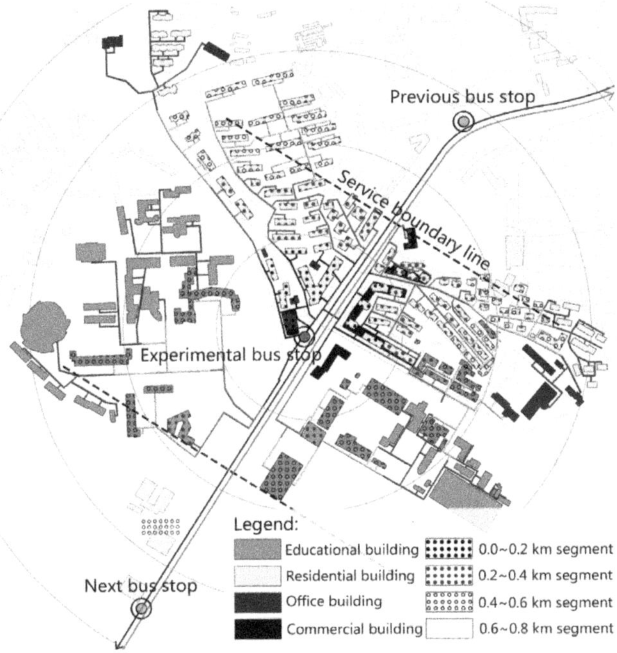

Figure 7. Example of the generation and distance impedance of the walking from-and-to models.

3.3.2. The Bus Travel from-and-to Models

The bus travel from-and-to models correspond to the gravity set associated with the generation and distance impedance of land use within the upstream and downstream collections of bus stop TAZs. Considering that it is only necessary to simulate the total gravity of the gravity set, in the generation estimation using the area-based origin unit method, it may be feasible to treat educational, residential, office, and commercial land uses, collectively referred to as "generation land use", within the service area of upstream and downstream collections of bus stop TAZs as homogeneous (fuzzy land use types, intensities, structure and spatial distributions without considering variations in generation parameters), and to take the area of "generation land use" to aggregate the production and attraction volumes of bus travel within the upstream and downstream collections of bus stop TAZs.

The distance impedance in the bus travel from-and-to models comprises two components: walking distance impedance and bus travel distance impedance within the upstream and downstream collections of bus stop TAZs. In the distance impedance estimation using the probability density method, the bus travel distance is evenly segmented, and a buffer is applied to the walking form and to distance. By combining the probabilities of both components, it becomes possible to estimate the spatial distance impedance within the bus travel from-and-to distance segment. This paper constructs the bus travel from ($P^{f}{}_{s}$) and to ($A^{t}{}_{s}$) models based on the actual conditions of land use and bus route system within the upstream and downstream collections of bus stop TAZs. The calculation formula and a model example (Figure 8) are as follows:

$$P^{f}{}_{s} = \Sigma^{n}{}_{i}\Sigma^{m}{}_{j}[S_i{\cdot}R_j{\cdot}(h_1F^{f}{}_{ij} + h_2N^{f}{}_{ij})], \quad (8)$$

$$A^{t}_{s} = \Sigma^{n}_{i}\Sigma^{m}_{j}[S_i \cdot R_j \cdot (h_1 F^{t}_{ij} + h_2 N^{t}_{ij})], \tag{9}$$

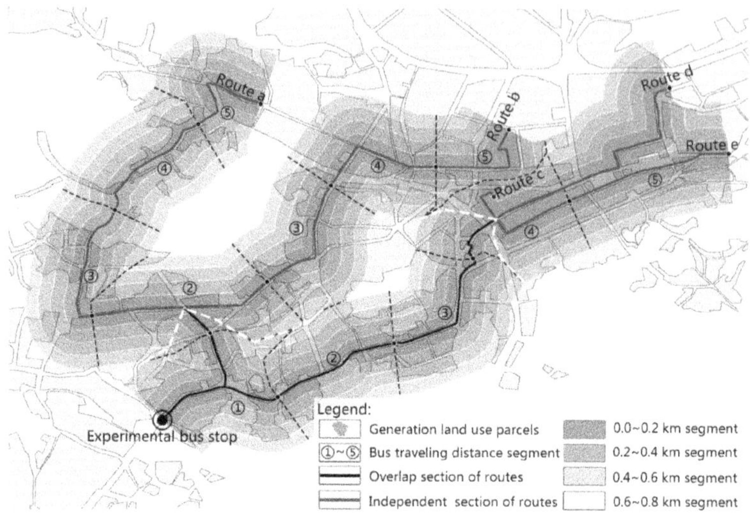

Figure 8. Example of the generation and distance impedance of the bus travel from-and-to models.

In Equations (8) and (9), R_j represents the distance decay probability corresponding to the bus travel from-and-to distance segments, where j ranges from 1 to 5; F^{f}_{ij}, N^{f}_{ij}, F^{t}_{ij}, and N^{t}_{ij} represent the area of "generation land use" within the independent and overlapping regions (some experimental bus stops serve multiple bus routes, resulting in overlapping land use areas) within the buffer zones i and j, measured in square kilometers (km^2); and h_1 and h_2 represent land use coefficients in these regions, set empirically to 1 and 2, respectively. This indicates that overlapping areas served by multiple routes are calculated at twice their per-unit area, measured in hectares per unit area (km^{-2}).

4. Results

4.1. The Basic Estimation Models' Parameters

Based on the construction of two basic estimation models, data from 38 experimental bus stops, corresponding to 68 groups (including 8 first or last bus stops with unidirectional data only), were investigated and collected, as shown in Tables A1 and A2 of Appendix A. The decay probabilities of walking from-and-to distance segments, as well as bus travel from-and-to distance segments obtained from Figure 6, are presented in Tables 1 and 2. By fitting the surveyed values of passenger flow volumes from and to 38 experimental bus stops with the basic estimation models, the relative production and attraction parameters of four types of land use (the unit building area) in the study area were determined, and the stepwise regression results are shown in Tables 3 and 4.

Table 1. The decay probabilities of walking from-and-to distance segments.

Parameter Symbol	S_1	S_2	S_3	S_4
Distance segment (km)	(0.0, 0.2)	(0.2, 0.4)	(0.4, 0.6)	(0.6, 0.8)
Estimated value (%)	59.46	21.02	7.43	2.63

Table 2. The decay probabilities of bus travel from-and-to distance segments.

Parameter Symbol	R_1	R_2	R_3	R_4	R_5
Distance segment (km)	(0.0, 2.0)	(2.0, 4.0)	(4.0, 6.0)	(6.0, 8.0)	(8.0, 10.0)
Estimated value (%)	65.98	28.72	12.50	5.44	2.37

Table 3. The relative production parameters of four types of land use (the unit building area).

Land Use Type	Educational	Residential	Office	Commercial
Parameter symbol (person·ha^{-1}·h^{-1})	O^e	O^r	O^o	O^c
Fitted relative value	13.2948	4.0310	52.6219	14.1178

Table 4. The relative attraction parameters of four types of land use (the unit building area).

Land Use Type	Educational	Residential	Office	Commercial
Parameter symbol (person·ha^{-1}·h^{-1})	T^e	T^r	T^o	T^c
Fitted relative value	15.9049	3.8221	26.9189	17.2681

4.2. Fit Validation of the Estimation Method

To test the explanatory power of the basic estimation models and validate the feasibility of the estimation method of passenger flow volumes from and to bus stops based on land use elements, a linear regression analysis was conducted between the surveyed values of passenger flow volumes of the 68 groups and the models' estimated values (the related data are shown in Table A2 of Appendix A), which can demonstrate the correlation between the two sets of values. The results are shown in Figure 9.

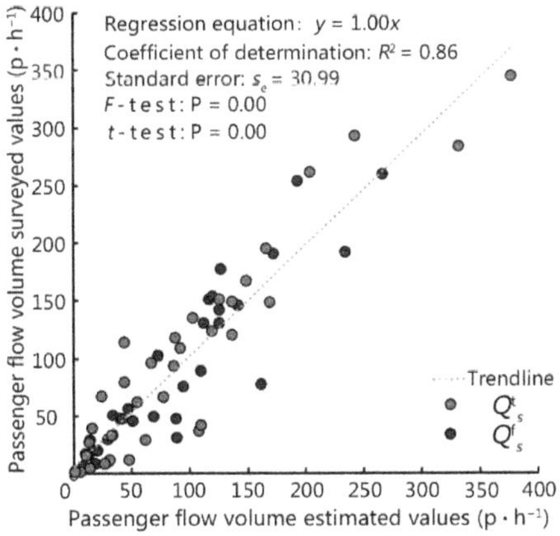

Figure 9. Regression of the surveyed values and estimated values of passenger flow volumes from and to bus stops.

The correlation analysis in Figure 9 shows a strong linear relationship between estimated and surveyed values, represented by the equation $y = 1.00x$. With an R^2 of 0.86, the model provides a good fit, indicating that estimated values have a high explanatory power for surveyed values. Additionally, both the F-test ($F = 1007.25$, $p < 0.05$) and t-test ($t = 31.74$, $p < 0.05$) confirm the validity of the linear regression model. These regression tests suggest the feasibility of the estimation method of passenger flow volumes from and to bus stops based on land use elements proposed in this paper. In essence, there exists a general relationship between bus travel and land use around bus stops and along bus routes, and passenger flow volumes from and to bus stops can be understood based on logical relationships between land use types, intensities, and spatial distributions. Furthermore, the combination of area-based origin unit method and distance impedance of probability density method effectively transforms this relationship.

5. Discussion

The construction of the gravitational logic of bus travel at the land use level using the bus travel generation aggregation of area-based origin unit method and the bus travel distance impedance of probability density method offers a theoretical foundation and methodological support for fostering fair [48,49], orderly [50], efficient, sustainable [51], and intensified demand organization in TOD.

The bus travel relative generation parameters (corresponding to generation efficiency) of different land use types in unit area and the bus travel relative distance decay law (corresponding to distance impedance, related to bus travel distance service levels), provide an orderly framework for coordinated development between land use and bus travel. In station area TOD, the basic unit order logic includes, the land use type layout should be organized based on generation efficiency to meet the bus travel intensity, and the same type of land use should follow a fair layout to reduce the impact of accessibility differences on travel efficiency. In citywide TOD, the order between basic units includes, the urban spatial functions should be relatively balanced to reduce pendulum-like commuting and the separation of residential and workplace areas, and the urban spatial structures should align with the operational characteristics (volume, speed, distance traveled, etc.) of different transportation systems to adapt to the structured upgrading of demand organization.

The construction of the estimation method of passenger flow volumes from and to bus stops provides actionable ideas for the straightforward analysis and optimization of TOD-related issues, as shown in Figure 10.

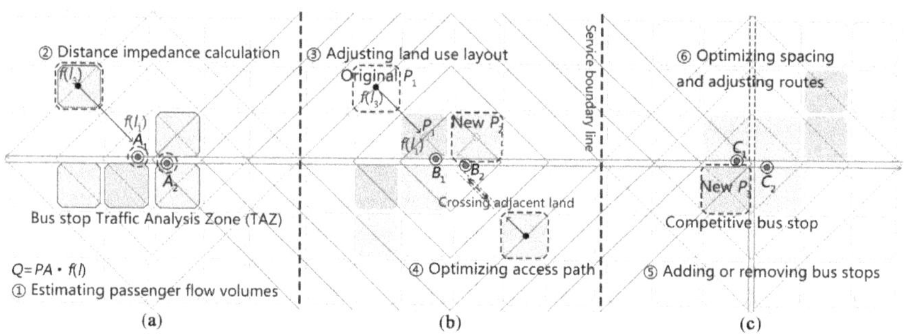

Figure 10. Application scenarios of the estimation method.

As depicted in Figure 10, within the constraints of the study area and time period and with the calibration of parameters such as production, attraction, and distance impedance in the basic estimation models, the application scenarios of the estimation method may include three main categories, as follows.

(a) Estimating passenger flow volumes within bus stop TAZs: When the relationship between land use around bus stops and along bus routes is known, but passenger flow volumes from and to bus stops are unknown, this estimation method can be used to estimate these volumes and diagnose potential deficiencies in passenger flow in terms of land use types, intensities, and spatial distributions.

(b) Updating land use within bus stop TAZs for station area TOD: For bus stops where the relationship between land use around bus stops and along bus routes is known, this estimation method can not only assess walking inhibition from and to the bus stops (due to inconvenient paths, unreasonable layout relationship, etc.) under the current land use layout but also evaluate improvements in passenger flow volumes after adjusting land use layouts and optimizing access paths. This facilitates organic land use renewal within bus stop TAZs from a bus travel perspective.

(c) Updating land use within upstream and downstream bus stops TAZs for citywide TOD: For bus routes where the land use generation and distance impedance are known, the estimation method can not only assess the benefits in passenger flow volumes from scenarios such as adding or removing bus stops, optimizing spacing, and adjusting routes but also consider the redistribution of passenger flow volumes for integrated development in the service area and structural upgrades to meet increased demand.

6. Conclusions

To unravel the general relationship between bus travel and land use around bus stops and along bus routes and to foster their coordinated development, this paper explores the estimation method of passenger flow volumes from and to bus stops (serving as a link between the bus system and its service area) based on land use elements. Through the analysis of the gravity model principle and method of estimating passenger flow volumes, the construction of basic estimation models corresponding to the gravity set based on land use elements and empirical testing of the estimation method, the main conclusions obtained are as follows.

Regression results demonstrate the feasibility of estimating passenger flow volumes based on land use elements. In other words, passenger flow volumes from and to bus stops can be derived from the land use types, intensities, and spatial distributions on both sides connected by these volumes by treating the land use elements as the gravity set, which is related to the supply and demand of passenger flow volumes. Decoding this general relationship provides a fair and orderly methodological framework for the coordinated development and strategic decision-making on bus travel and land use around bus stops and along bus routes, particularly in the context of TOD planning updates.

In contrast to existing complex and non-transferable analytic predictions concerning bus network relationships, this paper introduces the concept of passenger flow volumes from and to bus stops, deconstructing the point–line units corresponding to these volumes from bus network relationships to simplify the estimation of passenger flow volumes at the land use level. By using the area-based origin unit method, which hypothesizes the bus travel generation is closely correlated with land use types and their corresponding area scales, and the bus travel distance impedance of probability density method, which deciphers the decay probability of different bus travel distances under certain conditions, a portable estimation method based on gravity relationships is constructed and empirically verified.

While the function model has been tested and confirmed with relatively high confidence, ensuring the reliability of the estimation approach and method, there are limitations in its application and extension. The estimation model primarily simulates passenger flow volumes from and to bus stops during the morning peak period in the experimental area, so future research should expand the study area and time period for broader verification of the estimation method. Additionally, the simplified basic estimation models used for experimentation may suffer from issues such as poor representativeness of parameter values, inadequate transfer considerations, and insufficient prediction accuracy. To improve the model, it is necessary to minimize the interference of random factors in estimation of passenger flow volumes from and to bus stops and refine the aggregation of generation and distance impedance while improving the aggregation of transfer passenger flow volumes of bus travel.

Author Contributions: Conceptualization, J.C.; methodology, software, and validation, J.Z.; formal analysis, J.Z. and M.W.; investigation, resources, and data curation, J.Z., J.C. and W.Z.; writing—original draft preparation, J.Z., W.Z. and M.W.; writing—review and editing, J.Z. and J.C.; visualization, J.Z.; supervision, project administration, and funding acquisition, J.Z. and J.C. All authors have read and agreed to the published version of the manuscript.

Funding: This research was funded by the National Natural Science Foundation of China, grant 52278048, project Research on Theory and Intelligent Method of Road Network Quality Improving Planning from the Perspective of Dual Motive of City and Block. The APC was funded by the National Natural Science Foundation of China.

Data Availability Statement: The data presented in this study are available upon request from the first author due to the complexity and privacy concerns associated with the raw data. Summary data supporting the study results can be found in Appendix A.

Acknowledgments: The authors sincerely thank the editors and reviewers for their valuable comments on this manuscript.

Conflicts of Interest: The authors declare no conflicts of interest.

Appendix A

Table A1. Statistical data on building areas for four types of land use within four walking distance segments (km).

No.	M^e_i (ha)				M^r_i (ha)				M^o_i (ha)				M^c_i (ha)			
	(0.0, 0.2)	(0.2, 0.4)	(0.4, 0.6)	(0.6, 0.8)	(0.0, 0.2)	(0.2, 0.4)	(0.4, 0.6)	(0.6, 0.8)	(0.0, 0.2)	(0.2, 0.4)	(0.4, 0.6)	(0.6, 0.8)	(0.0, 0.2)	(0.2, 0.4)	(0.4, 0.6)	(0.6, 0.8)
1	0.00	0.00	1.77	0.00	2.24	2.43	3.39	2.64	0.00	0.34	0.11	0.27	0.00	0.16	0.00	0.00
2	0.00	0.00	0.00	0.00	2.64	2.06	2.41	3.25	0.06	1.57	0.09	0.15	0.13	0.426	0.00	0.00
3	0.00	1.60	4.29	8.95	9.81	10.62	12.95	0.00	1.64	0.01	2.58	0.18	2.52	0.30	0.00	2.48
4	0.00	1.70	9.31	13.14	6.30	10.91	15.23	4.10	1.64	0.01	1.05	2.76	2.31	0.54	0.04	0.00
5	0.96	11.93	1.18	5.11	5.91	6.23	11.59	7.73	0.99	0.00	1.14	0.00	0.04	0.17	0.00	0.00
6	2.51	7.13	4.27	0.61	1.90	8.21	14.84	4.17	0.99	0.00	1.14	0.00	0.00	0.11	0.10	0.32
7	0.00	5.82	2.54	0.00	5.08	23.00	9.90	7.29	1.04	1.99	0.96	0.11	1.32	0.76	0.24	0.07
8	0.00	0.79	12.35	12.76	0.00	14.21	26.22	14.52	0.29	2.76	0.00	0.962	3.31	1.89	1.71	0.18
9	0.00	0.79	12.35	12.76	0.00	14.21	26.22	14.52	0.29	2.76	0.00	0.96	3.31	1.89	1.71	0.18
10	0.00	5.82	2.54	0.00	5.08	23.00	9.90	7.29	1.04	1.99	0.96	0.11	1.32	0.76	0.24	0.07
11	0.00	0.79	12.35	12.76	0.00	14.21	26.22	14.52	0.29	2.76	0.00	0.96	3.31	1.89	1.71	0.18
12	0.00	0.79	12.35	12.76	0.00	14.21	26.22	14.52	0.29	2.76	0.00	0.96	3.31	1.89	1.71	0.18
13	1.42	5.37	10.12	15.25	2.22	9.92	22.37	12.44	0.00	0.20	2.74	0.26	3.76	2.86	0.29	0.49
14	1.42	5.37	10.12	15.25	2.22	9.92	22.37	12.44	0.00	0.20	2.74	0.26	3.76	2.86	0.29	0.49
15	1.42	5.37	10.12	15.25	2.22	9.92	22.37	12.44	0.00	0.20	2.74	0.26	3.76	2.86	0.29	0.49
16	2.21	9.56	13.39	14.13	6.45	8.69	4.71	0.00	0.00	0.10	0.00	0.00	2.42	0.89	0.18	0.00
17	1.47	6.75	11.45	11.94	6.08	9.40	4.38	1.06	0.00	0.22	0.00	1.83	1.37	1.85	0.21	0.00
18	4.16	7.41	8.75	0.66	2.88	9.01	3.75	4.55	0.00	0.10	0.00	3.97	1.39	4.28	1.95	0.38
19	2.83	8.10	2.47	1.62	6.95	9.77	5.74	2.44	0.00	0.10	0.00	2.047	2.84	2.77	1.95	0.38
20	0.00	7.85	13.47	4.14	10.27	10.02	9.09	14.94	0.00	0.00	0.00	2.05	0.18	2.80	3.43	0.73
21	0.00	7.85	13.47	4.14	10.27	10.02	9.09	14.94	0.00	0.00	0.00	2.05	0.18	2.80	3.43	0.73
22	1.02	5.00	3.25	7.83	6.76	5.33	13.39	3.87	0.00	0.00	0.16	2.46	0.18	0.00	0.19	0.11
23	1.02	5.00	3.25	7.83	6.76	5.33	13.39	3.87	0.00	0.00	0.16	2.46	0.18	0.00	0.19	0.11
24	4.63	1.11	0.00	0.00	0.86	6.35	2.26	5.65	0.00	0.04	0.45	1.04	0.06	0.32	0.11	0.05
25	3.41	1.92	4.73	6.02	4.38	11.47	6.62	5.78	0.04	0.04	2.67	0.58	0.06	0.71	0.08	0.05
26	0.00	0.00	3.53	2.41	5.18	2.11	0.00	0.00	0.20	0.19	0.00	0.00	0.09	0.39	0.00	0.00
27	0.00	0.00	3.53	2.41	5.18	2.11	0.00	0.00	0.20	0.19	0.00	0.00	0.09	0.39	0.00	0.00
28	0.00	1.77	2.21	1.12	4.38	4.01	0.92	3.98	0.10	0.97	0.06	1.60	0.05	0.17	1.92	2.55
29	0.00	1.77	2.21	1.12	4.38	4.01	0.92	3.98	0.10	0.97	0.06	1.60	0.05	0.17	1.92	2.55
30	0.14	9.90	15.06	15.25	3.66	11.93	5.64	9.48	0.04	0.14	1.17	0.22	0.71	0.11	0.00	0.12
31	0.14	9.90	15.06	15.25	3.66	11.93	5.64	9.48	0.04	0.14	1.17	0.22	0.71	0.11	0.00	0.12
32	1.32	1.62	0.07	0.00	6.21	11.03	9.97	8.79	1.37	1.64	0.57	1.73	0.20	0.26	0.25	1.74
33	1.32	1.62	0.07	0.00	6.21	11.03	9.97	8.79	1.37	1.64	0.57	1.73	0.20	0.26	0.25	1.74
34	0.00	3.76	0.38	0.00	8.11	18.58	23.48	18.89	2.54	0.00	8.79	6.83	0.60	2.28	1.17	0.20
35	1.39	2.75	0.00	0.00	8.84	26.21	19.23	5.42	2.54	3.36	11.26	1.53	1.71	1.75	0.35	0.11
36	1.39	2.75	0.00	0.00	8.84	26.21	19.23	5.42	2.54	3.36	11.26	1.53	1.71	1.75	0.35	0.11
37	0.00	3.01	1.13	0.00	5.99	20.29	16.50	15.82	2.54	0.00	8.38	4.98	0.60	2.69	0.53	0.09
38	0.00	3.01	1.13	0.00	5.99	20.29	16.50	15.82	2.54	0.00	8.38	4.98	0.60	2.69	0.53	0.09

Table A2. Surveyed values of passenger flow volumes and models' estimated values (including aggregate data from two basic models).

No.	$M^{e}_{i}\cdot S_i$ (ha)	$M^{r}_{i}\cdot S_i$ (ha)	$M^{o}_{i}\cdot S_i$ (ha)	$M^{c}_{i}\cdot S_i$ (ha)	A^{f}_{s} ($P\cdot h^{-1}$)	P^{f}_{s}	$A^{f}_{s}\cdot P^{f}_{s}$ ($P\cdot h^{-1}$)	Q^{f}_{s} ($P\cdot h^{-1}$)	P^{t}_{s} ($P\cdot h^{-1}$)	A^{t}_{s}	$P^{t}_{s}\cdot A^{t}_{s}$ ($P\cdot h^{-1}$)	Q^{t}_{s} ($P\cdot h^{-1}$)
1	0.13	2.16	0.09	0.03	13.26	0.99	13.2	17.0	15.49	0.86	13.3	27.5
2	0.00	2.27	0.37	0.17	21.62	0.86	18.5	8.5	31.19	0.99	31.0	12.0
3	0.89	9.03	1.17	1.63	108.39	1.59	172.7	191.0	133.00	1.82	242.5	293.0
4	1.40	7.28	1.13	1.49	106.09	1.82	193.4	254.0	128.29	1.59	204.3	262.0
5	3.30	5.89	0.67	0.06	94.19	1.72	161.9	78.0	103.89	1.63	169.0	149.0
6	3.33	4.07	0.67	0.04	87.22	1.63	141.8	146.0	96.53	1.72	165.9	195.0
7	1.41	8.78	1.11	0.97	102.64	0.87	89.0	31.0	126.33	0.86	108.2	37.0
8	1.42	5.32	0.78	2.49	106.91	1.17	125.5	131.0	116.44	1.03	119.6	124.0
9	1.42	5.32	0.78	2.49	106.91	1.18	125.7	142.0	116.44	0.88	103.0	135.0
10	1.41	8.78	1.11	0.97	102.64	0.86	87.9	48.0	126.33	0.87	109.5	42.0
11	1.42	5.32	0.78	2.49	106.91	1.03	109.8	89.0	116.44	1.17	136.7	149.5
12	1.42	5.32	0.778	2.49	106.91	0.88	94.5	76.0	116.44	1.18	136.9	120.5
13	3.12	5.40	0.25	2.87	126.75	0.57	72.5	103.0	—	—	—	—
14	3.12	5.40	0.25	2.87	126.75	0.92	117.1	151.0	117.18	1.07	125.7	151.0
15	3.12	5.40	0.25	2.87	—	—	—	—	117.18	0.57	67.0	96.5
16	4.69	6.02	0.02	1.64	126.52	0.89	112.2	130.5	110.89	0.70	77.3	67.0
17	3.46	5.95	0.09	1.22	101.28	1.26	127.1	177.5	92.08	1.61	148.6	167.5
18	4.70	4.01	0.12	1.88	125.90	0.55	68.8	49.5	111.75	0.24	26.6	8.5
19	3.61	6.68	0.07	2.43	126.87	0.24	30.2	10.5	113.10	0.55	61.8	29.0
20	2.76	9.28	0.05	0.97	—	—	—	—	90.61	0.53	47.6	12.0
21	2.76	9.28	0.05	0.97	97.55	0.53	51.3	46.0	—	—	—	—
22	2.11	6.24	0.08	0.12	61.53	0.18	11.0	15.5	58.91	0.55	32.3	33.0
23	2.11	6.24	0.08	0.12	61.53	0.55	33.7	34.0	58.91	0.18	10.5	15.5
24	2.99	2.16	0.07	0.11	59.57	0.57	34.1	51.0	53.65	0.10	5.4	2.0
25	2.94	5.66	0.25	0.19	78.32	0.10	7.9	7.0	77.52	0.57	44.3	80.0
26	0.33	3.52	0.16	0.14	25.21	0.04	1.0	2.0	28.70	0.56	16.1	39.5
27	0.33	3.52	0.16	0.14	25.21	0.56	14.1	29.0	28.70	0.04	1.1	2.0
28	0.57	3.62	0.31	0.28	—	—	—	—	42.40	0.57	24.1	67.5
29	0.57	3.62	0.31	0.28	35.97	0.57	20.4	20.5	—	—	—	—
30	3.68	5.35	0.14	0.45	—	—	—	—	84.40	0.52	44.2	114.0
31	3.68	5.35	0.14	0.45	90.62	0.52	47.4	56.5	—	—	—	—
32	1.13	6.98	1.25	0.24	82.36	0.49	40.3	47.5	112.21	0.12	13.9	5.0
33	1.13	6.98	1.25	0.24	82.36	0.12	10.2	17.0	112.21	0.49	54.9	62.5
34	0.82	10.97	2.34	0.93	134.10	0.22	29.4	30.0	191.55	0.48	91.7	109.0
35	1.41	12.34	3.09	1.41	177.19	1.50	266.5	260.0	251.15	1.32	332.3	284.0
36	1.41	12.34	3.09	1.41	177.19	1.32	234.5	192.0	251.15	1.50	377.8	344.5
37	0.72	9.47	2.26	0.97	125.23	0.34	42.3	48.0	180.50	0.48	86.5	94.0
38	0.72	9.47	2.26	0.97	125.23	0.96	119.9	154.0	180.50	0.49	87.6	118.0

References

1. Berawi, M.A.; Saroji, G.; Iskandar, F.A.; Ibrahim, B.E.; Miraj, P.; Sari, M. Optimizing Land Use Allocation of Transit-Oriented Development (TOD) to Generate Maximum Ridership. *Sustainability* **2020**, *12*, 3798. [CrossRef]
2. Hasibuan, H.S.; Mulyani, M. Transit-Oriented Development: Towards Achieving Sustainable Transport and Urban Development in Jakarta Metropolitan, Indonesia. *Sustainability* **2022**, *14*, 5244. [CrossRef]
3. Woo, J.H. Classification of TOD Typologies Based on Pedestrian Behavior for Sustainable and Active Urban Growth in Seoul. *Sustainability* **2021**, *13*, 3047. [CrossRef]
4. Stojanovski, T. Urban design and public transportation—Public spaces, visual proximity and Transit-Oriented Development (TOD). *J. Urban Des.* **2020**, *25*, 134–154.
5. Su, S.; Zhang, J.; He, S.; Zhang, H.; Hu, L.; Kang, M. Unraveling the impact of TOD on housing rental prices and implications on spatial planning: A comparative analysis of five Chinese megacities. *Habitat Int.* **2021**, *107*, 102309. [CrossRef]
6. Gan, Z.; Yang, M.; Feng, T.; Timmermans, H.P. Examining the relationship between built environment and metro ridership at station-to-station level. *Transp. Res. D Transp. Environ.* **2020**, *82*, 102332. [CrossRef]
7. Newman, P.; Davies-Slate, S.; Conley, D.; Hargroves, K.; Mouritz, M. From TOD to TAC: Why and How Transport and Urban Policy Needs to Shift to Regenerating Main Road Corridors with New Transit Systems. *Urban Sci.* **2021**, *5*, 52. [CrossRef]

8. Wang, B.; de Jong, M.; van Bueren, E.; Ersoy, A.; Meng, Y. Transit-Oriented Development in China: A Comparative Content Analysis of the Spatial Plans of High-Speed Railway Station Areas. *Land* **2023**, *12*, 1818. [CrossRef]
9. Carlton, I. Transit Planners' Transit-Oriented Development-Related Practices and Theories. *J. Plan. Educ. Res.* **2019**, *39*, 508–519. [CrossRef]
10. Ibrahim, S.M.; Ayad, H.M.; Turki, E.A.; Saadallah, D.M. Measuring Transit-Oriented Development (TOD) levels: Prioritize potential areas for TOD in Alexandria, Egypt using GIS-Spatial Multi-Criteria based model. *Alex. Eng. J.* **2023**, *67*, 241–255. [CrossRef]
11. Ibraeva, A.; de Almeida Correia, G.H.; Silva, C.; Antunes, A.P. Transit-oriented development: A review of research achievements and challenges. *Transp. Res. Part A Policy Pract.* **2020**, *132*, 110–130. [CrossRef]
12. Su, S.; Zhao, C.; Zhou, H.; Li, B.; Kang, M. Unraveling the relative contribution of TOD structural factors to metro ridership: A novel localized modeling approach with implications on spatial planning. *J. Transp. Geogr.* **2022**, *100*, 103308. [CrossRef]
13. Currie, G. Bus Transit Oriented Development—Strengths and Challenges Relative to Rail. *J. Public Transp.* **2006**, *9*, 1–21. [CrossRef]
14. Shen, Q.; Xu, S.; Lin, J. Effects of bus transit-oriented development (BTOD) on single-family property value in Seattle metropolitan area. *Urban Stud.* **2018**, *55*, 2960–2979. [CrossRef]
15. Pan, H.; Li, J.; Shen, Q.; Shi, C. What determines rail transit passenger volume? Implications for transit oriented development planning. *Transp. Res. Part D Transp. Environ.* **2017**, *57*, 52–63. [CrossRef]
16. Huang, Y.; Zhang, Z.; Xu, Q.; Dai, S.; Chen, Y. Causality between multi-scale built environment and rail transit ridership in Beijing and Tokyo. *Transp. Res. Part D Trans. Environ.* **2024**, *130*, 104150. [CrossRef]
17. Wegener, M. Land-Use Transport Interaction Models. In *Handbook of Regional Science*; Fischer, M., Nijkamp, P., Eds.; Springer: Berlin/Heidelberg, Germany, 2021; pp. 229–246. [CrossRef]
18. Ren, F.; Zhang, J.; Yang, X. Study on the Effect of Job Accessibility and Residential Location on Housing Occupancy Rate: A Case Study of Xiamen, China. *Land* **2023**, *12*, 912. [CrossRef]
19. Derakhti, L.; Baeten, G. Contradictions of Transit-Oriented Development in Low-Income Neighborhoods: The Case Study of Rosengård in Malmö, Sweden. *Urban Sci.* **2020**, *4*, 20. [CrossRef]
20. Lin, C.; Wang, K.; Wu, D.; Gong, B. Passenger Flow Prediction Based on Land Use around Metro Stations: A Case Study. *Sustainability* **2020**, *12*, 6844. [CrossRef]
21. Zhai, H.; Tian, R.; Cui, L.; Xu, X.; Zhang, W. A Novel Hierarchical Hybrid Model for Short-Term Bus Passenger Flow Forecasting. *J. Adv. Transp.* **2020**, *2020*, 7917353. [CrossRef]
22. Zhang, X.; Yan, M.; Xie, B.; Yang, H.; Ma, H. An automatic real-time bus schedule redesign method based on bus arrival time prediction. *Adv. Eng. Inform.* **2021**, *48*, 101295. [CrossRef]
23. Li, W.; Sui, L.; Zhou, M.; Dong, H. Short-term passenger flow forecast for urban rail transit based on multi-source data. *EURASIP J. Wirel. Comm.* **2021**, *2021*, 9. [CrossRef]
24. Wen, K.; Zhao, G.; He, B.; Ma, J.; Zhang, H. A decomposition-based forecasting method with transfer learning for railway short-term passenger flow in holidays. *Expert Syst. Appl.* **2022**, *189*, 116102. [CrossRef]
25. Liu, W.; Tan, Q.; Wu, W. Forecast and Early Warning of Regional Bus Passenger Flow Based on Machine Learning. *Math. Probl. Eng.* **2020**, *2020*, 6625435. [CrossRef]
26. Baghbani, A.; Bouguila, N.; Patterson, Z. Short-Term Passenger Flow Prediction Using a Bus Network Graph Convolutional Long Short-Term Memory Neural Network Model. *Transp. Res. Rec.* **2023**, *2677*, 1331–1340. [CrossRef]
27. Liu, L.; Chen, R. A novel passenger flow prediction model using deep learning methods. *Transp. Res. Part C Emerg. Technol.* **2017**, *84*, 74–91. [CrossRef]
28. Halyal, S.; Mulangi, R.H.; Harsha, M.M. Forecasting public transit passenger demand: With neural networks using APC data. *Case Stud.* **2022**, *10*, 965–975. [CrossRef]
29. Arhin, S.; Manandhar, B.; Baba-Adam, H. Predicting Travel Times of Bus Transit in Washington, D.C. Using Artificial Neural Networks. *Civ. Eng. J.* **2020**, *6*, 2245–2261. [CrossRef]
30. Lv, W.; Lv, Y.; Ouyang, Q.; Ren, Y. A Bus Passenger Flow Prediction Model Fused with Point-of-Interest Data Based on Extreme Gradient Boosting. *Appl. Sci.* **2022**, *12*, 940. [CrossRef]
31. Deepa, L.; Pinjari, A.R.; Nirmale, S.K.; Srinivasan, K.K.; Rambha, T. A direct demand model for bus transit ridership in Bengaluru, India. *Transp. Res. Part A Policy Pract.* **2022**, *163*, 126–147. [CrossRef]
32. Klar, B.; Lee, J.; Long, J.A.; Diab, E. The impacts of accessibility measure choice on public transit project evaluation: A comparative study of cumulative, gravity-based, and hybrid approaches. *J. Transp. Geogr.* **2023**, *106*, 103508. [CrossRef]
33. Lee, J.H.; Kim, J.W.; Lee, K.; Choi, M.Y. Generalized maximal entropy argument for the gravity law in human mobility. *Europhys. Lett.* **2020**, *132*, 48001. [CrossRef]
34. Zhang, N.; Wang, Z.; Chen, F.; Song, J.; Wang, J.; Li, Y. Low-Carbon Impact of Urban Rail Transit Based on Passenger Demand Forecast in Baoji. *Energies* **2020**, *13*, 782. [CrossRef]
35. Ranceva, J.; Ušpalytė-Vitkūnienė, R. Specifics of Creating a Public Transport Demand Model for Low-Density Regions: Lithuanian Case. *Sustainability* **2024**, *16*, 1412. [CrossRef]
36. Bhatt, D.; Minal. GIS and Gravity Model-Based Accessibility Measure for Delhi Metro. *Iran. J. Sci. Technol. Trans Civ. Eng.* **2022**, *46*, 3411–3428. [CrossRef]

37. Mashrur, S.M.; Lavoie, B.; Wang, K.; Habib, K.N. A Regional Multimodal Network Microsimulation (GTASim) for a Comprehensive Utility maximizing System of Travel Options Modelling (CUSTOM) in the Greater Toronto and Hamilton Area. *Procedia Comput. Sci.* **2023**, *220*, 110–118. [CrossRef]
38. Ding, J.; Zhang, Y.; Li, L. Accessibility measure of bus transit networks. *IET Intell. Transp. Syst.* **2018**, *12*, 682–688. [CrossRef]
39. Liu, X.; Chen, X.; Potoglou, D.; Tian, M.; Fu, Y. Travel impedance, the built environment, and customized-bus ridership: A stop-to-stop level analysis. *Transp. Res. Part D Trans. Environ.* **2023**, *122*, 103889. [CrossRef]
40. Dai, P.; Han, S.; Yang, X.; Fu, H.; Wang, Y.; Liu, J. Analysis of the Factors Affecting the Construction of Subway Stations in Residential Areas. *Sustainability* **2022**, *14*, 13075. [CrossRef]
41. Gao, K.; Yang, Y.; Li, A.; Qu, X. Spatial heterogeneity in distance decay of using bike sharing: An empirical large-scale analysis in Shanghai. *Transp. Res. Part D Trans. Environ.* **2021**, *94*, 102814.
42. Wu, X.; Lu, Y.; Lin, Y.; Yang, Y. Measuring the Destination Accessibility of Cycling Transfer Trips in Metro Station Areas: A Big Data Approach. *Int. J. Environ. Res. Public Health* **2019**, *16*, 2641. [CrossRef] [PubMed]
43. Chen, Y. The distance-decay function of geographical gravity model: Power law or exponential law? *Chaos Solitons Fractals* **2015**, *77*, 174–189. [CrossRef]
44. Shoup, D.C.; Shi, F.; Luo, Y.; Zhu, W.; Zhang, G. Truth in Transportation Planning. *Urban Transp. China.* **2022**, *20*, 49–59.
45. Rahman, M.; Yasmin, S.; Faghih-Imani, A.; Eluru, N. Examining the Bus Ridership Demand: Application of Spatio-Temporal Panel Models. *J. Adv. Transp.* **2021**, *2021*, 8844743. [CrossRef]
46. Zhang, J.; Cai, J.; Qi, M. Research on the spatiotemporal organization order of supply and demand based on the travel time/distance distribution law. *Urban Transp. China* 2024, *in press*.
47. Cai, J.; Liu, K.; Liu, L. Bus OD matrix estimation by VISUM model: Case of Xining of China. *J. Transp. Syst. Eng. Inf. Technol.* **2013**, *13*, 49–56.
48. Appleyard, B.S.; Frost, A.R.; Allen, C. Are all transit stations equal and equitable? Calculating sustainability, livability, health, & equity performance of smart growth & transit-oriented-development (TOD). *J. Transp. Health.* **2019**, *14*, 100584.
49. Turbay, A.L.B.; Pereira, R.H.M.; Firmino, R. The equity implications of TOD in Curitiba. *Case Stud.* **2024**, *16*, 101211. [CrossRef]
50. Knowles, R.D.; Ferbrache, F.; Nikitas, A. Transport's historical, contemporary and future role in shaping urban development: Re-evaluating transit oriented development. *Cities* **2020**, *99*, 102607. [CrossRef]
51. Liang, Y.; Du, M.; Wang, X.; Xu, X. Planning for urban life: A new approach of sustainable land use plan based on transit-oriented development. *Eval. Program Plann.* **2020**, *80*, 101811. [CrossRef]

Disclaimer/Publisher's Note: The statements, opinions and data contained in all publications are solely those of the individual author(s) and contributor(s) and not of MDPI and/or the editor(s). MDPI and/or the editor(s) disclaim responsibility for any injury to people or property resulting from any ideas, methods, instructions or products referred to in the content.

land

Article

Nature-Based Solutions Scenario Planning for Climate Change Adaptation in Arid and Semi-Arid Regions

Rifat Olgun [1,2,*], Chingwen Cheng [3] and Paul Coseo [2]

[1] Vocational School of Serik G-S. Sural, Akdeniz University, Antalya 07500, Türkiye
[2] The Design School, Arizona State University, Tempe, AZ 85287, USA; paul.coseo@asu.edu
[3] Stuckeman School, Pennsylvania State University, University Park, PA 16802, USA; cvc6428@psu.edu
* Correspondence: rifatolgun@akdeniz.edu.tr

Abstract: Extreme climatic conditions cause a decrease in ecosystem services, the disruption of the ecological balance, and damage to human populations, especially in areas with socially vulnerable groups. Nature-based solutions applying blue-green infrastructure (BGI) against these negative impacts of climate change have an important role in planning sustainable cities. This study aims to identify priority areas and develop scenarios and strategies for spatial planning to understand the tradeoffs in approaches and to maximize the benefits of ecosystem services provided by BGI in cities with arid and semi-arid climates, using Phoenix, Arizona, a swiftly urbanizing city in the Sonoran Desert, as the study area. Using GIS-based multi-criteria decision-making techniques and the Green Infrastructure Spatial Planning model integrated with the city's existing water structures, this study is conducted at the US census scale. The hotspots for BGI are mapped from the combined GIS-based multi-criteria evaluation and expert stakeholder-driven weighting. In the hotspots where priority areas for BGI in Phoenix are identified, the city center area with a high density of impervious surfaces is identified as the highest priority area. It is revealed that social vulnerability and environmental risks (flooding, heat) have a positive correlation in Phoenix, and stormwater management and the urban heat island are the criteria that should be considered first in BGI planning.

Keywords: climate change; nature-based solutions; blue-green infrastructure; spatial planning; scenario planning

1. Introduction

Ecosystems provide many benefits to human and non-human life. These benefits, which occur for free as part of the natural processes within ecosystems, are collectively termed ecosystem services [1,2]. Regulatory ecosystem services are crucial in reducing the impacts of climate change on cities and increasing their ability to withstand these impacts [3,4], such as mitigating climate change-induced flooding hazards [5]. Among the regulating services, key services include managing waste, enhancing the air and soil quality, mitigating natural disasters like floods and landslides, regulating the climate, controlling diseases, facilitating pollination, purifying water, and controlling harmful species [6]. However, the worldwide population growth and the consequent expansion of urban settlements are causing great damage to ecosystems [7]. This makes cities vulnerable to the impacts of climate change [8] and contributes to the climate justicescape, where vulnerable communities with a lack of green infrastructure investment are facing inequitable consequences from climate change impacts [9,10].

Cities are the areas where climate change is felt the most intensely [11]. Therefore, many social [9], economic [12–14], and environmental problems [15] arise in cities due to climate change. An increased urban heat island (UHI) effect [16], an increase in the number of hot days [17], increased air pollution [18], water shortages [19], changes in precipitation regimes [20], coastal floods, and floods caused by rainwater are among these problems [21].

Nowadays, about 30% of the world's population is exposed to these adverse climatic conditions for at least 20 days per year [22]. For instance, in just two years, 2021 and 2022, the devastating flooding disasters in Australia, China, India, Malaysia, Nigeria, Pakistan, and Western Europe, as well as the forest fires in Algeria, Argentina, Canada, Greece, Turkey, and the USA, resulted in the loss of thousands of lives and the destruction of homes as a result of the negative effects of climate change [23,24].

To prevent these problems in cities, researchers, decision makers, and planners have put forward green infrastructure (GI) systems that provide nature-based solutions (NbS) as an alternative urban planning approach [25,26] to traditional command-and-control gray infrastructure approaches (e.g., concrete dams and channelized drainage for flood control) [27]. Cohen-Shacham et al. [28] define NbS as societal activities to safeguard, steward, and repair urban ecosystems, which, at the same time, both effectively and adaptively make progress toward other societal challenges (e.g., social justice). In this way, nature-based solutions, of which GI is an example, align human well-being with biodiversity benefits simultaneously. Accordingly, countries in the United States [5,29], the United Kingdom [30], Europe [31], China [32], and other countries in Asia [8] have developed strategies and policies for the planning and implementation of GI systems. Towards the end of the initial decade of the twenty-first century, the BGI planning paradigm emerged, which, unlike prevailing planning methodologies, aims to plan open green spaces, natural and semi-natural areas (cultural landscapes), rivers and valleys, and natural corridors within urban environments as a cohesive system, combining GI with water management and aiming for a nature-based water cycle, as well as the benefits of green spaces [33]. BGI systems are a holistic combination of sustainable ecological functions and cultural values. Scientific studies have shown that when BGI is integrated into the urban structure at different levels (micro, medium, and macro levels), it increases the resilience of cities in dealing with the negative impacts of climate change and provides significant environmental, social, and economic advantages compared to gray infrastructure [12,33].

The geographical structure of regions is important in developing different planning scenarios and strategies [34]. For this reason, BGI systems are planned for various ecosystem services in different regions [35]. For example, BGI planning strategies have been developed to improve the local climatic conditions in Stuttgart (Germany) and Melbourne (Australia); reduce runoff in Seattle (USA), Portland (USA), and Rotterdam (the Netherlands); and protect the existing water resources in Durban (South Africa) [36,37]. Generally, research on the planning and design of BGI systems has been carried out in areas with heavy rainfall. Studies to control excess rainwater [38], reduce runoff, reduce the burden of gray infrastructure [20], and provide ecological connectivity [39] between existing green spaces in these regions are seen in the literature and practice [40,41]. However, very few studies have demonstrated the planning and implementation of BGI systems in arid and semi-arid regions. These studies in the literature especially focus on the use of green spaces to reduce temperatures and mitigate the urban heat island effect [6,15,42]. However, we have not encountered a BGI study in the literature that evaluates various criteria in a holistic approach through GIS-based multi-criteria evaluation and expert stakeholder-driven weighting in desert cities with high drought levels.

This study was prompted by the following research question: How can we identify strategic hotspots to maximize the multifunctionality of BGI in cities in arid or semi-arid regions at a regional scale and where are the priority areas for BGI? Accordingly, the aim of this study is to identify priority areas and develop scenarios and strategies for spatial planning to understand the tradeoffs in approaches and to maximize the benefits of ecosystem services provided by BGI in cities that have arid and semi-arid climates. In particular, evaluating different scenarios can provide decision-makers with a better understanding of how to anticipate governance and plan for uncertainty [43]. In this context, it is necessary to focus on more benefits in line with the requirements of the region, rather than focusing on only one benefit of BGI. For this reason, a GIS-based model that can analyze many criteria and combine these multiple criteria is useful for decision-making [44].

In this study, we adopt the Green Infrastructure Spatial Planning (GISP) model, developed by Meerow and Newell [29] for this purpose and applied to the city of Detroit, Michigan and in follow-up research by Meerow [25,45] to the assessment of the GI of Manila (Philippines), Los Angeles (USA), and New York City (USA). In addition, Chang et al. [46] used GISP to assess the GI of the Yanshuei River Basin (Taiwan). In some studies, GI planning frameworks have been created by integrating various spatial factors with the GISP model [47,48].

The City of Phoenix, Arizona, USA was selected as the study area, where the negative impacts of climate change, such as extreme heat and drought conditions, are occurring more frequently and intensely [49], as well as increasing summertime temperatures combined with urban-induced heat [50], water availability issues [51], and monsoon flash floods [52]. Thus, officials are increasingly concerned about green space [53] and equitable access to high-quality green space to cope with climate change impacts. In addition, microbursts accompanying thunderstorms in the city and small-scale destructive winds damaging trees and other green infrastructure are also influential factors for the study area [54,55].

2. Materials and Methods

2.1. Study Area

The City of Phoenix, the capital of Arizona, USA, is located in the northeastern part of the Sonoran Desert. It is the fifth-largest city in the United States by population [50] and the country's most populous state capital, with a population of 1,608,139 and a population density of approximately 8039.3 people per square kilometer [56] (Figure 1). Agricultural activities were intensively developed from the city's foundation in the late 1800s until the second half of the twentieth century. The city had a relatively low population growth rate since its establishment for many years [57]. However, with the development of air conditioning, along with advancements in technology, it has experienced rapid population growth and urbanization in recent decades. Today, it is one of the largest and fastest-growing metropolitan areas in the United States [52].

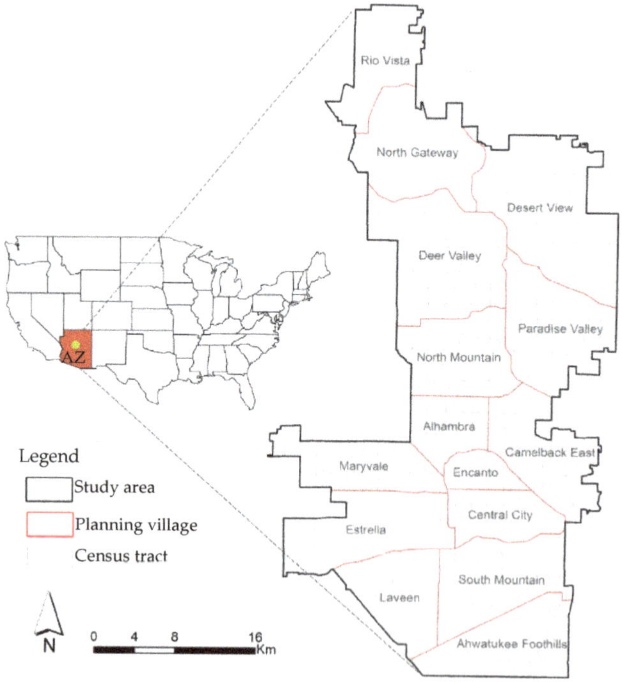

Figure 1. Map of the study area boundary of the City of Phoenix, AZ, USA.

Phoenix is one of the hottest cities in the United States due to its location in an arid and semi-arid region with mild winters and hot summers [49]. The 30-year average daily maximum temperature is above 38 °C for about 110 days of the year [53]. The average highest temperature is 41.3 °C in July, but the climate is rapidly shifting to a hotter regime. The 1991–2020 average summertime temperature (June–August) for Sky Harbor Airport, the official reporting station for Phoenix, was 27.9 °C. However, the summer of 2020 had an average temperature of 35.9 °C, shattering the previous hottest summer on record for 2015 with an average temperature of 35.1 °C [58]. By 2023, Phoenix broke the 2020 record with an average temperature of 36 °C (June–August). The winter months bring relief from the heat and the climate is mild, where the average daily high temperature is above 17.6 °C and the average daily low temperature is below 3.9 °C. Most of the rainfall occurs during the winter season and the monsoon season (15 June to 30 September). The average annual precipitation for the last two decades (2003–2023) is 163.3 mm (6.43 in) for Phoenix [58]. Approximately half of the average annual precipitation falls during the monsoon period [59] (Suchy et al., 2020). Flash floods during the monsoon period cause damage to people and property [52,58,60]. At the same time, increasing impervious surfaces due to urbanization have increased the urban heat island effect in Phoenix, causing the average daily air temperature to rise by 3.1 °C and the night-time minimum temperature to rise by 5 °C over the last 52 years [50,61].

2.2. Methods

This study employed the GISP model to analyze spatial data (i.e., social, topography, land cover–land use, zoning plan, weather, air pollution, and drainage) and identify "hotspots" where multiple BGI benefits are required based on ecosystem services. In addition, the BGI scenarios' tradeoffs and synergies between planning priorities were evaluated using the US census tract as a unit of analysis. The GISP model utilizes GIS-based multi-criteria decision-making techniques with six planning criteria identified: managing stormwater, reducing social vulnerability, increasing access to park areas, reducing the urban heat island effect, improving the air quality, and increasing the landscape connectivity (Table 1).

Table 1. The criteria and data sources utilized for the GISP model. Modified from Meerow and Newell [29].

GISP Model Criterion	Abbreviation	Criterion Database	Description
1. Managing Stormwater	MS	Flood susceptibility mapping using GIS-AHP multi-criteria analysis - Topographic wetness index - Elevation - Slope - Precipitation - Global man-made impervious surface dataset - Normalized difference vegetation index (NDVI) - Distance from river - Distance from road	The analytic hierarchy process (AHP) method is used to assign weights to all data. The resulting data are then overlaid, and the values for each census tract are averaged.
2. Reducing Social Vulnerability	SoVI	Heat vulnerability index (HVI) Social vulnerability index (SoVI)	The SoVI with 27 social variables and the HVI with 10 social variables created by Wright et al. [62] from the 2012–2016 American Community Survey.
3. Increasing Access to Park Areas	PA	The existing public parks (open street map, zoning plan) Census data	Population-weighted distance to the nearest park [25,63].

Table 1. *Cont.*

GISP Model Criterion	Abbreviation	Criterion Database	Description
4. Reducing the Urban Heat Island Effect	UHI	Land surface temperature (LST) and UHI	Estimates of average LST and UHI for each census tract obtained from Landsat for the last 10 years (years 2013, 2018, and 2023).
5. Improving Air Quality	AQ	Air quality index (O_3, PM2.5, PM10, CO, SO_2, NO_2)	Air quality index between 2013 and 2022 is used [64].
6. Increasing Landscape Connectivity	LC	The physical continuity of vegetated areas	The high-resolution (1 m) Lidar image created by Zhang and Turner II [65] and the "Patch Cohesion Index" metric in the Fragstats 4.2 software [66].

2.2.1. Managing Stormwater

The determination of flood areas in stormwater management is important in the preparation of strategic urban plans. In this context, the "Flood Susceptibility Mapping using GIS-AHP Multi-Criteria Analysis" method was used to determine flood areas. A database for the management of stormwater was created by using ten different data belonging to the study area (Table 1). Alos Palsar's Digital Elevation Model (DEM) data with a 12.5 m spatial resolution were used to create the data infrastructure for the topographic structure of the land, such as the slope, elevation, aspect, and surface flow direction "https://asf.alaska.edu (accessed on 12 September 2023)". Road information for the study area was obtained from the United States government's open data website "https://data.gov/ (accessed on 21 October 2023)". Data on water structures were obtained from the AZGeo Data Hub developed by the Arizona Geographic Information Council and Arizona State Land Department "https://azgeo-data-hub-agic.hub.arcgis.com/ (accessed on 17 September 2023)". The monthly average precipitation data of the study area for the last 10 years (2014–2023) were obtained from measurements at 7 different stations (Phoenix Airport, Litchfield Park, Youngtown, Phoenix Deer Valley Municipal Airport, Carefree, Scottsdale Municipal Airport, Tempe ASU) located in and around the study area [58]. Using the precipitation data obtained, the precipitation map of the study area was created with the inverse distance weighted (IDW) technique, which is frequently used in the creation of precipitation maps. The equations and analysis steps used to generate the TWI, GMIS, NDVI, and drainage density are shown in Supplementary File S1. In the last stage, after the database was created, each piece of data was weighted in the database with the analytic hierarchy process, one of the multi-criteria decision-making methods. Then, flood areas were determined by overlaying the data according to the weight values with the ArcGIS 10.5 software.

2.2.2. Reducing Social Vulnerability

The most widely used method for this is the social vulnerability index (SoVI). In this study, the SoVI with 27 social variables and the heat vulnerability index (HVI) with 10 social variables (see Supplementary File S1), created by Wright et al. [62] from the 2012–2016 American Community Survey (ACS), were used to identify the areas with a high concentration of socially disadvantaged and vulnerable groups within the study area. Then, the correlation between these two indices was analyzed, and the social vulnerability value was calculated for each census tract.

2.2.3. Increasing Access to Park Areas

A population-weighted approach was used to calculate the service areas of park areas. Census tract data produced by the US Census Bureau for 2020 were used to determine the population data. First of all, the population density in the study area was evaluated with the kernel density method. The population data of the census tract were distributed equally to the existing buildings within the tract in the zoning plan. To calculate the walking distances of the users, the roads that provided access to green areas were determined. In this context, highways that did not allow walking were removed from the database, and streets and roads that were under 35 miles per hour (mph) were used. To calculate more accurate accessibility values, the entry points of all park areas were identified and entered into the road database. The accessibility/service areas of park areas were determined in the ArcGIS 10.5 software using the GIS-based network analysis modules. Finally, the population within the total service areas of the parks and the percentage of the population within the service areas of the parks were calculated. The equations used to generate the census tract park total service area population and percentage of population covered by the park service are shown in Supplementary File S1.

2.2.4. Reducing the Urban Heat Island Effect

BGI systems are important in cooling the region and reducing the UHI effect. Therefore, determining areas with high land surface temperature and UHI values in the planning of BGI ensures the success of the plans. In determining the UHI effect, firstly, the land surface temperature (LST) needs to be determined. An LST analysis was performed to determine the change in the LST over 10 years and to determine the current situation. For the analysis, Landsat 8 (18 June 2013 and 2 July 2018) and Landsat 9 (8 July 2023) satellite images were downloaded from the USGS. While downloading the satellite images, it was ensured that the cloudiness of the data was less than 10% and that the data belonged to the period when the temperature was high. After the pre-processing of the satellite images, the LST analysis was performed. In the last stage, the relationship between the LST and NDVI was analyzed by correlation analysis, and the stack profile of the UHI was derived (of which details are shown in Supplementary File S1).

2.2.5. Improving Air Quality

The 10-year air quality index (2013–2022) data from six stations, produced by the United States Environmental Protection Agency (EPA), were used to evaluate the air quality in the study area [64]. Since the air quality index (AQI) takes into account six different air pollutants measured within the geographical area, it is a general indicator of the air quality of the region. Air pollutants considered in the calculation of the AQI are ozone (ppm), PM2.5 ($\mu g/m^3$), PM10 ($\mu g/m^3$), CO (ppm), SO_2 (ppb), NO_2 (ppb) (Supplementary File S1). An air quality map of the study area was created with the IDW interpolation technique, which used a linearly weighted combination of data from these measurement stations.

2.2.6. Increasing Landscape Connectivity

In addition to the social and technical benefits of GI, ecological benefits are also important for urban areas. Urban green areas provide habitats for many species. However, with increasing urbanization, these areas become fragmented and decrease. The high-resolution (1 m) Lidar image prepared by Zhang and Turner II [65] was used to identify and classify the land cover in the study area. After pre-processing the Lidar image in the ArcGIS 10.5 software, the study area was classified into 8 land classes. The map contained eight land cover classes, including (1) buildings, (2) asphalt, (3) bare soil and concrete, (4) trees and shrubs, (5) grass, (6) water, (7) active cropland, and (8) fallow (these classes are defined in Supplementary File S1).

Different software programs measure both the structural and functional connectivity in a region. In this research, the Fragstats 4.2 software, which has been used in many scientific studies, and the validity of the data that it produces is scientifically accepted, is preferred. With the "Patch Cohesion Index" metric in the Fragstats 4.2 software, the physical connectivity of habitat patches in the research area was evaluated at the landscape scale.

To be able to compare the maps (for the six criteria) obtained as a result of the analysis of the GISP criteria, a linear scale transformation ("score range") was applied to the measurement scales [29,47,67]. Thus, all criteria scores ranged from zero to ten.

2.3. GIS-Based Multi-Criteria Decision-Making Analysis

A survey form was prepared to obtain the opinions of experts working on GI and BGI systems around the world on the prioritization of the GISP model criteria (Supplementary File S3). After ethics committee approval (approval date 28 November 2023), the online survey form was sent to approximately 81 experts, and 39 experts, who were city planners, professional planners, and landscape architects, responded. In the survey form, they were asked to perform a pairwise comparison and weight the relative importance of the 6 criteria of the GISP model using three different methods (rating, ranking, and pairwise comparison).

The data from the pairwise comparison survey were analyzed with an Excel-based AHP calculator [68] and the Expert Choice V11 software. Therefore, the weight value of each criterion in the model was determined. Then, a hotspot map for the planning of GI was developed using the weighted linear combination technique and the weights of the criteria produced by the experts.

Quantifying the correlations between the GISP model criteria is important in determining the criteria' potential relationships. In this context, Pearson's bivariate correlation was used to reveal the spatial tradeoffs and synergies between the model criteria for all census tracts in Phoenix. The IBM SPSS 23 Statistics and JMP Statistics software (https://www.jmp.com/en_us/software/data-analysis-software.html) were used to analyze and create the graphs.

3. Results

This section summarizes the results of the GISP model criteria analysis of Phoenix. Maps and tables for each criterion obtained from the analysis results are shown in Supplementary File S2.

3.1. Managing Stormwater

To determine the flood risk areas of Phoenix and its flood susceptibility, the identified criteria were analyzed. The Camelback East, Paradise Valley, and North Mountain planning villages of Phoenix have a higher vegetation density compared to other planning villages. At the same time, due to the agricultural areas in the west of the Laveen and the Estrella planning villages, the density of green areas is also high in these villages. However, it is seen that the percentage of impervious (PI) surfaces is high in densely populated areas. In the Escanto and Central City planning villages, where industry, trade, and Phoenix Sky Harbor International Airport are located, the drainage density is high due to the high PI values. The TWI values are also high in these areas, with low slopes and dense settlements. Phoenix's land elevation increases from south to north. However, South Mountain Park and Preserve in the south of the city has the highest elevation in the south. It is also observed that the average annual precipitation increases as one moves northward depending on the elevation. Phoenix does not have a very sloping land structure. Most of the area is in the 0–5% slope range. Phoenix has an extensive road network that provides connectivity in and around the city. Within the scope of the study, heavily used roads in the city's extensive road network were evaluated. In this context, the city center and the area where the airport is located are under the influence of roads. Phoenix has rivers and water channels of different widths and lengths, with various flow rates, both continuous and seasonal. In this regard, most of the city is located within a four-kilometer radius of rivers and canals. The soil structure of the study area is in the HSG-C group. The areas north of the Ahwatukee Foothills, east of Desert View, southwest of the North Gateway, and northwest of Deer Valley have soil structures in the HSG-B group. There is a group of soils with high surface runoff if drainage is not provided, especially in the area around the Salt River that runs through the Central City, South Mountain, Laveen, and Estrella planning villages.

The flood susceptibility of Phoenix is analyzed using GIS-based multi-criteria decision analysis. At this stage of the analysis, the flood susceptibility criteria and sub-criteria are weighted according to the information obtained from the literature. Then, all criteria are overlaid according to these weight values (which are shown in Supplementary File S2). According to the results of the analysis, the flood risk is high in areas with densely populated areas. The high PI values, low vegetation density, and high topographic wetness values increase the risk of flooding in these areas (Figure 2).

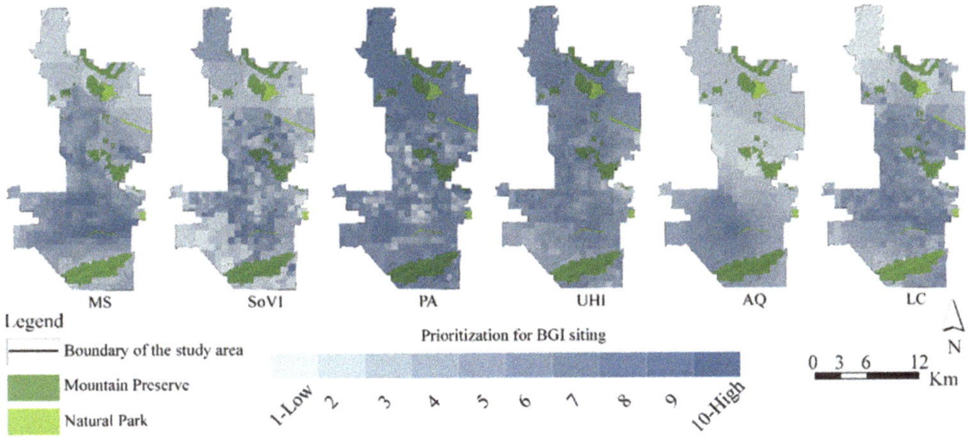

Figure 2. Relative prioritized maps of GISP model criteria for each census tract in Phoenix, AZ.

3.2. Reducing Social Vulnerability

Census tracts with high values of social vulnerability are present on both sides of the east–west (I-10) and north–south (I-17) interstate highways crossing through Phoenix. The North Mountain and Alhambra planning villages in the study area have high SoVI values. There is also a high density of socially vulnerable populations in the South Mountain region. In the Central City planning village, there are census tracts with high SoVI values around Phoenix Sky Harbor International Airport.

The HVI has higher values in the center of Phoenix, in areas with industrial and commercial centers. At the same time, the HVI is higher in areas with high percentages of ethnic minorities and elderly residents, low incomes, low vegetation density, and high surface temperatures. The HVI values are particularly high north of South Mountain, southwest of Central City, east of the Estrella and Laveen planning villages, and east of the Alhambra and North Mountain planning villages.

The relationship between the SoVI and HVI was evaluated by correlation analysis. According to the results of the analysis, a weakly positive (r = 0.359) and significant ($p < 0.05$) relationship was found between the SoVI and HVI. An increasing SoVI value increases the HVI value. Similarly, an increasing HVI value leads to an increasing SoVI value. Accordingly, the SoVI and HVI values of the census tracts were combined with the same impact ratio to determine the social vulnerability of the study area (see Supplementary File S2).

3.3. Increasing Access to Park Areas

It is observed that the population density (persons per km^2) is high in the Maryvale planning village. This planning village is followed by Alhambra, Central City, and the southern region of North Mountain. When the distribution of park areas is evaluated, the public park areas in Phoenix are not planned and implemented in an equal and fair manner throughout the city. Park areas are unevenly distributed across the city. This situation shows that individuals living in the city do not benefit from park areas equally. Not all of the population in the study area has access to park areas within a 10 min (800 m) walking distance. However, the Central City planning village has few residential areas due to the presence of industrial areas, commercial centers, and the airport. Therefore, approximately 92% of the existing residential areas in this region have access to park areas.

3.4. Reducing the Urban Heat Island Effect

The results of the analysis show that the LST of the study area varies between regions according to the vegetation structure/density of the city and the presence of impervious surfaces. Especially in areas with a dense urban texture, the surface temperature is quite high. It is seen that, due to the increase in urbanization and the expansion of settlement areas over time, the impervious surfaces have increased, and the land surface temperature values have also increased. Due to the ecological characteristics of Phoenix, which does not have dense natural vegetation cover (it has extensive bare soil), the surface temperature has high values in the regions outside the settlements. In addition, when the study area is evaluated in terms of the UHI effect, it is seen that the UHI is high in the city center.

In this context, the relationship between the land surface temperature and vegetation presence/density of Phoenix (years 2023, 2018, 2013) is evaluated by correlation analysis. According to the results of the analysis, it is concluded that there is a negative correlation (r: -0.466, $p < 0.001$ for 2023; r: -0.582, $p < 0.001$ for 2018; r: -0.523, $p < 0.001$ for 2013) between the vegetation structure and land surface temperature. In this case, as the presence/density of green areas increases, the land surface temperature decreases. When the LST and NDVI maps are evaluated together, the cooling effect of green areas on the land surface is realized not only at the location but also in its immediate surroundings (see Supplementary File S2).

3.5. Improving Air Quality

Phoenix's topography impacts the region's air pollution [69]. The region has a valley structure surrounded by mountains from the north and south. This situation prevents the polluted air from moving away from the region with the wind effect [70]. At the same time, the fact that the region has dry air with low humidity and little rainfall causes particles in the air to accumulate. Another cause of air pollution in Phoenix is temperature changes [42]. The reason for the high levels of air pollution, especially in the region around the Salt River, is that when the temperatures drop, cold air flows southward from the mountains to the lower reaches of the valley [71,72]. This moving air causes pollutants to accumulate in the Salt River zone in the south of the study area.

3.6. Increasing Landscape Connectivity

The patch cohesion index for vegetated land cover is calculated for each tract in the study area and it is assumed that these areas would provide habitats for a wide range of urban-dwelling species. In Phoenix, the ecological vulnerability is high in areas of dense urbanization, especially near the interstate highways and the airport. The patches in this region are smaller and poorly connected. These fragmented landscape patches are isolated and more vulnerable. Due to the Natural Park and Mountain Preserve areas to the north, south, and east of the research area, the habitat areas/ecosystem patches in these areas have a more holistic structure. The connectivity between the patches in this region is stronger than in residential areas. However, the areas of high ecological vulnerability around the rivers and canals in Phoenix contain flat, wide, and bare soils [67]. At the same time, these areas are located close to the toxic release inventory areas, which increases this sensitivity. In this context, wetlands, around rivers and canals, do not contribute much in terms of altering the pattern of ecological vulnerability [67] and do not provide any benefit in reducing the ecological fragility.

In the last stage, a linear scale transformation ("score range") was applied to the measurement scales to be able to compare the maps (for six indicators) obtained as a result of the analysis of the GISP criteria. In this way, all criteria scores ranged from 0 to 10. The six different criteria maps for each criterion are shown in Figure 2. Each map shows the relative prioritization of census tracts in Phoenix according to the criteria set for BGI planning. Darker-shaded tracts on the maps represent areas with higher priority for BGI development based on the model. According to the maps, the priority areas for each of the BGI planning criteria differ.

4. Discussion

4.1. Synergies and Tradeoffs between Planning Criteria

Assessing the correlations between the GISP model criteria is important in identifying spatial tradeoffs and synergies for such efforts as scenario planning. The findings of the correlation analysis are shown in Figure 3. The colors represent whether the correlation between the criteria is positive (blue) or negative (brown), while the saturation of the colors shows the strength of the association. The Pearson correlation coefficients are written above the dots. The presence of an X in a box indicates that there is no statistically significant relationship between the criteria. The figure also includes kurtosis–skewness values, statistical significance values, and criterion distributions.

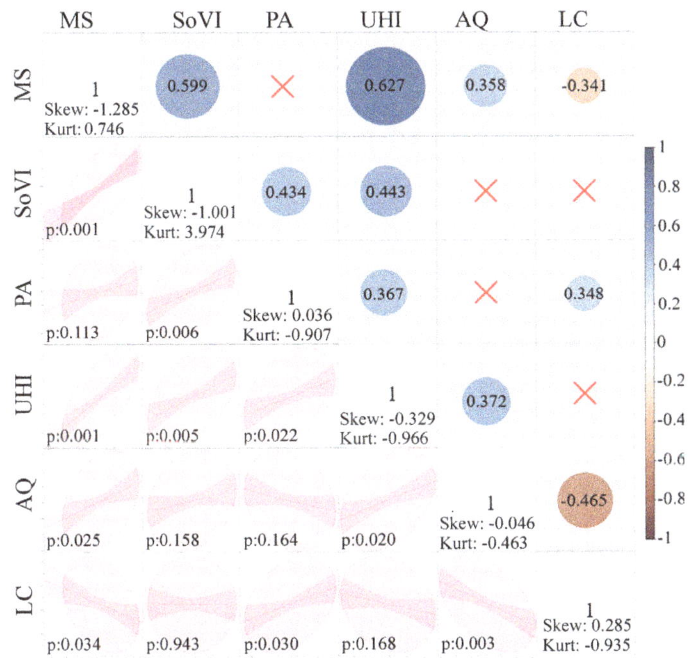

Figure 3. Statistical relationship between the planning criteria in the study.

Stormwater and the SoVI, UHI, and air quality have positive correlations (synergy) when the results of the correlation analysis between the criteria are considered. Considering that the problem of stormwater is experienced in regions with more residential areas and impervious surfaces, it is expected that the UHI value, air pollution, and SoVI value will be higher in these areas in Phoenix. There is a positive correlation between the SoVI, park area accessibility, and UHI. These three synergistic criteria show similarities within the urban structure. Areas with insufficient park area availability and limited park area accessibility have high UHI values and are socially vulnerable areas.

Decreasing the distance between park areas increases the connectivity between landscape patches. Thus, there is a positive correlation between the accessibility (or "availability") of park areas and landscape connectivity. The increase in landscape connectivity due to the availability of park areas mitigates the impact of air pollution. Therefore, there is a tradeoff between air pollution and landscape connectivity.

According to the survey results, there is no statistically significant relationship between the SoVI and air pollution. However, when the current structure is evaluated, there is a high population density of individuals with lower socioeconomic status and ethnic/racial minorities in the southern and western regions of Phoenix. Additionally, since the airport and industrial zone are located in this area, air pollution is higher in these areas than in the rest of the city. Grineski et al. [73], deBuys [54], and Kaus [74] emphasize that individuals living in these areas have a high rate of illness caused by exposure to air pollution. This fact shows that there is a synergy between air pollution and the SoVI.

4.2. Prioritization for BGI Planning

The expert stakeholders who participated in the survey rated the importance of the GISP criteria on a five-point scale from least to most important and ranked them in order of importance. Accordingly, in the rating of the planning criteria, stormwater management is seen as the most important criterion for BGI planning in cities. When the ranking of the planning criteria is evaluated, reducing the urban heat island effect is ranked first (Table 2).

Table 2. Results of the expert survey (N:39).

Criterion	Rating			Ranking			Pairwise Comparison Weights	
	Rank Order	Mean	Std. Deviation	Rank Order	Mean	Std. Deviation	Rank Order	Weight
MS	1	4.67	0.529	2	2.54	2.539	1	0.325
SoVI	3	4.43	0.552	3	3.31	1.217	4	0.112
PA	5	3.97	0.706	4	3.64	1.112	5	0.071
UHI	2	4.46	0.554	1	1.87	1.301	2	0.270
AQ	4	4.05	0.647	5	4.62	1.184	3	0.170
LC	6	3.71	0.825	6	4.79	1.852	6	0.052

MS: managing stormwater, SoVI: reducing social vulnerability, PA: increasing access to park areas, UHI: reducing the urban heat island effect, AQ: improving air quality, LC: increasing landscape connectivity.

In the survey, the experts performed pairwise comparisons of the planning criteria using the AHP technique. According to the results of the analysis of the pairwise comparison data, the criteria were weighted within themselves. In this context, stormwater management had the highest weight value, followed by UHI, AQ, SoVI, PA, and LC. Based on these weight values, the census tracts for each criterion were overlaid in the ArcGIS 10.5 software, resulting in the creation of a map identifying hotspots for BGI planning (Figure 4).

Being in a hot and arid climate and poor-draining soil conditions, Phoenix receives low annual precipitation, yet the amount of precipitation is not uniformly distributed at certain intervals, which contributes to the instantaneously high rainfall intensity, resulting in flash floods in urbanized areas. Therefore, nature-based structures (bioretention systems, bioswales, etc.) that will reduce stormwater runoff, harvest rainwater, and facilitate groundwater recharge should be planned in areas with high flood risks, especially in urban centers where the number of impervious surfaces is high. This will both reduce the flooding risk and provide a precaution against water scarcity [75], which will occur in the coming years due to climate change and increased urbanization and water consumption.

The rivers (Salt, Gila, and Verde) and canals in and around Phoenix, which have limited water resources, are important for the sustainability of the riparian ecosystems and ecological structure in the region. The wetlands and habitats around these rivers have shrunk and disappeared over time due to the impact of urbanization [76]. However, the Tres Rios wetland on the banks of the Salt River southwest of Phoenix provides ecosystem services, flood control, habitat restoration, recreation, and environmental education [77–79].

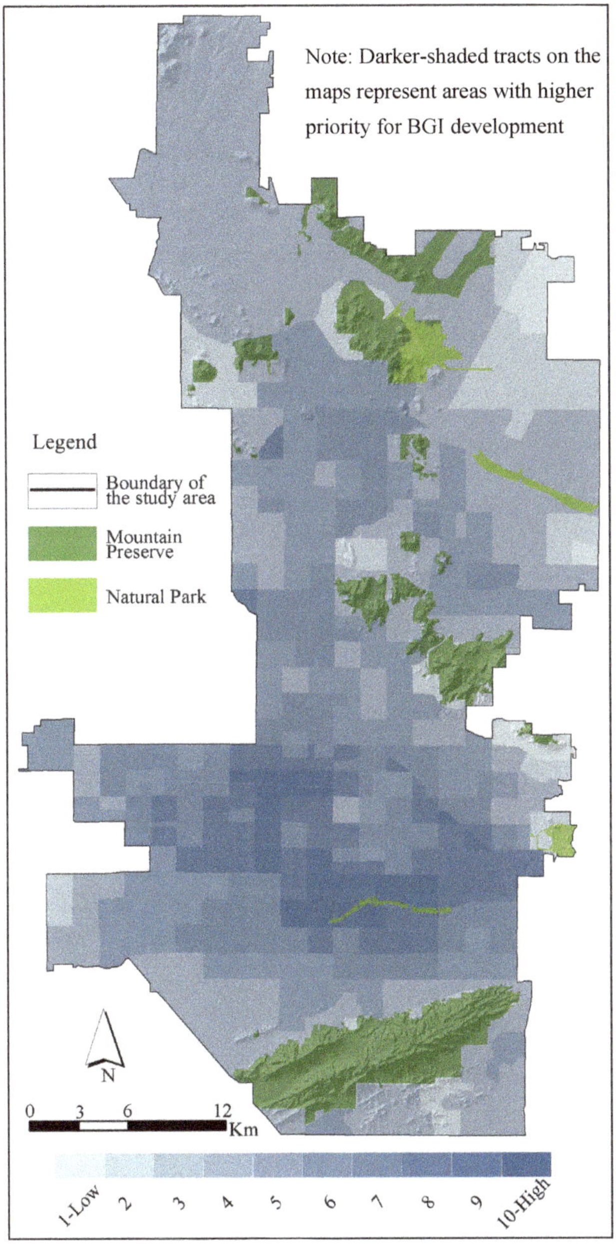

Figure 4. Map of hotspots for BGI planning in Phoenix, AZ.

Implementing urban infrastructure [80] and nature-based solutions needed to combat climate change for the entire city requires a holistic analysis of the geomorphology of the city and an understanding of the social needs and political will. For this reason, planning the infrastructure required to combat climate change in a way that ensures social and environmental justice in the most vulnerable parts of the city will reduce the impacts of climate change in these regions. Nonetheless, in Phoenix, there are fewer projects implemented to mitigate climate change problems in regions inhabited by socially vulnerable people [54,81]. As a result, air pollution and the UHI impact are high in these regions, the amount of green space is low, and water management is ineffective.

As seen in the hotspot map of the study area, the priority areas for BGI planning are the areas with a high urban density. The implementation of BGI in these areas will increase the resilience of the city against the effects of climate change and contribute to social and environmental justice.

4.3. NbS for Just and Healthy Urban Planning Implications

The United Nations World Urbanization Prospects estimate that, by 2050, approximately 68% of the world's population will live in urban areas [82]. Most of the projected urbanization due to population growth is foreseen to take place in arid areas, which cover 41% of the global land surface and have 35% of the world's largest cities [83]. In this context, the city of Phoenix, within the Phoenix Metropolitan Area, which has been called "the least sustainable city in the world" by Ross [84], is an important example of dryland urbanization. The transformation of Phoenix from an agricultural region to a metropolis with intense urban activity has led to extensive changes in land cover and land use. As a result of these land cover and land use changes, several interrelated environmental problems have emerged, such as deteriorating air quality [85], regional biodiversity loss [86], water scarcity [87], and an urban heat island [75]. This situation threatens the sustainability of the city.

The rapid increase and spread of urbanization in Phoenix has led to increased development in floodplains and increased volumes of impervious surfaces. This increases the risk of flash flooding in the city and the negative effects of the urban heat island [50,61]. Although the city of Phoenix does not receive heavy rainfall throughout the year, the area experiences flash floods, especially during the monsoon season. Therefore, reducing the impervious surfaces, increasing the vegetation density, and constructing stormwater retention facilities will contribute to the reduction of stormwater runoff and groundwater recharge.

The ecosystems of arid cities are different from those of cities with intense precipitation due to factors such as inadequate water resources, a low density of green areas, a high land surface temperature and urban heat island effect, irregular rainfall, and a diverse vegetation structure. Therefore, Phoenix has the potential to offer different BGI recommendations compared to wetter cities (Baltimore, Portland, Seattle, etc.).

The water demand for green infrastructure raises water management concerns, especially for cities in arid or semi-arid regions [75,88], because the water use in cities located in these regions could exceed local water availability [89]. The Phoenix region, which is arid and developing in terms of urbanization, is supplied by local rivers in the Salt–Verde basins, water from the Colorado River, groundwater, and wastewater [90]. However, with intense urbanization and population growth, the excessive water consumption in various areas (industrial productions, parks, nurseries, golf courses, lawns, etc.) leads to significant reductions in surface water and groundwater withdrawals in Phoenix [91]. Therefore, it is important to establish a relationship between urban green infrastructure and water resource management in Phoenix for the sustainability of the city.

Studies have shown that the preference for the arid landscape (xeriscape) design approach rather than traditional and common landscape designs in the planning and implementation of green spaces in arid regions contributes significantly to water saving [92,93]. However, two different studies in Phoenix have noted that xeriscape application increased the air temperatures and thermal discomfort. At the same time, this leads to excessive energy consumption for cooling systems and the risk of increased rates of heat-related morbidity and mortality [75,94]. For this reason, landscape areas should include trees that will provide spatial cooling through both shading and evapotranspiration, reduce the urban UHI effect, and increase the thermal comfort. In this context, xeriscape application in the landscapes of Phoenix with arid-tolerant tree species (low-water-use trees) will provide a cooling effect, reduce the UHI effect, and contribute significantly to water savings.

Increasing impervious surfaces due to urbanization increase the surface runoff and UHI. The study carried out by Flower et al. [95] in Salt Lake City (USA) revealed that the surface temperature of pervious concrete under the sun was equal to that of conventional concrete in the shade. It was also found to be, on average, 10 °C cooler than asphalt under the sun [96]. Therefore, the use of pervious surfaces in structural applications in Phoenix will contribute to reducing the surface runoff and temperature effects.

Since the boundaries of the study area determined are very wide, it is not possible to measure the surface temperature of each point at the same time. For this reason, satellite images of three different periods (2013, 2018, 2023) obtained by the remote sensing method were used in the study. The LST and UHI values of each period were calculated from these satellite images. The reason for considering LST values is that studies have shown that the LST is significantly correlated with the air temperature [53,97]. Phoenix is one of the hottest cities in the USA [98], and, with global temperatures rising due to climate change, extreme heat in the region is the most deadly threat [99]. In this city, characterized by a desert ecosystem, a master plan has been prepared that aims to increase the amount of green space to reduce the UHI [100]. This master plan aims to increase the tree canopy from 10% to 25% to reduce the city's heat. However, as emphasized by Zhang et al. [53], to make the most of the potential cooling benefits of green spaces, not only the quantity and spatial size of green spaces but also the density and distribution of green spaces need to be considered in planning and implementation. Since more than 45% of the land in Phoenix is used for residential areas and much of it is privately owned [101], this constrains the planning and implementation of green infrastructure in certain areas [60].

The findings of the study show that social vulnerability and environmental risks (flooding, heat) have a positive correlation in Phoenix, as in other cities [25,29,45]. In Phoenix, low-income, high-percentage renters and low-percentage English speakers are the variables that most increase the value of social vulnerability [67]. Zhang et al. [53] found that when the presence of green spaces and the social structure are considered together, neighborhoods with high social vulnerability have fewer green spaces. This conclusion is also valid for Phoenix. Prioritizing green infrastructure planning in areas with high social vulnerability will contribute to reducing the environmental risks in the region. In addition, increasing the accessibility of socially vulnerable groups to park areas within the green infrastructure will increase the level of utilization of the opportunities and benefits of these areas.

In the 20-year air quality report prepared by the American Lung Association, it is determined that 4 out of 10 Americans live in unhealthy air quality. In this report, Phoenix ranks seventh in ozone pollution [102]. When the AQI values (last 10 years) of Phoenix are examined, it is seen that air pollution is high, especially in the southern region of the city. The intensive industrial and commercial areas, highways, and the airport in this region [103] are thought to contribute to the high level of air pollution. Studies on reducing air pollution show that GI is an effective planning approach to reduce particulate matter in the air and improve the air quality [104,105]. For this reason, researchers propose street afforestation, roof gardens, facade greening, vertical gardens, and green corridor systems in the whole city as solutions to reduce air pollution in cities [106,107]. The City of Phoenix has prepared long-term plans to reduce air pollution (especially reducing ozone and particulate pollution) and improve the air quality. The City of Phoenix's objective is to achieve healthy air quality for people in Phoenix by 2050. This will be achieved by increasing the number of excellent and good days for air quality to 90% annually [51]. Furthermore, the "City of Phoenix Tree and Shade Master Plan", prepared by the City of Phoenix in 2010, aims to reduce the carbon footprint of the region, reduce stormwater runoff, increase bio-diversity, improve the air quality, reduce the urban heat island effect, and reduce energy costs. With the successful implementation of these plans, Phoenix expects to achieve many goals and policies in terms of urban sustainability. In this context, planning water structures and green areas in a holistic structure through natural and artificial corridors will contribute to the city's effective utilization of ecosystem services and will be effective in the implementation of nature-based solutions.

5. Conclusions

The empirical data provided by the GISP model criteria at the stage of determining priority areas for BGI will provide a great advantage for planners and decision-makers to develop planning scenarios and strategies that consider the urban ecosystem services provided in the interconnected social–ecological–technological systems [108]. The model, which has a flexible structure in creating the database [25,45], will support the development of anticipatory, participatory, and multifunctional BGI policies and strategies in the field of landscape architecture and planning. In addition, the fact that new criteria could be included in the model depending on the social, economic, cultural, and ecological structures of the cities will offer planning approaches according to the internal dynamics of the cities.

Climate change is causing the rapid warming of cities, where extreme heat and droughts are likely to be major challenges in the next few decades [109,110]. Thus, using BGI as NbS is an indispensable planning approach for arid and semi-arid regions. Being located in a region characterized by arid climatic conditions, Phoenix is adversely affected by extreme climatic conditions, which are rapidly changing due to climate change. This negative impact particularly damages fragile populations (children, the elderly, and people with disabilities) in areas where socially vulnerable groups are located. It also leads to a decrease in ecosystem services and the disruption of the ecological balance.

Nature-based solutions, such as BGI, buffer against the negative impacts of climate change, and they have an important role in planning and building sustainable and livable cities. In this context, integrating BGI into Phoenix's zoning planning processes and plans will contribute to mitigating many of the negative impacts associated with climate change and urbanization. Finally, our city-wide analysis precludes a higher level of detailed granularity and the recommendation of site-specific NbS BGI strategies at the city block level; however, we believe that subsequent studies will be able to focus on the identified hotspots.

Supplementary Materials: The following supporting information can be downloaded at: https://www.mdpi.com/article/10.3390/land13091464/s1, File S1: Formulas and flowcharts for the methodology of the study [62–64,111–120]; File S2: Maps and tables of sub-criteria analysis results [100,121]; File S3: Analytic hierarchy process survey.

Author Contributions: Conceptualization, R.O., P.C. and C.C.; methodology, R.O.; software, R.O.; formal analysis, R.O.; investigation, R.O.; writing—original draft preparation, R.O., P.C. and C.C.; writing—review and editing, R.O., P.C. and C.C.; visualization, R.O. All authors have read and agreed to the published version of the manuscript.

Funding: This research was funded by the Scientific and Technological Research Council of Turkey (TUBITAK) Science Fellowships and Grant Programme Directorate, grant number 1059B192201012, 2022/1.

Data Availability Statement: The raw data supporting the conclusions of this article will be made available by the authors on request.

Conflicts of Interest: The authors declare no conflicts of interest.

References

1. Langemeyer, J.; Wedgwood, D.; McPhearson, T.; Baró, F.; Madsen, A.L.; Barton, D.N. Creating urban green infrastructure where it is needed—A spatial ecosystem service-based decision analysis of green roofs in Barcelona. *Sci. Total Environ.* **2020**, *707*, 135487. [CrossRef] [PubMed]
2. García-Pardo, K.A.; Moreno-Rangel, D.; Domínguez-Amarillo, S.; García-Chávez, J.R. Remote sensing for the assessment of ecosystem services provided by urban vegetation: A review of the methods applied. *Urban For. Urban Green.* **2022**, *74*, 127636. [CrossRef]
3. Arnold, J.; Kleemann, J.; Fürst, C. A Differentiated Spatial Assessment of Urban Ecosystem Services Based on Land Use Data in Halle, Germany. *Land* **2018**, *7*, 101. [CrossRef]
4. Hansen, R.; Olafsson, A.S.; Van der Jagt, A.P.; Rall, E.; Pauleit, S. Planning multifunctional green infrastructure for compact cities: What is the state of practice? *Ecol. Indic.* **2019**, *96*, 99–110. [CrossRef]

5. Cheng, C.; Yang, Y.E.; Ryan, R.; Yu, Q.; Brabec, E. Assessing climate change-induced flooding mitigation for adaptation in Boston's Charles River watershed, USA. *Landsc. Urban Plan.* **2017**, *167*, 25–36. [CrossRef]
6. Kim, G.; Coseo, P. Urban park systems to support sustainability: The role of urban park systems in hot arid urban climates. *Forests* **2018**, *9*, 439. [CrossRef]
7. Biasin, A.; Masiero, M.; Amato, G.; Pettenella, D. Nature-based solutions modeling and cost-benefit analysis to face climate change risks in an urban area: The case of Turin (Italy). *Land* **2023**, *12*, 280. [CrossRef]
8. Lee, D.; Oh, K.; Suh, J. Diagnosis and prioritization of vulnerable areas of urban ecosystem regulation services. *Land* **2022**, *11*, 1804. [CrossRef]
9. Cheng, C. Spatial climate justice and green infrastructure assessment: A case study for the Huron River watershed, Michigan, USA. *J. Geogr. Inf. Sci.* **2016**, *1*, 176–190. [CrossRef]
10. Cheng, C. Climate justicescape and implications for urban resilience in American cities. In *The Routledge Handbook of Urban Resilience*, 1st ed.; Burayidi, M.A., Allen, A., Twigg, J., Wamsler, C., Eds.; Routledge: London, UK, 2019; pp. 83–96.
11. Kim, J.; Kang, J. Development of hazard capacity factor design model for net-zero: Evaluation of the flood adaptation effects considering green-gray infrastructure interaction. *Sustain. Cities Soc.* **2023**, *96*, 104625. [CrossRef]
12. Pour, S.H.; Wahab, A.K.A.; Shahid, S.; Asaduzzaman, M.; Dewan, A. Low impact development techniques to mitigate the impacts of climate-change-induced urban floods: Current trends, issues and challenges. *Sustain. Cities Soc.* **2020**, *62*, 102373. [CrossRef]
13. Gao, D.; Feng, H.; Cao, Y. The spatial spillover effect of innovative city policy on carbon efficiency: Evidence from China. *Econ. Surv. Singap.* **2024**, 1–23. [CrossRef]
14. Gao, D.; Zhou, X.; Wan, J. Unlocking sustainability potential: The impact of green finance reform on corporate ESG performance. *Corp. Soc. Responsib. Environ. Manag.* **2024**, *31*, 4211–4226. [CrossRef]
15. Alkhaled, S.; Coseo, P.; Brazel, A.; Cheng, C.; Sailor, D. Between aspiration and actuality: A systematic review of morphological heat mitigation strategies in hot urban deserts. *Urban Clim.* **2020**, *31*, 100570. [CrossRef]
16. Hu, X.; Cao, J.; Qian, Y.; Zhou, W.; Zheng, Z. Extreme heat events in mainland China from 1981 to 2015: Spatial patterns, temporal trends, and urbanization impacts. *Sustain. Cities Soc.* **2024**, *100*, 104999. [CrossRef]
17. Lipponen, A.H.; Mikkonen, S.; Kollanus, V.; Tiittanen, P.; Lanki, T. Increase in summertime ambient temperature is associated with decreased sick leave risk in Helsinki, Finland. *Environ. Res.* **2024**, *240*, 117396. [CrossRef]
18. Ravindiran, G.; Rajamanickam, S.; Kanagarathinam, K.; Hayder, G.; Janardhan, G.; Arunkumar, P.; Arunachalam, S.; AlObaid, A.A.; Warad, I.; Muniasamy, S.K. Impact of air pollutants on climate change and prediction of air quality index using machine learning models. *Environ. Res.* **2023**, *239*, 117354. [CrossRef]
19. Olmstead, S.M. Climate change adaptation and water resource management: A review of the literature. *Energy Econ.* **2014**, *46*, 500–509. [CrossRef]
20. Almaaitah, T.; Appleby, M.; Rosenblat, H.; Drake, J.; Joksimovic, D. The potential of blue-green infrastructure as a climate change adaptation strategy: A systematic literature review. *Blue-Green Syst.* **2021**, *3*, 223–248. [CrossRef]
21. Undurraga, R.; Vicuña, S.; Melo, O. Compensating water service interruptions to implement a safe-to-fail approach to climate change adaptation in urban water supply. *Water* **2020**, *12*, 1540. [CrossRef]
22. Lenaker, A.; Seymour, F. *Global Emissions and Local Deforestation Are Combining to Create Dangerous Levels of Heat Stress in the Tropics*; World Resources Institute: Washington, DC, USA, 2022. Available online: https://www.wri.org/ (accessed on 2 September 2023).
23. Hutley, N.; Dean, A.; Hart, N.; Daley, J. *Uninsurable Nation: Australia's Most Climate-Vulnerable Places*, 1st ed.; Climate Council: Sydney, Australia, 2022; pp. 1–29.
24. Ariyarathna, I.S.; Abeyrathna, W.P.; Jamei, E.; Chau, H. A Review of the application of blue–green infrastructure (BGI) as an effective urban flood mitigation strategy for livable and healthy cities in Australia. *Architecture* **2023**, *3*, 461–476. [CrossRef]
25. Meerow, S. The politics of multifunctional green infrastructure planning in New York City. *Cities* **2020**, *100*, 102621. [CrossRef]
26. Acosta, F.; Haroon, S. Memorial Parking Trees: Resilient Modular Design with Nature-Based Solutions in Vulnerable Urban Areas. *Land* **2021**, *10*, 298. [CrossRef]
27. Hamstead, Z.; Coseo, P. Building policies, plans, and cities to manage extreme weather events: Perspectives from urban planning and landscape architecture. In *Transforming Global Health*, 1st ed.; Smith, K.H., Ram, P.K., Eds.; Springer: Cham, Switzerland, 2020; Volume 1, pp. 261–283.
28. Cohen-Shacham, E.; Walters, G.; Janzen, C.; Maginnis, S. *Nature-Based Solutions to Address Global Societal Challenges*, 1st ed.; IUCN: Gland, Switzerland, 2016; 97p.
29. Meerow, S.; Newell, J.P. Spatial planning for multifunctional green infrastructure: Growing resilience in Detroit. *Landsc. Urban Plan.* **2017**, *159*, 62–75. [CrossRef]
30. Parker, J.; Zingoni de Baro, M.E. Green infrastructure in the urban environment: A systematic quantitative review. *Sustainability* **2019**, *11*, 3182. [CrossRef]
31. Slätmo, E.; Nilsson, K.; Turunen, E. Implementing green infrastructure in spatial planning in Europe. *Land* **2019**, *8*, 62. [CrossRef]
32. Wei, J.; Qian, J.; Tao, Y.; Hu, F.; Ou, W. Evaluating spatial priority of urban green infrastructure for urban sustainability in areas of rapid urbanization: A case study of Pukou in China. *Sustainability* **2018**, *10*, 327. [CrossRef]
33. Everett, G.; Adekola, O.; Lamond, J. Developing a blue-green infrastructure (BGI) community engagement framework template. *Urban Des. Int.* **2023**, *28*, 172–188. [CrossRef]

34. Vujadinovič, D.; Šabić, D. The importance of regions in geographical research. *Importance Reg. Geogr. Res.* **2017**, *913*, 195–208. [CrossRef]
35. O'Donnell, E.C.; Netusil, N.R.; Chan, F.K.S.; Dolman, N.J.; Gosling, S.N. International perceptions of urban blue-green infrastructure: A comparison across four cities. *Water* **2021**, *13*, 544. [CrossRef]
36. Dolman, N.; O'Donnell, H. 5 Lessons Learned from Blue-Green Infrastructure Delivery. Available online: https://www.ice.org.uk/news-and-insight/the-civil-engineer/july-2021/theory-and-practice-of-blue-green-infrastructure (accessed on 5 December 2023).
37. Valente de Macedo, L.S.; Barda Picavet, M.E.; Puppim de Oliveira, J.A.; Shih, W. Urban green and blue infrastructure: A critical analysis of research on developing countries. *J. Clean. Prod.* **2021**, *313*, 127898. [CrossRef]
38. Suleiman, L. Blue-green infrastructure, from niche to mainstream: Challenges and opportunities for planning in Stockholm. *Technol. Forecast. Soc. Chang.* **2021**, *166*, 120528. [CrossRef]
39. Molné, F.; Donati, G.F.; Bolliger, J.; Fischer, M.; Maurer, M.; Bach, P.M. Supporting the planning of urban blue-green infrastructure for biodiversity: A multi-scale prioritisation framework. *J. Environ. Manag.* **2023**, *342*, 118069. [CrossRef] [PubMed]
40. Kozak, D.; Henderson, H.; Rotbart, D.; Aradas, R. Blue-green infrastructure (BGI) in dense urban watersheds. The case of the Medrano Stream Basin (MSB) in Buenos Aires. *Sustainability* **2020**, *12*, 2163. [CrossRef]
41. Gomes, M.V.R.; Battemarco, B.P.; Guimarães, L.F.; Oliveira, A.K.B.D.; Rutigliani, V.D.A.; Cabral, F.M.; Bezerra, R.D.O.P.; Lourenço, I.B.; Rezende, O.M.; Magalhães, P.C.D.; et al. The use of blue-green infrastructure as a multifunctional approach to watersheds with socio-environmental vulnerability. *Blue-Green Syst.* **2021**, *3*, 281–297. [CrossRef]
42. Ortega-Rosas, C.I.; Enciso-Miranda, C.A.; Macías-Duarte, A.; Morales-Romero, D.; Villarruel-Sahagún, L. Urban vegetation cover correlates with environmental variables in a desert city: Insights of mitigation measures to climate change. *Urban Ecosyst.* **2020**, *23*, 1191–1207. [CrossRef]
43. Quay, R. Anticipatory governance: A tool for climate change adaptation. *J. Am. Plan. Assoc.* **2010**, *76*, 496–511. [CrossRef]
44. Sharma, A.; Woodruff, S.; Budhathoki, M.; Hamlet, A.; Chen, F.; Fernando, H. Role of green roofs in reducing heat stress in vulnerable urban communities a multidisciplinary approach. *Environ. Res. Lett.* **2018**, *13*, 194011. [CrossRef]
45. Meerow, S. A green infrastructure spatial planning model for evaluating ecosystem service tradeoffs and synergies across three coastal megacities. *Environ. Res. Lett.* **2019**, *14*, 125011. [CrossRef]
46. Chang, H.-S.; Lin, Z.-H.; Hsu, Y.-Y. Planning for green infrastructure and mapping synergies and trade-offs: A case study in the Yanshuei River Basin, Taiwan. *Urban For. Urban Green.* **2021**, *65*, 127325. [CrossRef]
47. Tran, T.J.; Helmus, M.R.; Behm, J.E. Green infrastructure space and traits (GIST) model: Integrating green infrastructure spatial placement and plant traits to maximize multifunctionality. *Urban For. Urban Green.* **2020**, *49*, 126635. [CrossRef]
48. Dong, X.; Ye, Y.; Yang, R.; Li, X. Planning for green infrastructure based on integration of multi-driving factors: A case study in pilot site of sponge city. *Sustain. Cities Soc.* **2023**, *93*, 104549. [CrossRef]
49. Arizona State Climate Office. Available online: https://azclimate.asu.edu/ (accessed on 17 October 2023).
50. Cilek, A.; Unal, M.; Middel, A. The effects of 2-D and 3-D urban landscape metrics on mean radiant temperature in hot-arid Phoenix and Tempe, Arizona, USA. *Sustain. Cities Soc.* **2024**, *101*, 105116. [CrossRef]
51. The City of Phoenix. Drought Information and Clean Air. 2023. Available online: https://www.phoenix.gov/ (accessed on 10 September 2023).
52. Helmrich, A.; Kuhn, A.; Roque, A.; Santibanez, A.; Kim, Y.; Grimm, N.B.; Chester, M. Interdependence of social-ecological-technological systems in Phoenix, Arizona: Consequences of an extreme precipitation event. *J. Infrastruct. Preserv. Resil.* **2023**, *4*, 1–13. [CrossRef]
53. Zhang, Y.; Murray, A.T.; Turner, B. Optimizing green space locations to reduce daytime and nighttime urban heat island effects in Phoenix, Arizona. *Landsc. Urban Plan.* **2017**, *165*, 162–171. [CrossRef]
54. deBuys, W. The Least Sustainable City: Phoenix as a Harbinger for Our Hot Future. 2013. Available online: https://grist.org/climate-energy/the-least-sustainable-city-phoenix-as-a-harbinger-for-our-hot-future/ (accessed on 11 December 2023).
55. Romantic, D.; Taszarek, M.; Brooks, H. Convective environments leading to microburst, macroburst and downburst events across the United States. *Weather Clim. Extrem.* **2022**, *37*, 100474. [CrossRef]
56. Census Demographic and Housing Characteristics File. 2020. Available online: https://data.census.gov/ (accessed on 7 July 2023).
57. Aragon, N.U.; Stuhlmacher, M.; Smith, J.P.; Clinton, N.; Georgescu, M. Urban agriculture's bounty: Contributions to Phoenix's sustainability goals. *Environ. Res. Lett.* **2019**, *14*, 105001. [CrossRef]
58. National Weather Service. Annual and Monthly Record data for Phoenix. 2023. Available online: https://www.weather.gov/psr/ (accessed on 1 January 2024).
59. Suchy, A.K.; Palta, M.M.; Stromberg, J.C.; Childers, D.L. High potential nitrate removal by urban accidental wetlands in a desert city: Limitations and spatiotemporal patterns. *Ecosystems* **2020**, *23*, 1227–1242. [CrossRef]
60. Meerow, S.; Helmrich, A.M.; Andrade, R.; Larson, K.L. How do heat and flood risk drive residential green infrastructure implementation in Phoenix, Arizona? *Urban Ecosyst.* **2021**, *24*, 989–1000. [CrossRef]
61. Kane, K.; Connors, J.P.; Galletti, C.S. Beyond fragmentation at the fringe: A path-dependent, high-resolution analysis of urban land cover in Phoenix, Arizona. *Appl. Geogr.* **2014**, *52*, 123–134. [CrossRef]

62. Wright, M.; Watkins, L.; Hondula, D.; Kurtz, L.; Chakalian, P.; Harlan, S.; Declet-Barreto, J. Social and Heat Vulnerability Indices in Phoenix, Arizona. Environmental Data Initiative. 2019. Available online: https://scholar.google.com/citations?view_op=view_citation&hl=en&user=vMd8LmIAAAAJ&sortby=pubdate&citation_for_view=vMd8LmIAAAAJ:qjMakFHDy7sC (accessed on 17 September 2023).
63. Logan, T.; Williams, T.; Nisbet, A.; Liberman, K.; Zuo, C.; Guikema, S. Evaluating urban accessibility: Leveraging open-source data and analytics to overcome existing limitations. *Environ. Plan. B Urban Anal. City Sci.* **2019**, *46*, 897–913. [CrossRef]
64. Environmental Protection Agency. 2023. Available online: https://www.epa.gov/outdoor-air-quality-data/download-daily-data (accessed on 6 September 2023).
65. Zhang, Y.; Turner II, B.L. Land-Cover Mapping of the Central Arizona Region Based on 2015 National Agriculture Imagery Program (NAIP) Imagery. 2020. Available online: https://sustainability-innovation.asu.edu/caplter/data/view/knb-lter-cap.685.1/ (accessed on 8 September 2023).
66. McGarigal, K.; Cushman, S.A.; Ene, E. Fragstats v4: Spatial Pattern Analysis Program for Categorical Maps. Available online: https://www.fragstats.org (accessed on 12 October 2023).
67. Chang, H.; Pallathadka, A.; Sauer, J.; Grimm, N.B.; Zimmerman, R.; Cheng, C.; Iwaniec, D.M.; Kim, Y.; Lloyd, R.; McPhearson, T.; et al. Assessment of urban flood vulnerability using the social-ecological-technological systems framework in six US cities. *Sustain. Cities Soc.* **2021**, *68*, 102786. [CrossRef]
68. Goepel, K.D. Implementation of an online software tool for the analytic hierarchy process (AHP-OS). *Int. J. Anal. Hierarchy Process.* **2018**, *10*, 469–487.
69. Wang, G.; Ostoja-Starzewski, M. Influence of topography on the Phoenix CO_2 dome: A computational study. *Atmos. Sci. Lett.* **2004**, *5*, 103–107. [CrossRef]
70. Pope, R.; Wu, J. Characterizing air pollution patterns on multiple time scales in urban areas: A landscape ecological approach. *Urban Ecosyst.* **2014**, *17*, 855–874. [CrossRef]
71. Stone, E. Winter Air Really Is Worse in South, West Phoenix. Here's Why. Azcentral. 2019. Available online: https://www.azcentral.com/ (accessed on 7 September 2023).
72. IQAir. What Is the Air Quality in Phoenix Now? Available online: https://www.iqair.com/us/ (accessed on 12 November 2023).
73. Grineski, S.; Bolin, B.; Boone, C. Criteria air pollution and marginalized populations: Environmental inequality in metropolitan Phoenix, Arizona. *Soc. Sci. Q.* **2007**, *88*, 535–554. [CrossRef]
74. Kaus, E. The Importance of Green Infrastructures: The Case of Phoenix. 2020. Available online: https://upe2020.wordpress.com/2020/11/21/green-infrastructure-phoenix/ (accessed on 2 January 2024).
75. Yang, J.; Wang, Z. Planning for a sustainable desert city: The potential water buffering capacity of urban green infrastructure. *Landsc. Urban Plan.* **2017**, *167*, 339–347. [CrossRef]
76. Whittley, K. Tres Rios. Available online: http://www.azheritagewaters.nau.edu/loc_TresRios.html (accessed on 10 December 2023).
77. Economides, C. Green Infrastructure: Sustainable Solutions in 11 Cities across the United States. 2014. Available online: https://water.columbia.edu/ (accessed on 23 October 2023).
78. Stefanakis, A.I. The role of constructed wetlands as green infrastructure for sustainable urban water management. *Sustainability* **2019**, *11*, 6981. [CrossRef]
79. The City of Phoenix. History of Tres Rios. Available online: https://www.phoenix.gov/waterservices/tresrios/wetlandsinfo (accessed on 3 January 2024).
80. Kim, G.; Miller, P.A. The impact of green infrastructure on human health and well-being: The example of the Huckleberry Trail and the Heritage Community Park and Natural Area in Blacksburg, Virginia. *Sustain. Cities Soc.* **2019**, *48*, 101562. [CrossRef]
81. Depietri, Y.; McPhearson, T. Integrating the grey, green, and blue in cities: Nature-based solutions for climate change adaptation and risk reduction. In *Theory and Practice of Urban Sustainability Transitions*, 1st ed.; Kabisch, N., Korn, H., Stadler, J., Bonn, A., Eds.; Springer: Cham, Switzerland, 2017; Volume 1, pp. 91–109.
82. Ritchie, H.; Roser, M. Urbanization. 2019. Available online: https://ourworldindata.org/urbanization (accessed on 3 October 2023).
83. Gilman, J.; Wu, J. Identifying broken linkages coupling water availability and dryland urbanization for sustainability: The case of the Phoenix metropolitan region, USA. *J. Environ. Manag.* **2024**, *352*, 120097. [CrossRef]
84. Ross, A. *Bird on Fire: Lessons from the World's Least Sustainable City*, 1st ed.; Oxford University Press: Oxford, UK, 2011.
85. Lee, S.M.; Fernando, H.J.S.; Grossman-Clarke, S. Mm5-Smoke-Cmaq as a modeling tool for 8-h ozone regulatory enforcement: Application to the state of Arizona. *Environ. Model. Assess.* **2007**, *12*, 63–74. [CrossRef]
86. Roggema, R.; Tillie, N. Realizing emergent ecologies: Nature-based solutions from design to implementation. *Land* **2022**, *11*, 1972. [CrossRef]
87. Wentz, E.A.; Gober, P. Determinants of small-area water consumption for the City of Phoenix, Arizona. Sustain. *Water Resour. Manag.* **2007**, *21*, 1849–1863. [CrossRef]
88. Shashua-Bar, L.; Pearlmutter, D.; Erell, E. The influence of trees and grass on outdoor thermal comfort in a hot-arid environment. *Int. J. Climatol.* **2011**, *31*, 1498–1506. [CrossRef]
89. Houdeshel, C.D.; Hultine, K.R.; Johnson, N.C.; Pomeroy, C.A. Evaluation of three vegetation treatments in bioretention gardens in a semi-arid climate. *Landsc. Urban Plan.* **2015**, *135*, 62–72. [CrossRef]

90. Larson, K.L.; Wiek, A.; Withycombe Keeler, L. A comprehensive sustainability appraisal of water governance in Phoenix, AZ. *J. Environ. Manag.* **2013**, *116*, 58–71. [CrossRef]
91. Gober, P.; Kirkwood, C.W. Vulnerability assessment of climate-induced water shortage in Phoenix. *Proc. Natl. Acad. Sci. USA* **2010**, *107*, 21295–21299. [CrossRef]
92. DeNicola, E.; Aburizaiza, O.S.; Siddique, A.; Khwaja, H.; Carpenter, D.O. Climate change and water scarcity: The case of Saudi Arabia. *Ann. Glob. Health* **2015**, *81*, 342–353. [CrossRef]
93. Ismaeil, E.M.H.; Sobaih, A.E.E. Assessing xeriscaping as a retrofit sustainable water consumption approach for a Desert University campus. *Water* **2022**, *14*, 1681. [CrossRef]
94. Chow, W.T.; Brazel, A.J. Assessing xeriscaping as a sustainable heat island mitigation approach for a desert city. *Build. Environ.* **2011**, *47*, 170–181. [CrossRef]
95. Flower, W.; Burian, S.J.; Pomeroy, C.A.; Pardyjak, E.R. Surface Temperature and Heat Exchange Differences between Pervious Concrete and Traditional Concrete and Asphalt Pavements. In Proceedings of the Low Impact Development Conference, San Francisco, CA, USA, 11–14 April 2010.
96. Meerow, S.; Natarajan, M.; Krantz, D. Green infrastructure performance in arid and semi-arid urban environments. *Urban Water J.* **2021**, *18*, 275–285. [CrossRef]
97. Klok, L.; Zwart, S.; Verhagen, H.; Mauri, E. The surface heat island of Rotterdam and its relationship with urban surface characteristics. *Resour. Conserv. Recycl.* **2012**, *64*, 23–29. [CrossRef]
98. Zheng, B.; Myint, S.W.; Fan, C. Spatial configuration of anthropogenic land cover impacts on urban warming. *Landsc. Urban Plan.* **2014**, *130*, 104–111. [CrossRef]
99. Hondula, D.M.; Davis, R.E.; Saha, M.V.; Wegner, C.R.; Veazey, L.M. Geographic dimensions of heat-related mortality in seven U.S. cities. *Environ. Res.* **2015**, *138*, 439–452. [CrossRef]
100. The City of Phoenix. The City of Phoenix Tree and Shade Master Plan (2010). Available online: https://www.phoenix.gov/parkssite/Documents/071957.pdf (accessed on 7 October 2023).
101. Keys, E.; Wentz, E.A.; Redman, C.L. The spatial structure of land use from 1970–2000 in the Phoenix, Arizona, metropolitan area. *Prof. Geogr.* **2007**, *59*, 131–147. [CrossRef]
102. American Lung Association. State of the Air. 2019. Available online: https://www.lung.org/ (accessed on 23 October 2023).
103. Balsas, C.J.L. Improving land-use-based urbanism through award-winning planning practice and scholarship: A North American case study. *Socio. Ecol. Pract. Res.* **2023**, *5*, 275–291. [CrossRef]
104. Ferrari, B.; Quatrini, V.; Barbati, A.; Corona, P.; Masini, E.; Russo, D. Conservation and enhancement of the green infrastructure as a nature-based solution for Rome's sustainable development. *Urban Ecosyst.* **2019**, *22*, 865–878. [CrossRef]
105. Seok, Y.; Ji, E.; Lee, J. How to manage urban disturbances: Focused on social-ecological vulnerability to fine dust pollution. *Environ. Impact Assess. Rev.* **2024**, *104*, 107317. [CrossRef]
106. Chen, L.; Liu, C.; Zhang, L.; Zou, R.; Zhang, Z. Variation in tree species ability to capture and retain airborne fine particulate matter ($PM_{2.5}$). *Sci. Rep.* **2017**, *7*, 3206. [CrossRef]
107. Jia, Y.; Lu, K.; Zheng, T.; Li, X.; Liu, X.; Peng, Z.; He, H. Effects of roadside green infrastructure on particle exposure: A focus on cyclists and pedestrians on pathways between urban roads and vegetative barriers. *Atmos. Pollut. Res.* **2021**, *12*, 1–12. [CrossRef]
108. McPhearson, T.; Cook, E.M.; Berbés-Blázquez, M.; Cheng, C.; Grimm, N.B.; Andersson, E.; Barbosa, O.; Chandler, D.G.; Chang, H.; Chester, M.V.; et al. A social-ecological-technological systems framework for urban ecosystem services. *One Earth* **2022**, *5*, 505–518. [CrossRef]
109. Li, X.; Stringer, L.C.; Dallimer, M. The impacts of urbanisation and climate change on the urban thermal environment in Africa. *Climate* **2022**, *10*, 164. [CrossRef]
110. Wang, T.; Sun, F. Integrated drought vulnerability and risk assessment for future scenarios: An indicator based analysis. *Sci. Total Environ.* **2023**, *900*, 165591. [CrossRef] [PubMed]
111. Kopecký, M.; Macek, M.; Wild, J. Topographic Wetness Index calculation guidelines based on measured soil moisture and plant species composition. *Sci. Total Environ.* **2021**, *757*, 143785. [CrossRef] [PubMed]
112. Brown de Colstoun, E.C.; Huang, C.; Wang, P.; Tilton, J.C.; Tan, B.; Phillips, J.; Niemczura, S.; Ling, P.Y.; Wolfe, R.E. Global Man-Made Impervious Surface (GMIS) Dataset from Landsat. 2017. Available online: https://sedac.ciesin.columbia.edu/data/set/ulandsat-gmis-v1 (accessed on 25 October 2023).
113. Aslan, N. Detection of Urban Heat Islands from Landsat Imageries: A Case Study of West Mediterranean Region. Master's Thesis, Akdeniz University, Antalya, Turkey, 2016.
114. Artis, D.A.; Carnahan, W.H. Survey of emissivity variability in thermography of urban areas. *Remote Sens. Environ.* **1982**, *12*, 313–329. [CrossRef]
115. Sobrino, J.A.; Jiménez-Muñoz, J.C.; Paolini, L. Land surface temperature retrieval from Landsat TM 5. *Remote Sens. Environ.* **2004**, *90*, 434–440. [CrossRef]
116. Olgun, R.; Yılmaz, T. Determination of ecologically suitable settlement areas by using GIS-based multi-criteria decision making analysis: The case of Nigde province. *Fresen. Environ. Bull.* **2019**, *28*, 3768–3777.
117. Görener, Ö. A swot-ahp approach for assessment of medical tourism sector in Turkey. *Alphanumeric J.* **2016**, *4*, 159–170. [CrossRef]
118. Wind, Y.; Saaty, T.L. Marketing applications of the analytic hierarchy process. *Manag. Sci.* **1980**, *26*, 641–658. [CrossRef]

119. Özdemir Işık, B.; Demir, S. Integrated multi-criteria decision-making methods for the sustainability of historical–cultural structures on the Trabzon coastline. *Sustainability* **2017**, *9*, 2114. [CrossRef]
120. Zare, K.; Mehri-Tekmeh, J.; Karimi, S. A swot framework for analyzing the electricity supply chain using an integrated AHP methodology combined with fuzzy-topsis. *Int. Strat. Manag. Rev.* **2015**, *3*, 66–80. [CrossRef]
121. Fao-Unesco. The Digital Soil Map of the World. Available online: https://www.fao.org/soils-portal/data-hub/soil-maps-and-databases/ (accessed on 21 October 2023).

Disclaimer/Publisher's Note: The statements, opinions and data contained in all publications are solely those of the individual author(s) and contributor(s) and not of MDPI and/or the editor(s). MDPI and/or the editor(s) disclaim responsibility for any injury to people or property resulting from any ideas, methods, instructions or products referred to in the content.

land

MDPI

Article

Eco-Spatial Indices as an Effective Tool for Climate Change Adaptation in Residential Neighbourhoods—Comparative Study

Renata Giedych, Gabriela Maksymiuk * and Agata Cieszewska

Department of Landscape Architecture, Institute of Environmental Engineering, Warsaw University of Life Sciences, 159 Nowoursynowska Street, 02-776 Warsaw, Poland; renata_giedych@sggw.edu.pl (R.G.); agata_cieszewska@sggw.edu.pl (A.C.)
* Correspondence: gabriela_maksymiuk@sggw.edu.pl (G.M.)

Abstract: Eco-spatial indices are commonly used tools to improve the quality of the environment in cities. Initially modelled on the Berlin BAF, indices have evolved to address current challenges, particularly climate change adaptation. The Ratio of Biologically Vital Areas (RBVA), introduced in Poland in the mid-1990s, is an early planning tool for implementing Nature-based Solutions (NbSs) at the site level. This research aimed to assess the effectiveness of the RBVA in Poland compared to its counterparts in Oslo and Malmö. The study employed a serious simulation game developed under the Norwegian-funded CoAdapt project, testing six development scenarios, varied in terms of applied NbSs, for a typical multi-family housing estate. The adaptive potential of the tested scenarios was assessed based on the values of five environmental parameters calculated in the game; that is, air temperature, oxygen production, CO_2 sequestration, rainwater harvesting, and biodiversity. The findings revealed that the RBVA, in its current form, has limited effectiveness in supporting climate adaptation. Its two-dimensional nature makes it less effective than the more comprehensive Green Factors used in Oslo and Malmö. The research presented in the article proves that better-constructed indices result in the efficiency of applied NbSs and consequently better adaptation to climate change.

Keywords: nature-based solutions (NbSs); green space factor; serious game; planning tool

1. Introduction

The year 2023 has been identified as the warmest on record since observations began in 1850, according to the Copernicus Climate Change Service [1]. In urban areas, which house the majority of the global population, the effects of climate change are particularly severe. Numerous studies highlight the impact of changes in radiation balance due to the high proportion of impervious surfaces [2], reduced vegetation, and limited water resources, which in turn affect oxygen production and exacerbate air pollution [3,4]. Additionally, there has been an increased frequency of extreme events, including recurring droughts and floods. These factors not only threaten the health of residents but also degrade the overall quality of daily life [5]. The above reasons clearly indicate the need to implement climate change adaptation solutions in cities to mitigate the impacts of climate change on the economy, society, and the environment. Examples of adaptation measures include spatial planning tools aimed to implement various forms of green infrastructure, ranging from regional-scale ecological networks, through urban systems of parks and green spaces, to specific Nature-based Solutions (NbSs) like green roofs or sustainable drainage systems. Basic planning tools for implementing NbSs are green space factors [6], which are also known as eco-spatial indices [7] or green area indicators [8]. These measures are designed to introduce various types of green and blue elements in densely urbanised areas [9]. The first of its kind in the world was the Biotope Area Factor (BAF), introduced in Berlin in the early 1990s [7,10,11]. It is expressed by the share of the area occupied by different eco-friendly elements (from vegetation growing on the ground, green roofs and green walls, to various types of permeable surfaces) to the total land area. The adopted system of weighting factors adequate to the ecological effectiveness of each element allows for a very flexible and site-specific design [7,12].

Over the past thirty years, eco-spatial indices have been implemented in many places around the world, e.g., in Poland [7], Malmö [10], Stockholm [13], Helsinki [13], London [14], Vienna [15], Seattle [16], Melbourne [17], Seoul [16], and São Paulo [18]. Most of them originally were modelled on the Berlin BAF [7,16], but consequently, they have been creatively developed by adding new components to meet today's challenges. Initially, the priority was to improve the functioning of urban ecosystems, and now more attention is focused on issues related to adaptation to climate change, e.g., flash floods [19] and microclimate regulation [12]. It should also be added that over time, qualitative elements began to be added in addition to quantitative ones related solely to the area occupied by each element. For example, in Helsinki, it is related to ensuring habitat connectivity and the sustainable management of rainwater, and, e.g., in the case of Malmö and Seattle, the vertical structure of vegetation is considered [16,19]. As a result, the scope of NbSs used in urban planning is constantly expanding, evolving with advances in eco-spatial indices.

Nature-based Solutions (NbSs), as defined by Bulkeley et al. [20], are approaches that draw inspiration from and are supported by natural processes. These solutions are characterised by their economic cost-efficiency, the range of environmental, economic, and social benefits they provide, and their role in aiding climate change adaptation [21]. According to the International Union for Conservation of Nature (IUCN) [22], NbSs are efforts to conserve, sustainably manage, and restore natural and modified ecosystems in ways that effectively and adaptively address societal challenges to ensure both human well-being and biodiversity benefits. Thus, NbSs can be treated as an overarching concept encompassing other significant ideas and discussions, such as ecosystem-based adaptation, green infrastructure, and nature's contributions to people [23,24]. NbSs are now broadly recognised as a crucial foundation for creating resilient and liveable cities. These include green infrastructure interventions that can be implemented at various scales, from site-specific solutions like green roofs and green walls to citywide measures such as parks and urban forests [25]. At the neighbourhood scale, the NbSs refer to diverse vegetation, social gardens, rain gardens, bioswales, green roofs, etc.

As Chausson et al. [12] pointed out in their review paper on the effectiveness of Nature-based Solutions in adapting to climate change and minimising hydrometeorological risks to humans, most authors described positive impacts of NbS use on climate change adaptation. Moreover, NbS measures are equally or even more effective than traditional grey infrastructure [26,27]. Many articles detailing specific efficiency results regarding NbS functionality focus on extensive green roofs. These studies highlight the roofs' ability to retain rainwater, mitigate the effects of heavy rains (such as reducing peak flow), and lower air and indoor temperatures, which subsequently reduces energy costs for cooling or heating. Green roofs reflect 20–30% of solar radiation and absorb up to 60% through photosynthesis, transferring less than 20% of heat to the ground. Depending on the climate zone, they reduce heat flow to the ground by 70–90% in summer and 10–30% in winter. The higher albedo of green roofs compared to bituminous roofing (0.7–0.85 vs. 0.1–0.2) helps mitigate the urban heat island effect [28]. Green roofs can lower surface temperatures by 7.3 °C and air temperatures above them by 0.5 °C [29] and reduce mortality during heat waves [30]. The large-scale, citywide use of green roofs can reduce air temperature by 0.3 °C to 3 °C [28]. According to research by Liu et al. [31] average annual roof retention is higher in dry continental climates (73.8%, Lanzhou) than in humid climates (48.9%, Beijing). On the other hand, empirical studies of the assessment of the retention capacity of a 6 cm roof in a Mediterranean climate have shown that the retention ranges from 32 to 54% of precipitation, and interestingly, no significant improvement in retention was observed when increasing the substrate layer (up to 15 cm), which leads to the conclusion that a roof of 6 cm is sufficient in this climate [32].

Studies on green roofs as a strategic solution, like Zölch et al. [33] for Munich, show that covering 50% of the city's land with green roofs could reduce surface runoff by 80.9%. The effectiveness of NbSs is influenced by city-specific factors, as shown in comprehensive research like the study by Cortinovis et al. [34], which modelled various NbS scenarios and their climate impacts in cities such as Barcelona, Utrecht, and Malmö. For example, densely built Barcelona has a high potential for green roofs, while Malmö, with more open spaces, benefits more from street and park plantings. The best adaptation results from a combination of NbSs, with green roofs reducing surface runoff and tall vegetation in parks mitigating heat waves. However, unsealing car parks showed relatively small benefits. These findings align with Majidi et al. [35], who tested NbSs in Bangkok to reduce flood risk and improve thermal comfort. Among the five scenarios, green roofs were most effective at reducing surface runoff and precipitation peak, while rain gardens with trees best improved thermal comfort. In turn, Vojinovic et al. [36] found that large-scale solutions like retention reservoirs and canal reconstruction, when combined with traditional grey infrastructure, offer the highest level of safety during heavy rains. Single NbSs, such as green roofs or bioretention basins, are less effective on their own. The best flood protection results from integrating both small-scale and large-scale NbSs with traditional flood protection measures. American research [37] on green infrastructure solutions to mitigate flooding in residential areas confirms the effectiveness of comprehensive implementations. Using a system of combined bioretention basins, artificial wetlands, and green roofs, surface runoff was reduced by approximately 36.4% over a 2-year period. Additionally, the size of the rainfall event was the strongest predictor of retention for all systems.

Another frequently studied issue is the impact of green and blue infrastructure (GBI) on lowering city air temperature. A notable study by Zuvela-Aloise et al. [38] simulated the effects of GBI on urban cooling in Vienna. Adding about 350 hectares of parks, either centrally or on the outskirts, was analysed, and parks in central areas had a more significant cooling effect, impacting over 8000 hectares around them, while parks on the outskirts had a lesser impact on surrounding temperatures. Additionally, water reservoirs (at 18 °C) were found to cool the environment more effectively than parks of the same size and location, with green infrastructure being more efficient in windless conditions. These results are consistent with the results of studies on the effect of vegetation cover on the temperature of the ground surface, in built-up areas and on the outskirts of the city, in conditions of extreme summer heat, obtained by Ossola et al. [39]. In the city of Adelaide (AU), tree canopies, and to a lesser extent grass and low vegetation, reduced ground temperatures by 6 °C during the day, though not at night. In built-up areas away from the coast, even small patches of vegetation in courtyards and gardens lowered ground temperatures within 30 m. Also, Marando et al. [40] highlight the benefits of tree canopy coverage in reducing urban temperatures, with an average decrease of 1.07 °C and up to 2.9 °C in some cases. Their study, involving spatial, mathematical, and statistical modelling for 601 European cities with populations over 50,000, found that at least 16% tree canopy coverage is needed to achieve a 1 °C reduction in temperature. In contrast, empirical studies on the effect of linear parks on thermal comfort found a temperature reduction of only 0.5 °C compared to nearby built-up areas, with a minimal impact on the urban heat island effect. However, user surveys revealed a significantly higher level of satisfaction, with respondents reporting a noticeable improvement in thermal comfort within the park [41].

The examples indicated above show the many benefits that can be derived from using NbSs. And while individual NBSs are being studied, there is an observed scarcity of studies that evaluate them in the context of eco-spatial index effectiveness. Although the literature on comparisons of the indices themselves is quite rich [12,42–46], the studies conducted usually involve a comparison of the selection of components that make up the index and the weighting factors assigned to each component.

To fill this knowledge gap in our research, we focus on comparing the effectiveness of three eco-spatial indices: the Ratio of Biologically Vital Area (RBVA) used in Poland [7,47] and two indices used in Scandinavian cities, i.e., Malmö Green Factor (MGF) [48] and Oslo Green Factor (OGF) [49].

The eco-spatial index that was introduced in Poland during the mid-1990s is one of the first of its kind used in the world just after Berlin's Biotope Area Factor (BAF) [7]. In contrast to the abovementioned Scandinavian eco-spatial indices, RBVA is used in the whole of Poland and is defined in national law [47]. It should also be emphasised that of the three indices examined, it is the least complex index, considering only the quantitative aspects of NbSs and not the qualitative ones.

The research presented here aims to assess the environmental effectiveness of RBVA in comparison with its counterparts applied in Oslo and Malmö. This comparison will make it possible to assess to what extent the current index in Poland effectively enforces planning provisions favourable to residential areas from the point of view of their climate change adaptation. This assessment is of great practical importance since it seems that the existing index in Poland is less effective due to its quantitative character than the Scandinavian ones.

To evaluate the environmental effectiveness of studied eco-spatial indices, we applied a serious simulation game, "Neighbourhood with Climate", developed as part of a Norwegian-funded CoAdapt project. Using a typical multi-family housing estate as an example study area, we tested different development scenarios, each based on the feasible NbS specific to each of the indices analysed.

2. Materials and Methods

2.1. Research Background

The research on eco-spatial indices was performed as a part of the Norwegian-funded project CoAdapt–Communities for Climate Change Action that dealt with the possibilities of the adaptation of residential areas to climate change with the application of NbSs. Therefore, the selection of indices, the tool for evaluating adaptation potential and the test area are linked to this project. To evaluate the effectiveness of eco-spatial indices as a measure for developing the adaptive potential of multi-family housing estates, three indices were selected—one quantitative and two qualitative, discussed in detail in Section 2.2. To determine the adaptation potential of the spatial solutions (including various NbSs) resulting from the application of each index, the serious computer game "Neighbourhood with Climate" (described in Section 2.3) was used. To fulfil the research aim, we selected a multi-family housing estate in Warsaw (characterised in Section 2.5), which served as the tested area.

2.2. Eco-Spatial Indices under Consideration

A key objective of eco-spatial indices (ESIs) is to set a minimum standard for the proportion of blue and green elements that must be incorporated into a developed area [8]. Although ESIs may differ from each other, the way they are calculated is similar and is expressed as a ratio of the area covered by greenery, open water, permeable paving, stormwater infiltration, etc., in relation to the total site area [43]. Therefore, the following general formula for their calculation can be given:

$$Eco - spatial\ index\ value = \frac{areas\ of\ green\ and\ blue\ elements \times weighting\ factors}{total\ site\ area}$$

What makes them different is the number of elements taken into account and the size of the weighting factor attributed to each element (see Table 1).

RBVA was introduced into Polish planning practice in the mid-1990s and Polish law in 2002 as a mandatory element of local spatial development plans [7]. From 2023 onwards, national standards for a minimum value of the index were introduced [50]. The calculation of the index considers areas covered by vegetation, water, and roof gardens, giving the first two a weighting of 1, and for roof gardens half the value of vegetation on the ground, i.e., 0.5.

Table 1. Comparison of studied eco-spatial indices.

Characteristics	RBVA	MGF	OGF
		Weighting Factors	
Components of the index calculated by area occupied			
Vegetation on the ground	1.0	1.0	1.0–1.4
Vegetation on underground constructions	-	0.6–0.8	-
Green roofs	0.5	0.8	0.3–0.9
Green walls	-	0.7	0.3–0.6
Open water	1.0	1.0	3.0
Retention basins	-	-	1.0
Sealed surfaces runoff to green areas	-	0.2	0.2
Permeable pavements	-	0.2–0.4	0.2–0.4
Components of the index calculated in area per piece			
Trees	-	0.4 [1]	1.0 [2]
Shrubs	-	0.2 [3]	-
Constant variable counted, if any, for the area			
Vegetation and water management	-	-	0.05–0.15
Minimal target value	0.2–0.5	0.5–0.6	0.7–0.8

[1] The maximum area of 25 m^2 for each tree; [2] the maximum area of 20–70 m^2 for each tree (depending on existing and planned trees); [3] the maximum area of 5 m^2 for each shrub.

In Malmö, the green factor was introduced in 2001 in connection with the Bo01, the international housing exhibition fair. The positive results led to the instrument now being used in the city's new construction projects, primarily via its inclusion in the green building programme [51]. Principles for calculating MGF are set out in the municipal guidelines [48]. MGF takes into account vegetation on the ground, vegetation on underground constructions (where the weighting factor depends on the depth of the soil), areas covered with climbing plants, drainage from sealed surfaces to green areas, and permeable and semipermeable pavements. In addition, the variation in vegetation cover is considered by adding an additional area for each tree and shrub.

The most complex of the indices studied is the OGF. The OGF was first introduced in 2014 and was designed as a tool to facilitate the integration of blue and green structures in building and landscape architectural design at the property level [52]. The latest guidance for calculating the Oslo index from 2023 [49] besides components of the index calculated by area occupied and by area per piece, takes into consideration additional fixed variables related to vegetation and water management. This last component concerns the physical expansion of the existing blue–green structure, the restoration or establishment of new habitats for biological diversity, the collection of stormwater for irrigation and other reuse, the coordination of measures with adjacent areas and/or owners of neighbouring lands, and the reopening of closed waterways, streams, and rivers in pipes. In the case of the OGF, it is noteworthy that existing vegetation is given prominence in the calculation of the index; this applies to existing vegetation on the ground and existing trees. An additional characteristic of the OGF is the division of trees into small, large, and medium-sized and considering retention basins where stormwater can be diverted and infiltrated through a permeable surface.

With regard to the target minimum value of the index in the case of RBVA and MGF, it depends on the land use, and in the case of OGF, on the zone in which the site is located. For residential areas, the index values are RBVA—0.3 and MGF—0.6. For Oslo—in the central part—0.7, and in the peripheral part of the city, 0.8.

2.3. *Simulation Tool*

Serious computer games are increasingly being used as tools to test new, often experimental solutions in land use as well [53,54]. Such a game is the computer game "Neighbourhood with Climate" (https://coadapt.pl/en/game/, accessed on 9 August 2024) [55]. It is a strategy simulation game in which players adapt a selected neighbourhood to climate change by introducing Nature-based Solutions (NbSs). The game is based on the Open Street Map combined with an elevation model. The menu of the game includes 37 diverse NbSs that have been parameterised, e.g., trees, shrubs, flower meadows, orchards, green roofs and walls, permeable pavements, and raingardens, and create different adaptation scenarios.

Parameterisation algorithms were based on literature indicators [56–60] including published environmental calculators [61] and research conducted within the CoAdapt project. The game allows to learn about the effectiveness of NbSs and the benefits of adaptation to climate change, which is included in the information layer and the post-game summary.

Parameterisation also allows to estimate the overall benefits that result from the NbS introduced in the game: environmental (climate, hydrological, biodiversity), social, and economic.

In the conducted study, the game was used as a tool to assess the potential of the settlement to adapt to climate change by introducing various NbSs with reference to eco-spatial indices. Five environmental parameters were analysed: air temperature, oxygen production, CO_2 sequestration, rainwater harvesting, and biodiversity. These parameters indicate how the tested area is prepared for heat waves or heavy rains, which are the most crucial problems related to climate threats in Poland.

2.4. *Research Framework*

2.4.1. Calculations of Eco-Spatial Indices

The first step of the research was calculating the values of three eco-spatial indices, i.e., RBVA, MGF, and OGF, for the existing state of the studied housing estate, according to the rules specific to each of them. All data necessary for the studied index calculations were compiled on the basis of topographic data retrieved from the National Database of Topographic Objects (in Polish, abbreviated as BDOT) [62], orthophotos [63], and fieldwork performed in July 2024. The inventory comprised all development elements, including architectural objects and vegetation, i.e., buildings, pavements, existing trees, hedgerows, groups of shrubs, and lawns. The calculations of the area occupied by buildings, pavements, and vegetation were made in ArcGIS Pro 3.3.1 software. Next, we determined whether the calculated RBVA met the minimum amount of biologically active area required for residential development in Warsaw and whether it would meet the required ratio for such development in Malmö (MGF) and Oslo (OGF). When calculating MGF and OGF values, we use conversion factors for existing inventoried development elements in agreement with the practice for the specific index [48,49]. According to the planning guidelines for Malmö and Oslo, the assumed minimum value of the indices was 0.6 for MGF (in regards to residential areas) and 0.8 for OGF (in regards to peripheral urban zones).

2.4.2. Scenarios Building

The second step of the research was the development of variant versions of scenarios for the spatial development of the test housing estate in order to meet the required minimum level of MGF and OGF. Altogether, six different scenarios (that is, three for each index) were developed, characterised by using a different array of NbSs (e.g., vegetation elements on the ground only, solutions on architectural surfaces, and a more diverse combination of NbSs). For the correctness of calculations, the area of climbing plants or vertical gardens was counted only up to the 3rd floor (up to 9 m of facade) and reduced by the area of both windows and doors, and the area of green roofs accounted for 0.7 of the total roof area, due to existing chimneys and structural elements.

2.4.3. Simulation Phase

The third step was a simulation phase, in which the "Neighbourhood with Climate" game was applied with the purpose of assessing the adaptation potential of the test housing estate, both in the existing way it is developed and for the 6 studied scenarios. The adaptive potential was assessed based on the results of the values of the 5 environmental parameters calculated in the game; that is, air temperature, oxygen production, CO_2 sequestration, rainwater harvesting, and biodiversity. The application of the "Neighbourhood with Climate" game as a simulation tool first required, from the NbS menu, the selection of the elements that are variables of the analysed eco-spatial indices. Table 2 shows a relationship between the NbSs used in the game and the variables applied in the indices studied. It is worth noting that MGF takes into account trees with a stem girth of more than 35 cm, calculated for the maximum area of 25 m^2 for each tree, which is compatible with a "medium tree" in the "Neighbourhood with Climate" game. Also treated differently are shrubs, which are only considered for MGF if they exceed 3 m in height, while when calculating OGF, shrubs up to 2 m are relevant.

Table 2. Relationship between NbSs used in the game and the variables applied in the indices studied.

NbS Categories	"Neighbourhood with Climate" Game Menu	RBVA	MGF	OGF
Areas covered by vegetation on the ground		□	□	□
Large trees Medium trees Small trees	□ □ □		 □ [1] 	□ □ □
Shrubs	□		□ [2]	□ [3]
Unmown lawn Mown lawn Groundcovers	□ □ □			□ □ □
Retention pond Infiltration basin	□ □	□ 	□ 	□ □
Rainwater harvesting	□			□
Green facades Green walls Green roofs [4]	□ □ □	 □	□ □	□ □ □
Permeable pavement [5] Reinforced lawn	□ □		□ 	□ □

[1] Taking into account the tree parameters indicated in the calculation of the index in Malmö, trees were classified as medium. [2] Shrubs higher than three meters: calculated for the maximum area of 5 m^2 for each shrub. [3] Shrubs up to 2 m. [4] In the case of Oslo, the weighting factor for extensive green roofs is taken into account (due to the actual feasibility of such a solution in the test housing estate). [5] For Malmö and Oslo, surfaces with the highest weighting factor are taken into account.

2.5. Case Study

The area of analysis is the Klobucka housing estate, located in a residential district in the southern part of Warsaw. The rationale for choosing this particular housing estate is that it is a typical example of contemporary multi-family housing in Warsaw, where the vast majority (88 per cent) of residents live in multi-family buildings [64]. In addition, it is a social housing development that needs to improve the quality of space and adapt to climate change. The case study represents a relatively new social housing estate that was built in 2011, covering an area of approximately 1.99 ha. The estate comprises 9 buildings ranging from 4 to 7 floors with 265 apartments. The number of inhabitants, as approximately estimated based on the calculation of the average number of people per apartment according to the indicators of the Ministry of Development and Technology for 2022 [65], is

640 people (2.42 people/apartment). The development layout is radial, and all buildings are arranged in a fan shape (see Figure 1). A good pedestrian communication system of concrete sidewalks connects the buildings, and parking lots and access roads are located at the edge of the estate. The housing estate has poor vegetation structure, mown lawns, locally planted hedgerows, and predominant young trees.

Figure 1. Existing condition of the Klobucka housing estate showing the spatial layout and distribution of buildings, permeable surfaces, and vegetation structure. Diagonal aerial photo up to date for 2022: [66].

As it concerns the environmental conditions, the estate is located in an area rated as moderately exposed to the urban heat island (UHI) [67,68]. Flooding has not been reported here, although in the vicinity, in the valley of a small stream called Potok Sluzewski, a problem with rainwater pouring out of the riverbed and flooding one of Warsaw's main streets has been observed for years [69].

3. Results

3.1. Calculations of Eco-Spatial Indices of Klobucka Housing Estate

The inventory included those development elements and surfaces that are necessary for calculating the studied eco-spatial indices. The share of individual land cover categories is shown in Table 3.

Table 3. Share of various surface types in the studied housing estate.

Land Cover Categories	Area (m^2)	Share in a Total Housing Estate Area (%)
Buildings	4441.57	22.37
Impervious surfaces	7026.34	35.39
Greenery	8387.65	42.24
Total area	19,855.56	100.00

As it concerns greenery, almost the entire housing estate area, besides buildings, two playgrounds, and sidewalks, is covered with mown lawns, making the greenery cover equal to 8387.65m^2. Reinforced lawns have been used at the entrances to the buildings, but their area is very insignificant at about 88m^2. Within the estate, there are 38 very small trees (mainly lindens and globe maples, which, even as adults, would not reach crown sizes of more than 2 m). In addition, there are a few low hedges and individual shrubs in the estate, with a total area of about 1413 m^2. Due to the predominance of very young trees, a tree canopy cover (TCC) in the Klobucka estate, is only about 4.23%.

To calculate the RBVA, the proportion of land covered by impervious surfaces, i.e., under buildings, paved sidewalks and playgrounds, was subtracted from the total area of the housing estate. In total, impervious surfaces occupy 57.78%, so the RBVA is 42.22%. The Klobucka neighbourhood does not have any areas arranged as green roofs or any vegetation placed on other architectural surfaces, so this did not affect the value of this indicator.

Among the variables affecting the MGF score in relation to the studied housing estate, only the biologically active area was applicable, characterised in MGF guidelines as an "area where the plant roots have direct contact with deeper soil layers, and water can freely percolate to groundwater level". Other variables, such as existing trees, did not affect the value of the index, as their size is too negligible and relevant to MGF are trees with a trunk girth of more than 35 cm. The value of the MGF for the estate's existing state is 0.42.

Regarding the OGF, more variables could be taken into account due to other methods of index calculations, i.e., permeable ground covered by grass, moss or lichen, layer of shrubs up to 2 metres high, and also reinforced lawn (that is, a lawn prepared for higher intensity of use). The OGF for the existing state of the studied housing estate is 0.6, based on the proportion of permeable surfaces covered with grass, including reinforced lawns and shrubs. The results of the three calculated indices for the Klobucka housing estate's existing state are presented in Table 4.

Table 4. Values of studied indices for the existing state of Klobucka housing estate.

	RBVA	MGF	OGF
Spatial planning requirements	Min. 30%	Min. 0.6	Min. 0.8
Existing state	42.24%	0.42	0.6

The obtained values of the examined indices were compared with planning requirements, which are political decisions in each country. According to the spatial planning regulations in force in Poland, each residential development must have at least 30% of the biologically active area, hence the obtained RBVA result of 42.22% is sufficient from the point of view of the regulations. However, as for the other two indices are concerned, they are not sufficient. According to current regulations in Malmö, residential areas must have an MGF of at least 0.6. Also in Oslo, residential areas located on the urban periphery (that is, like the Klobucka estate in the context of Warsaw) must meet an OGF of at least 0.8.

3.2. Scenarios for MGF and OGF

Since the analysed housing estate does not meet the minimum values of the MGF and OGF indices, a series of six development scenarios was elaborated (three for each index), which, using different NbSs scored in the examined indices, would allow them to achieve their minimum values. The scenarios tested were differentiated in terms of the solutions used, as schematically presented in Figure 2 and described in Table 5. The MGF1 and OGF1 scenarios were based on the widest possible selection of NbSs, so they are called miscellaneous scenarios. In the second version, the scenarios were supposed to comprise only vegetation on the ground, i.e., trees and shrubs as in OGF2, while permeable pavement was added because it was not possible to achieve the desired minimum index value with the introduced vegetation elements only in MGF2. In the third version of scenarios, the choice of NbSs was limited only to those that can be applied to architectural surfaces; that is, roofs and building walls. All the developed scenarios meet the assumed minimum values of the eco-spatial indices studied.

Figure 2. Idea diagrams of the tested scenarios.

Table 5. Characteristics of MGF and OGF eco-spatial index scenarios developed for the Klobucka housing estate as compared to the existing state.

"Neighbourhood with Climate" Game Menu	Existing State	MGF1 (Miscellaneous)	MGF2 (Vegetation on the Ground, Pavements)	MGF3 (Vegetation on Buildings)	OGF1 (Miscellaneous)	OGF2 (Vegetation on Ground)	OGF3 (Vegetation on Buildings)
Large trees	-	NA	NA	NA	8	89	-
Medium trees	-	12	89	-	-	-	-
Small trees	38	NA	NA	NA	38	38	38
Shrubs	1413 m^2	-	1000 m^2	-	1413 m^2	2413 m^2	1413 m^2
Mown lawn	8300 m^2	NA	NA	NA	8300 m^2	8300 m^2	8300 m^2
Retention pond	-	30 m^2	-	-	-	-	-
Rainwater harvesting	-	NA	NA	NA	20 barrels	-	-
Green facades	-	4896 m^2 (on 8 buildings)	-	1836 m^2 (on 3 buildings)	2448 m^2 (on 4 buildings)	-	-
Green walls	-	NA	NA	NA	2448 m^2 (on 4 buildings)	-	4896 m^2 (on 8 buildings)
Green roofs	-	-		3200 m^2 (on 8 buildings)	1200 m^2 (on 3 buildings)	-	3900 m^2 (on 9 buildings)
Permeable pavement	-	-	4000 m^2	-	-	-	-
Reinforced lawn	88 m^2	NA	NA	NA	88 m^2	88 m^2	88 m^2
Index value		0.6	0.6	0.61	0.8	0.8	0.8

In the MGF1 scenario, an additional 12 medium-sized trees (i.e., those with a trunk circumference of at least 35 cm and a canopy cover of 25 m^2), a 30 m^2 open retention pond, and climbers on the facades of four buildings were introduced in the studied estate. On the other hand, in the MGF2 scenario, only vegetation elements, i.e., trees and shrubs, were used as a base, but the addition of as many as 89 trees and 1000 m^2 of shrubs (which is the maximum value in the Klobucka neighbourhood due to area shortage) did not allow the achievement of the minimum MGF index. Therefore, in this scenario, 4000 m^2 of concrete pavements were also replaced with permeable sidewalks. This solution made it possible to achieve the required minimum MGF index value of 0.6. In the third scenario developed for Malmö (MGF3), only NbSs on architectural surfaces were used; that is, green facades on three buildings and extensive green roofs on eight buildings.

As the OGF is also partly based on qualitative variables, such as those related to restoring connectivity between green spaces or collecting rainwater for reuse, 20 rain barrels could possibly be incorporated into the OGF1 scenario. In addition, this scenario introduced eight more large trees, climbing plants (i.e., green facades on four buildings), and vertical gardens on another four buildings.

On the other hand, in the OGF2 scenario, only plant solutions were used, i.e., 89 large trees were introduced, which was already enough to achieve the required value of the index. In the last scenario OGF3, as in the case of MGF3, NbSs were introduced only on the facades and roofs of buildings. In order to achieve the index value of 0.8, it was necessary to introduce green facades on eight buildings and green extensive roofs on all nine buildings located in the studied neighbourhood.

Data on the current development of the studied neighbourhood and input data describing all six scenarios were entered into the serious game "Neighbourhood with Climate". Sample visualisations of the existing state and individual scenarios are shown in Figure 3.

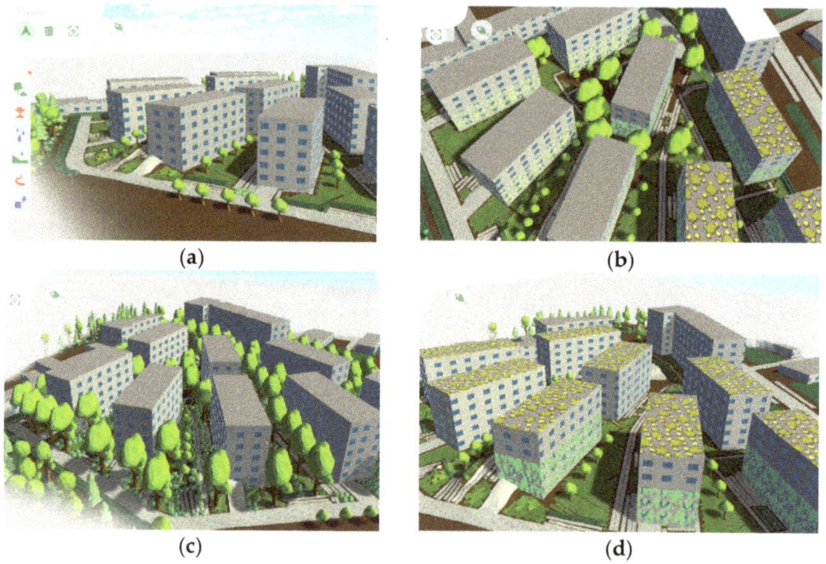

Figure 3. The Klobucka housing estate as visualised in the "Neighbourhood with Climate" serious game: (**a**) the existing state of the Klobucka housing estate; (**b**) OGF1 scenario visualisation; (**c**) OGF2 scenario visualisation; (**d**) MGF3 scenario visualisation.

3.3. *Adaptive Potential of Studied Scenarios*

The adaptation potential for the estate's current land-use state and the six scenarios studied was calculated using the "Neighbourhood with Climate" simulation game. The results obtained for environmental parameters such as temperature, oxygen production, carbon sequestration, rainwater harvesting, and biodiversity are shown in Table 6.

Table 6. Environmental parameters of tested scenarios (based on the "Neighbourhood with Climate" game).

	Existing State	MGF1	MGF2	MGF3	OGF1	OGF2	OGF3
temperature (°C)	20.15	20.04	19.92	19.84	19.8	19.08	19.76
oxygen production (kg)	770,480.5	843,024.4	2,094,981	964,531.7	1,009,858	2,871,481	1,045,389
CO_2 sequestration (kg)	91,649.79	93,588.55	166,669.8	99,139.92	102,353	217,709.8	100,828
water harvesting (l)	0	267.84	0	793.93	2860.35	0	1007.46
biodiversity	1.52	1.8	1.74	1.85	1.83	1.88	1.83

The four-degree green shading scale—the darker the colour, the better the result achieved.

The air temperature was reduced the most in the OGF2 scenario (a decrease of 1.07 °C), which confirms the very clear impact of large trees, of which as many as 89 were used in this setting (see Figure 4). In the other scenarios, the temperature reduction was less significant but noticeable. A similar level of temperature reduction was observed in the OGF3 and OGF1 scenarios (0.39 °C and 0.35 °C reduction, respectively). In the case of the OGF3 scenario, green roofs placed on all nine buildings and vertical gardens (green walls) placed on the facades of eight buildings of the test estate are significant in reducing the temperature. On the other hand, a similar result obtained in the OGF1 scenario is due to the use of a combination of climbing plants, vertical gardens, green roofs, and several large trees. Among the scenarios developed for the MGF index, the most favourable result in terms of temperature reduction was obtained in the MGF3 scenario (−0.31 °C), in which NbSs were introduced only on architectural surfaces. This shows that a significant share of green roofs (here located on as many as eight buildings) and climbing plants (placed on the facades of three buildings) in terms of lowering air temperature is a more effective solution than the use of a significant number of trees, but smaller ones. In addition, the applied water reservoir in the MGF1 scenario did not substantially affect the temperature reduction, which is due to its insignificant area (only 30 m^2). At the same time, it should be noted that the possibility of introducing specific NbSs is always dictated by the field possibilities in a given space. When comparing temperature reduction results, all three OGF scenarios outperformed MGF, indicating that the higher index value required in Oslo (0.8) compared to Malmö (0.6) has a positive impact.

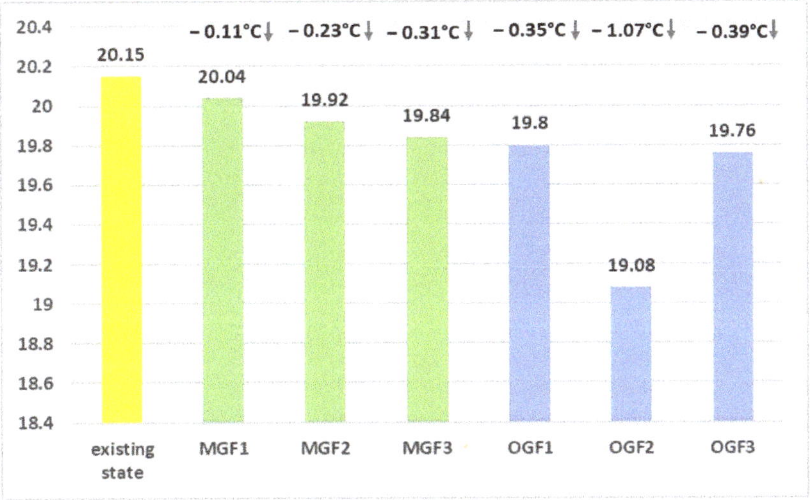

Figure 4. Impact of scenarios on the air temperature reduction.

Oxygen production and carbon sequestration are processes closely related to plant photosynthesis, and their value is influenced by the amount of biomass. The results for oxygen production and carbon sequestration are shown in Figure 5.

The highest values in this regard were obtained in the OGF2 scenario, where the increase in oxygen production is +272.69% and carbon sequestration is +137.55%. Slightly lower, but still significant results were obtained for the MGF2 scenario, where the increase was +171.91% and +81.86% for oxygen production and CO_2 sequestration, respectively. In both scenarios, such a significant increase in the studied parameters is due to the introduction of a large number of large or medium-sized trees in the estate. In the other analysed scenarios, the increase in these two parameters was at a similar level, with a slight advantage for the OGF variants.

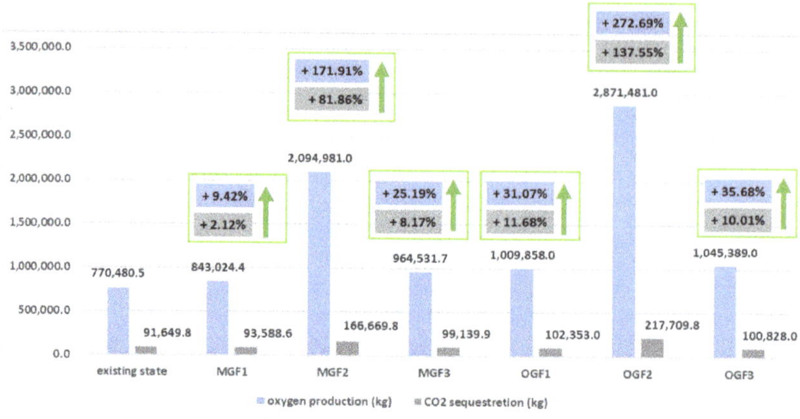

Figure 5. Impact of the scenarios on oxygen production and CO_2 sequestration.

As for rainwater harvesting, there are no solutions for doing so in the current state of development of the Klobucka housing estate. The largest amount of rainwater is allowed to be collected by the OGF1 scenario (that is, 2860.35 L), which uses 20 barrels at the gutters and extensive green roofs on four buildings (see Figure 6). The application of extensive green roofs alone on all buildings of the studied housing estate, as proposed in the OGF3 scenario, allows for the collection of 1007.46 L of rainwater. The lower value of this parameter obtained in the MGF3 scenario (i.e., 793.93 L) is due to the use of extensive green roofs not on all buildings. Scenarios based only on the introduction of vegetation, i.e., MGF2 and OGF2, do not provide any benefit in terms of water harvesting.

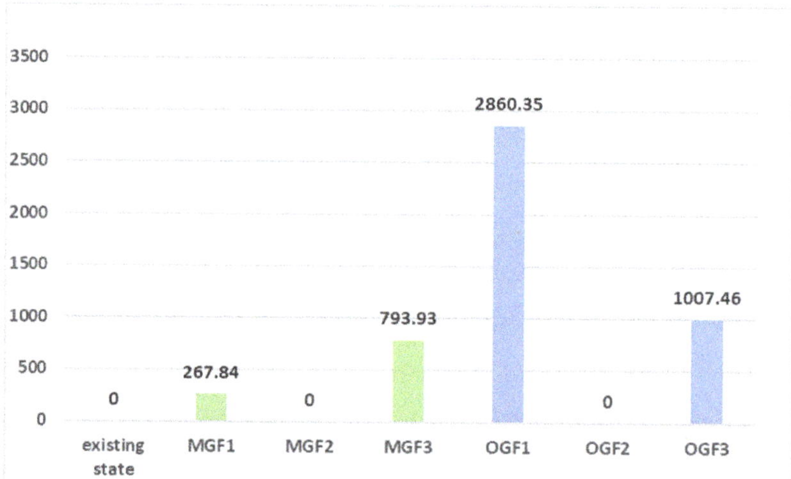

Figure 6. Impact of scenarios on water harvesting.

The last parameter examined was the effect of the applied land-use scenarios and the introduced NbSs on the level of biodiversity. All analysed scenarios increase the level of biodiversity to a moderate degree, and the most favourable in this regard is the OGF2 variant, in which this parameter increased to a value of 1.88, which is 23.7%.

4. Discussion

Our research is embedded in the trend of research on evaluating the effectiveness of eco-spatial indices (in various approaches, from improving microclimatic conditions, in the context of the impact of the use of these indicators to ensuring a certain degree of ecosystem service delivery or improving the quality of space from the point of view of landscape issues). Over the past several years, the use of various indices has been gaining popularity, precisely because of the convenience of application, flexibility, and transferability potential.

Articles comparing eco-spatial indices have appeared since the introduction of the Berlin Biotope Area Factor (BAF). In many cities, BAF was later adapted and implemented, e.g., in Malmö, Helsinki and Padua in Europe, as well as in Seattle in the USA [43,70]. These indices have also been tested in other countries, including Greece [12], Turkey [42] and Finland [43]. Among the articles comparing indices, the study by Song and Kil [16] stands out, where they compared a modified index used in Korea with the original index from Berlin, Malmö, and Seattle. The comparison primarily focused on guidelines available on their respective homepages, particularly on the categories and their associated values. These studies were preceded by others aimed at improving the index used in Korea [44,45, 71], which focused on costs, evaluations, and weights used as criteria for certifying green buildings and ecological environments.

A different approach was presented in an article by Silva et al. [46], comparing four Environmental Quotas (EQs) from São Paulo with factors from Berlin, Malmö, and Seattle. The study aimed to contribute to a better understanding of environmental quality issues, ultimately improving the EQ index in São Paulo. The research compared model solutions for the indexes on a hypothetical plot of 500 m^2 with a single building, analysing selected Nature-based Solutions (NbSs) such as small trees, green roofs, green walls, and bioretention areas. This study mainly focused on the types of solutions and the area they covered, without considering the environmental impacts of their implementation. However, it provides insight into the varying effects that the application of these indexes can have.

The discussed articles, however, did not concern applying research in real spaces; they did not take into account the entire neighbourhood area but only a single building. Specific environmental values achieved through the application of the tested indexes were also not provided. Additionally, an important contribution to the discussion on comparative studies of eco-spatial indexes is the work by Ring et al. [15], which highlighted the limitations of these indexes, including an excessive focus on environmental aspects without considering socio-cultural factors, as well as the lack of consideration for floor area ratio. This omission results in a focus on the relationship between built-up area and plot size, without considering the size of the building.

This last element is particularly emphasised in the article, where a comparison of the indices from Malmö and Oslo with the Polish index, based solely on the relationship between built-up areas and areas covered with vegetation and water, shows how different spatial, environmental, and socio-cultural impacts can result from the use of indexes based on various components. One of the major limitations in using eco-spatial indices lies in how they are calculated. For instance, the Polish RBVA indicator considers only the area covered by vegetation or water, without accounting for vegetation structure features that are included in the indices for Oslo and Malmö, such as variations in tree weight depending on whether they are planted or naturally occurring, or the circumference of the trunk. The RBVA is limited by its narrow selection of NbS elements and the omission of these important qualitative factors. The application of different scenarios for the OGF and MGF indicators illustrates how varying the complexity of an index can lead to significantly different adaptation outcomes for the same area. It is crucial to note that the weighting of individual elements ultimately determines the results achieved.

The method by which each indicator is calculated is a strategic and political choice, which in turn influences the type of NbS that is promoted. For example, in Oslo, the high weighting of large trees significantly impacts developers' decisions regarding the NbS they implement, as they align their projects with these scoring priorities.

5. Conclusions

In summary, the conducted analyses underscore that the structure of the applied Nature-based Solutions (NbSs) is a critical aspect when comparing different spatial indices and working on improving urban planning tools aimed at enhancing the quality of life in cities. The characteristics of the NbS structure, such as the type of solutions used and their assigned weight, can lead to significantly different environmental outcomes. This is particularly evident in the rather flat, two-dimensional index used in Poland, where an appropriate proportion of lawn often suffices to meet environmental requirements. In contrast, comparisons with more complex indices like those in Malmö or Oslo clearly show that environmental benefits, including improved thermal conditions and rainwater harvesting, are achieved when a more sophisticated NbS structure is considered. This is highlighted by the scenarios applied, such as focusing exclusively on elements integrated into the architectural structures of buildings (as preferred in some indices like Malmö) or placing greater emphasis on preserving large trees (as in Oslo). The research presented in the article demonstrates that well-constructed indices lead to more efficient Nature-based Solutions (NbSs), thereby enhancing adaptation to climate change. A review of only three eco-spatial indices reveals significant differences in the outcomes of their application. Expanding the analysis to include the effectiveness of other eco-spatial indices would undoubtedly be of great interest, particularly in the context of improving urban environments and, consequently, the quality of life—a key objective of the Sustainable Development Goals 2030.

Author Contributions: Conceptualisation, R.G., G.M. and A.C.; methodology, G.M.; software, G.M.; validation, G.M., R.G. and A.C.; formal analysis, G.M., R.G. and A.C.; investigation, R.G., G.M. and A.C.; resources, G.M., R.G. and A.C.; data curation, G.M.; writing—original draft preparation, G.M., A.C. and R.G.; writing—review and editing, G.M., A.C. and R.G.; visualisation, A.C. and G.M.; funding acquisition, A.C. All authors have read and agreed to the published version of the manuscript.

Funding: This research was funded by CoAdapt: Communities for Climate Change Action, grant number Adapt/0002/2020-00-Communities for Climate Change Action.

Data Availability Statement: The raw data supporting the conclusions of this article will be made available by the authors on request.

Acknowledgments: The authors would like to thank all members of the CoAdapt team.

Conflicts of Interest: The authors declare no conflicts of interest.

References

1. Copernicus: 2023 Is the Hottest Year on Record, with Global Temperatures close to the 1.5 °C Limit. Copernicus. Available online: https://climate.copernicus.eu/copernicus-2023-hottest-year-record (accessed on 9 August 2024).
2. Liu, R.; Han, Z.; Wu, J.; Hu, Y.; Li, J. The impacts of urban surface characteristics on radiation balance and meteorological variables in the boundary layer around Beijing in summertime. *Atmos. Res.* **2017**, *197*, 167–176. [CrossRef]
3. Junior, D.P.M.; Bueno, C.; da Silva, C.M. The Effect of Urban Green Spaces on Reduction of Particulate Matter Concentration. *Bull. Environ. Contam. Toxicol.* **2022**, *108*, 1104–1110. [CrossRef] [PubMed]
4. Ramyar, R.; Ackerman, A.; Johnston, D.M. Adapting cities for climate change through urban green infrastructure planning. *Cities* **2021**, *117*, 103316. [CrossRef]
5. Girard, L.F.; Nocca, F. Climate Change and Health Impacts in Urban Areas: Towards Hybrid Evaluation Tools for New Governance. *Atmosphere* **2020**, *11*, 1344. [CrossRef]
6. Perera, A.; Chathurika, S.; Davies, P.; Graham, P. A global review of urban blue-green planning tools. *Land Use Policy* **2024**, *140*, 107093. [CrossRef]
7. Szulczewska, B.; Giedych, R.; Borowski, J.; Kuchcik, M.; Sikorski, P.; Mazurkiewicz, A.; Stańczyk, T. How much green is needed for a vital neighbourhood? In search for empirical evidence. *Land Use Policy* **2014**, *38*, 330–345. [CrossRef]
8. Stange, E.E.; Barton, D.N.; Andersson, E.; Haase, D. Comparing the implicit valuation of ecosystem services from nature-based solutions in performance-based green area indicators across three European cities. *Landsc. Urban Plan.* **2022**, *219*, 104310. [CrossRef]

9. Reinwald, F.; Brandenburg, C.; Gabor, A.; Hinterkörner, P.; Kainz, A.; Kraus, F.; Ring, Z.; Scharf, B.; Tötzer, T.; Damyanovic, D. Multi-Level Toolset for Steering Urban Green Infrastructure to Support the Development of Climate-Proofed Cities. *Sustainability* **2021**, *13*, 12111. [CrossRef]
10. Pristeri, G.; Peroni, F.; Pappalardo, S.E.; Codato, D.; Castaldo, A.G.; Masi, A.; De Marchi, M. Mapping and Assessing Soil Sealing in Padua Municipality through Biotope Area Factor Index. *Sustainability* **2020**, *12*, 5167. [CrossRef]
11. Slätmo, E.; Nilsson, K.; Turunen, E. Implementing Green Infrastructure in Spatial Planning in Europe. *Land* **2019**, *8*, 62. [CrossRef]
12. Vartholomaios, A.; Kalogirou, N.; Athanassiou, E.; Papadopoulou, M. The green space factor as a tool for regulating the urban microclimate in vegetation-deprived Greek cities. In Proceedings of the International Conference on "Changing Cities": Spatial, Morphological, Formal & Socio-Economic Dimensions, Skiathos, Greece, 18–21 June 2013.
13. Khoshkar, S.; Hammer, M.; Borgström, S.; Balfors, B. Ways Forward for Advancing Ecosystem Services in Municipal Planning—Experiences from Stockholm County. *Land* **2020**, *9*, 296. [CrossRef]
14. Schrade, S.N.; Collomb, N.; Coleman, A. *Making GI Count—Tackling Climate Justice and Improving Delivery Using Urban Green Space Factors*; Town and Country Planning Association: London, UK, 2021; pp. 385–391.
15. Ring, Z.; Damyanovic, D.; Reinwald, F. Green and open space factor Vienna: A steering and evaluation tool for urban green infrastructure. *Urban For. Urban Green.* **2021**, *62*, 127131. [CrossRef]
16. Song, B.-H.; Kil, S.-H. Research Trends for Improvement of Biotope Area Index. *J. People Plants Environ.* **2018**, *21*, 103–116. [CrossRef]
17. Bush, J.; Ashley, G.; Foster, B.; Hall, G. Integrating Green Infrastructure into Urban Planning: Developing Melbourne's Green Factor Tool. *Urban Plan.* **2021**, *6*, 20–31. [CrossRef]
18. Caetano, P.M.D.; Pereira, H.M.S.B.; Figueiredo, L.C.R.; Sepe, P.M.; Giatti, L.L. The City of São Paulo's Environmental Quota: A Policy to Embrace Urban Environmental Services and Green Infrastructure Inequalities in the Global South. *Front. Sustain. Cities* **2021**, *3*, 685875. [CrossRef]
19. Podhajska, E.; Burszta-Adamiak, E.; Drzeniecka-Osiadacz, A.; Zienowicz, M.; Podhajski, B.; Sawiński, T.; Jasińska, A. Sustainability as a Function of an Area: Application of Multi-Criteria Evaluation in Assessing the Effectiveness of Nature-Based Solutions. *Atmosphere* **2021**, *12*, 1464. [CrossRef]
20. Bulkeley, H.; Naumann, S.; Vojinovic, Z.; Calfapietra, C.; Whiteoak, K.; Freitas, T.; Vandewoestijne, S.; Wild, T. *Nature-Based Solutions: State of the Art in EU Funded Projects*; Publications Office of the European Union: Luxembourg, 2020; ISBN 978-92-76-17334-2. Available online: https://data.europa.eu/doi/10.2777/236007 (accessed on 9 June 2023).
21. Seddon, N.; Chausson, A.; Berry, P.; Girardin, C.A.J.; Smith, A.; Turner, B. Understanding the value and limits of nature-based solutions to climate change and other global challenges. *Philos. Trans. R. Soc. B Biol. Sci.* **2020**, *375*, 20190120. [CrossRef]
22. Nature-Based Solutions. IUCN. Available online: https://iucn.org/our-work/nature-based-solutions (accessed on 9 August 2024).
23. Nesshöver, C.; Assmuth, T.; Irvine, K.N.; Rusch, G.M.; Waylen, K.A.; Delbaere, B.; Haase, D.; Jones-Walters, L.; Keune, H.; Kovacs, E.; et al. The science, policy and practice of nature-based solutions: An interdisciplinary perspective. *Sci. Total Environ.* **2017**, *579*, 1215–1227. [CrossRef]
24. Pauleit, S.; Zölch, T.; Hansen, R.; Randrup, T.; Konijnendijk, C. Nature-Based Solutions and Climate Change—Four Shades of Green. In *Nature-Based Solutions to Climate Change Adaptation in Urban Areas. Theory and Practice of Urban Sustainability Transitions*; Springer: Cham, Switzerland, 2017; pp. 29–49. ISBN 978-3-319-56091-5.
25. McPhearson, T.; Andersson, E.; Elmqvist, T.; Frantzeskaki, N. Resilience of and through urban ecosystem services. *Ecosyst. Serv.* **2015**, *12*, 152–156. [CrossRef]
26. Chausson, A.; Turner, B.; Seddon, D.; Chabaneix, N.; Girardin, C.A.J.; Kapos, V.; Key, I.; Roe, D.; Smith, A.; Woroniecki, S.; et al. Mapping the effectiveness of nature-based solutions for climate change adaptation. *Glob. Chang. Biol.* **2020**, *26*, 6134–6155. [CrossRef]
27. Pugliese, F.; Caroppi, G.; Zingraff-Hamed, A.; Lupp, G.; Gerundo, C. Assessment of NBSs effectiveness for flood risk management: The Isar River case study. *AQUA J. Water Supply Res. Technol.* **2021**, *71*, 42–61. [CrossRef]
28. Berardi, U.; GhaffarianHoseini, A.; GhaffarianHoseini, A. State-of-the-art analysis of the environmental benefits of green roofs. *Appl. Energy* **2014**, *115*, 411–428. [CrossRef]
29. Qin, X. A Green Roof Test Bed for Stormwater Management and Reduction of Urban Heat Island Effect in Singapore. *Br. J. Environ. Clim. Change* **2013**, *2*, 410–420. [CrossRef] [PubMed]
30. Marvuglia, A.; Koppelaar, R.; Rugani, B. The effect of green roofs on the reduction of mortality due to heatwaves: Results from the application of a spatial microsimulation model to four European cities. *Ecol. Model.* **2020**, *438*, 109351. [CrossRef]
31. Liu, W.; Engel, B.A.; Feng, Q.; Li, R. Simulating annual runoff retention performance of extensive green roofs: A comparison of four climatic regions in China. *J. Hydrol.* **2022**, *610*, 127871. [CrossRef]
32. Palermo, S.A.; Turco, M.; Principato, F.; Piro, P. Hydrological Effectiveness of an Extensive Green Roof in Mediterranean Climate. *Water* **2019**, *11*, 1378. [CrossRef]
33. Zölch, T.; Henze, L.; Keilholz, P.; Pauleit, S. Regulating urban surface runoff through nature-based solutions—An assessment at the micro-scale. *Environ. Res.* **2017**, *157*, 135–144. [CrossRef]
34. Cortinovis, C.; Olsson, P.; Boke-Olén, N.; Hedlund, K. Scaling up nature-based solutions for climate-change adaptation: Potential and benefits in three European cities. *Urban For. Urban Green.* **2022**, *67*, 127450. [CrossRef]

35. Majidi, A.N.; Vojinovic, Z.; Alves, A.; Weesakul, S.; Sanchez, A.; Boogaard, F.; Kluck, J. Planning Nature-Based Solutions for Urban Flood Reduction and Thermal Comfort Enhancement. *Sustainability* **2019**, *11*, 6361. [CrossRef]
36. Vojinovic, Z.; Alves, A.; Gómez, J.P.; Weesakul, S.; Keerakamolchai, W.; Meesuk, V.; Sanchez, A. Effectiveness of small- and large-scale Nature-Based Solutions for flood mitigation: The case of Ayutthaya, Thailand. *Sci. Total Environ.* **2021**, *789*, 147725. [CrossRef]
37. Thompson, R.; Tjaden, S.; Tilley, D. Stormwater Retention of an In-Series System Composed of a Green Roof, Constructed Wetland, and Bioretention Cell for a Single-Family Home. *J. Sustain. Water Built Environ.* **2022**, *8*, 04021023. [CrossRef]
38. Zuvela-Aloise, M.; Koch, R.; Buchholz, S.; Früh, B. Modelling the potential of green and blue infrastructure to reduce urban heat load in the city of Vienna. *Clim. Chang.* **2016**, *135*, 425–438. [CrossRef]
39. Ossola, A.; Jenerette, G.D.; McGrath, A.; Chow, W.; Hughes, L.; Leishman, M.R. Small vegetated patches greatly reduce urban surface temperature during a summer heatwave in Adelaide, Australia. *Landsc. Urban Plan.* **2021**, *209*, 104046. [CrossRef]
40. Marando, F.; Heris, M.P.; Zulian, G.; Udías, A.; Mentaschi, L.; Chrysoulakis, N.; Parastatidis, D.; Maes, J. Urban heat island mitigation by green infrastructure in European Functional Urban Areas. *Sustain. Cities Soc.* **2022**, *77*, 103564. [CrossRef]
41. Giannakis, E.; Bruggeman, A.; Poulou, D.; Zoumides, C.; Eliades, M. Linear Parks along Urban Rivers: Perceptions of Thermal Comfort and Climate Change Adaptation in Cyprus. *Sustainability* **2016**, *8*, 1023. [CrossRef]
42. Kacmaz, G. The Biotope Area Factor Method for Sustainable Urban Landscapes: The Cases of the Bornova and Bayraklı Districts, İzmir. *J. Int. Environ. Appl. Sci.* **2021**, *16*, 82–90.
43. Juhola, S. Planning for a green city: The Green Factor tool. *Urban For. Urban Green.* **2018**, *34*, 254–258. [CrossRef]
44. Jang, D.; Kim, H.; Kang, B.K. A basic study on Improvement for Biotops Area Ratio through the Post Evaluation Plan for Outdoor Space of Apartment Housings. *KIEAE J.* **2010**, *10*, 91–96.
45. Yeom, D.; Lee, K.-I. Study on the Improvement of G-SEED Management System. *J. Archit. Inst. Korea Plan. Des.* **2015**, *31*, 13–22. [CrossRef]
46. Da Silva, P.W.S.; Benites, H.S.; Monteiro, L.M.; Duarte, D.H.S. INSTRUMENTOS URBANÍSTICOS PARA INCREMENTO DE VEGETAÇÃO EM ÁREAS URBANAS: Análise comparada a partir da quota ambiental do município de São Paulo. *Cad. Zygmunt Bauman* **2019**, *8*, 167–187. Available online: https://periodicoseletronicos.ufma.br/index.php/bauman/article/view/10227 (accessed on 9 August 2024).
47. Szulczewska, B.; Giedych, R.; Blaszke, M.; Wójcik-Gront, E.; Nowak, M.; Legutko-Kobus, P. Ratio of Biologically Vital Area in Local Spatial Plans as an Instrument of Green Infrastructure Creation in Single—And Multi-Family Residential in Small and Medium-Sized Towns in Poland. *Teka Kom. Urban. Archit. Oddziału Pol. Akad. Nauk Krakowie* **2022**, *L*, 189–201.
48. Riktlinjer för Grönytefaktor (Guidelines for the Green Space Factor), Malmö Stad, 2014, Stadsbyggnadskontoret Godkänd av Stadsbyggnadsnämnden 2014. Available online: https://malmo.se/download/18.2b036ae717c5447e582895b/1637596358956/Gr%C3%B6nytefaktordec+2014.pdf (accessed on 15 July 2024).
49. *Norm for Vegetasjon og Vannhåndtering—Blågrønn Factor (Norm for Vegetation and Water Management—Blue-Green Factor), Oslo, 2023*; Plan-og Bygningsetaten: Oslo, Norway, 2023.
50. Rozporządzenie Ministra Rozwoju i Technologii z dnia 8 Grudnia 2023 r. w Sprawie Projektu Planu Ogólnego Gminy, Dokumentowania Prac Planistycznych w Zakresie tego Planu oraz Wydawania z Niego Wypisów i Wyrysów (Ordinance of the Minister of Development and Technology of 8 December 2023 on the Draft General Plan of a Municipality, the Documentation of Planning Work for This Plan and the Issuing of Extracts from It). 2023. Available online: https://isap.sejm.gov.pl/isap.nsf/download.xsp/WDU20230002758/O/D20232758.pdf (accessed on 15 July 2024).
51. Lenhart, J.; Fitzgerald, J. Eco-Districts as a Transition Pathway to Low-Carbon Cities. In *Creating Low Carbon Cities*; Dhakal, S., Ruth, M., Eds.; Springer International Publishing: Cham, Switzerland, 2017; pp. 187–199. ISBN 978-3-319-49730-3.
52. Horvath, P.; Barton, D.N.; Hauglin, E.A.; Ellefsen, H.W. *Blue-Green Factor (BGF) Mapping in QGIS. User Guide and Documentation*; Norsk Institutt for Naturforskning (NINA): Trondheim, Norway, 2017; ISBN 978-82-426-3176-3.
53. Irsalina, N.; Gamal, A. Urban Planning Simulation Game and the Development of Spatial Competence Urban Planning Simulation Game and the Development of Spatial Competence. *IOP Conf. Ser. Earth Environ. Sci.* **2019**, *396*, 012016.
54. Sousa, M.; Antunes, A.P.; Pinto, N.; Zagalo, N. Serious Games in Spatial Planning: Strengths, Limitations and Support Frameworks. *Int. J. Serious Games* **2022**, *9*, 115–133. [CrossRef]
55. Game | Projekt CoAdapt—Communities for Climate Change Action. Available online: https://coadapt.pl/en/game/ (accessed on 9 August 2024).
56. Kántor, N.; Kovács, A.; Takács, Á. Small-scale human-biometeorological impacts of shading by a large tree. *Open Geosci.* **2016**, *8*, 231–245. [CrossRef]
57. Srivanit, M.; Hokao, K. Evaluating the cooling effects of greening for improving the outdoor thermal environment at an institutional campus in the summer. *Build. Environ.* **2013**, *66*, 158–172. [CrossRef]
58. Gillner, S.; Vogt, J.; Tharang, A.; Dettmann, S.; Roloff, A. Role of street trees in mitigating effects of heat and drought at highly sealed urban sites. *Landsc. Urban Plan.* **2015**, *143*, 33–42. [CrossRef]
59. Armson, D.; Stringer, P.; Ennos, A.R. The effect of tree shade and grass on surface and globe temperatures in an urban area. *Urban For. Urban Green.* **2012**, *11*, 245–255. [CrossRef]
60. Edmondson, J.L.; Stott, I.; Davies, Z.G.; Gaston, K.J.; Leake, J.R. Soil surface temperatures reveal moderation of the urban heat island effect by trees and shrubs. *Sci. Rep.* **2016**, *6*, 33708. [CrossRef]

61. Kalkulator Korzyści z Drzew (Trees Benefits Calculator). Available online: https://www.omnicalculator.com/pl/ekologia/korzysci-z-drzew (accessed on 10 June 2024).
62. National Database of Topographic Objects. Available online: https://bdot10k.geoportal.gov.pl/ (accessed on 15 June 2024).
63. Warsaw City Map Service. Available online: https://mapa.um.warszawa.pl/en/index.html (accessed on 15 June 2024).
64. Ilu Nas Jest? A Ilu Będzie? Raporty z Planowania WARSZAWY (How Many of Us Are There, and How Many Will There Be? Warsaw Planning Reports). Available online: https://architektura.um.warszawa.pl/documents/12025039/21022710/RAPORT_Z_PROJEKTOWANIA_WARSZAWY_CZ1_Ilu_nas_jest_a_Ilu_nas_b%C4%99dzie.pdf/68b51821-af6f-ec1c-586e-279eea2fc9ce?t=1619802121397 (accessed on 15 July 2024).
65. Podstawowe Informacje o Mieszkalnictwie w Polsce—Ministerstwo Rozwoju i Technologii—Portal Gov.pl. Available online: https://www.gov.pl/web/rozwoj-technologia/podstawowe-informacje-o-mieszkalnictwie-w-polsce (accessed on 9 August 2024).
66. Warszawa Ukośna. Available online: https://ukosne.um.warszawa.pl/ (accessed on 9 August 2024).
67. Błażejczyk, A.; Błażejczyk, K.; Degórska, B.; Dudek, W.; Kręcisz, B.; Kuchcik, M.; Milewski, P.; Pałczyński, C.; Szmyd, J. (Eds.) *Miejska Wyspa Ciepła w Warszawie: Uwarunkowania Klimatyczne i Urbanistyczne*; Wydawnictwo Akademickie Sedno: Instytut Geografii i Przestrzennego Zagospodarowania PAN: Warszawa, Poland, 2014; ISBN 978-83-7963-018-9.
68. Kuchcik, M.; Milewski, P. Miejska wyspa ciepła w Warszawie—Próba oceny z wykorzystaniem Local Climate Zones. *Acta Geogr. Lodz.* **2016**, *104*, 21–33.
69. Barszcz, M. Zastosowanie modelu SWMM do prognozy przepływów prawdopodobnych w zlewni miejskiej. *Przegląd Nauk. Inż. Kształt. Śr.* **2015**, *2015*, 209–223.
70. Peroni, F.; Pristeri, G.; Codato, D.; Pappalardo, S.E.; De Marchi, M. Biotope Area Factor: An Ecological Urban Index to Geovisualize Soil Sealing in Padua, Italy. *Sustainability* **2020**, *12*, 150. [CrossRef]
71. Son, D.P. The preliminary study for the introduction of biotope area ratio in renewal project area: Focused on cost-effectiveness analysis. *J. Korea Plan. Assoc.* **2015**, *50*, 109–122. [CrossRef]

Disclaimer/Publisher's Note: The statements, opinions and data contained in all publications are solely those of the individual author(s) and contributor(s) and not of MDPI and/or the editor(s). MDPI and/or the editor(s) disclaim responsibility for any injury to people or property resulting from any ideas, methods, instructions or products referred to in the content.

land

MDPI

Article

Exploring the Impact of Urban Amenities on Business Circle Vitality Using Multi-Source Big Data

Yi Ji [1,*], Zilong Wang [1] and Dan Zhu [2]

[1] College of Economics and Management, Nanjing University of Aeronautics and Astronautics, Nanjing 211106, China; wangzilong@nuaa.edu.cn

[2] School of Economics and Management, Tianjin Chengjian University, Tianjin 300384, China; 2024010@tcu.edu.cn

* Correspondence: jiyibx2009511@nuaa.edu.cn

Abstract: Urban business circles are important locations for economic and social activities. Improving the vitality of urban business circles is conducive to stimulating the potential of the consumer market and promoting sustainable economic development. However, targeted research on the factors influencing business circle vitality is lacking. Therefore, in this study, we aimed to quantitatively examine the impact of the number and diversity of urban amenities on business circle vitality at the street block level using open-source geospatial big data from 32 Chinese metropolises. We found that the number of residential, transportation, educational, cultural, and recreational amenities and the diversity of catering and retail amenities had significant positive impacts on business circle vitality. Catering and retail diversity were the two most critical factors, followed by the number of transportation, cultural, and recreational amenities. However, the effect of urban amenities on business circle vitality varied considerably across different cities and business districts. The results of this study contribute to a holistic understanding of how to improve the vitality of business circles by optimizing urban amenities at the street block level.

Keywords: urban amenities; business circle vitality; big data

1. Introduction

As urban development gradually moves from the industrial to the post-industrial era, the contribution of traditional resources or capital investments to economic growth diminishes [1]. Modern consumption-based economies have emerged as important sources of economic growth. Thus, the importance of cities increasingly depends on their role as consumption centers [2]. As show by major cities such as New York, Tokyo and Shanghai, consumption vitality has increasingly become a new driving force of urban development. Since the 21st century, the importance of many major cities has increasingly depended on their role as centers of consumption rather than production [2]. Stimulating the vitality of the consumer market and promoting the development of urban commerce are conducive to maintaining the quality of life of residents and supporting the sustainable development of economies. Furthermore, urban consumption, leisure, and entertainment have become important attractions for young talent and are key to improving cities' competitiveness. Improving urban commercial vitality and creating a favorable consumer environment have gradually become the focuses of local governments in China [3]. The urban business circle is a radiation area where stores expand in a certain direction and distance to attract customers [4]. As a promoter of urban commercial activities, it fulfills important functions such as meeting shopping needs and providing leisure and entertainment. Therefore, exploring the impact of urban amenities on business circle vitality at the street block level is of great practical importance for meeting residents' consumption demands and contributing to sustainable urban development.

However, existing research on business circle vitality and its influencing factors is inadequate. In terms of business circle delimitation, previous researchers have assumed that a business circle radiates outward from a business center while ignoring the fact that people's social and economic activities occur via street networks [2]. Moreover, some studies have delimited business circles based on equidistant grids, which may not reflect the actual spatial structure of a city [4,5].

Furthermore, collecting large-scale data on business circle vitality from fine spatiotemporal units is challenging. In general, business circle vitality can be measured based on the volume or density of pedestrian flow in the business circle [6]. Traditional studies have relied heavily on field survey data to estimate population dynamics. This data collection is cumbersome and time consuming. In recent years, with the rapid development of communication technology, the increasing number of mobile phone users, and wide use of Web2.0 applications, the connection between cyberspace and the geographical location of human activities has been realized [7–9]. Various spatial and temporal marker data, including global positioning system (GPS) data, mobile phone signals, and social media location records, are gradually making it possible to measure urban vitality accurately [10,11]. However, previous studies have examined the vitality of the city as a whole, ignoring the categorical refinement of functions within the city [12–14]. On this basis, scholars have studied the impact of factors such as mixed land use [15,16], the built environment [17–20], the urban landscape [21,22], urban design [23,24], urban traffic planning [25] and urban infrastructure [1] on urban vitality. However, these studies are not well targeted because the functions and social attributes of different areas within the city are different.

The urban business circle is the urban core functional area, with intensive land use and active commercial activities. It is also an important place in terms of enhancing urban consumption vitality, affecting the economic development of the entire city. Therefore, it is very important to conduct in-depth research on business circle vitality in order to propose targeted optimization paths. Research on the impact of urban amenities on business circle vitality is limited. Additionally, there is a lack of research on the heterogeneity of this effect across cities and business circles.

To address the gaps in the existing research, we used multi-source big data and the temporal distance to delimit urban business circles based on the identification of street blocks. Additionally, rather than focusing on urban vitality at the city scale, we studied more refined and targeted business circles and aimed to examine the impact of the number and diversity of urban amenities on business circle vitality at the block level. Moreover, we expanded the study of urban business circle vitality from a single city to 32 cities in China and analyzed heterogeneity across different cities and business circles.

The remainder of this paper follows. Section 2 reviews and summarizes the literature on urban business circle vitality. Section 3 details the data and methodology. Section 4 presents the results of this study. Section 5 discusses the results and Section 6 concludes this paper.

2. Literature Review

2.1. Delimitation of Urban Business Circles

In terms of the delimitation of urban business circles, existing studies are mainly based on the qualitative identification of business centers or the quantitative delimitation of business circles based on point of interest (POI) data. For example, Shi et al. [2] unified the study area as core business circles with a uniform radius of 500 m when analyzing the spatial characteristics of the number of takeouts in Shanghai's business circles. Shi et al. [5] used POI data from shopping centers to visualize the spatial distribution and structural characteristics of Shanghai's business circles through hot spot analysis. These methods typically assume that geographical space is a continuous uniform plane and that the distance between two points can be measured using the Euclidean distance. In reality, geographical spaces are heterogeneous, with various geographical elements along a network of streets [26]. Therefore, scholars have proposed replacing the plane with road network space, and replacing plane kernel density estimation (KDE) with network KDE [27]. However, both planar and network KDE only change the similarity measurement method for commercial POIs and do not consider the characteristics of road network density in business districts. Based on this, Yang et al. [28] proposed a commercial intersection KDE method that combines road intersections with KDE and used it to identify the central business district of Nanjing using POI data and road networks. In contrast to previous studies, the current study proposes a method to delimit business circles using temporal distance by combining multi-source big data on factors such as POIs and road networks and from sources such as location-based services (LBSs). When customers decide which store to visit, they prefer those where visits take less time. Therefore, temporal distance, which is closer to psychological distance, is more valuable in decision making and can delimit urban business circles with high precision, but has been rarely addressed in previous research [26].

2.2. Urban Business Circle Vitality

With the widespread use of geotagged data, such as mobile phone locations, social media check-ins, and transit smart cards, researchers can capture the concentration of people in cities in a timely and efficient manner [7,9]. This provides new opportunities to measure urban business circle vitality quantitatively. However, only a few quantitative analyses of the factors that influence the vitality of urban business circles have been conducted. Relevant studies have mainly focused on the overall city level [12–14]. Gao et al. [29] found that POI density, building density, and road intersection density had a significant positive effect on urban vitality, using Munich as an example. Mouratidis and Poortinga [30] used geospatial data from the Oslo metropolitan area to establish that urban vitality was mainly shaped by neighborhood density and land use mix. Huang et al. [31] found that urban functions, accessibility, and walkability were associated with urban vitality. Wu et al. [32] argued that a vibrant urban environment relies on effectively developed urban functions catering to residents' needs. Wang et al. [16] measured the urban vitality of Beijing using social media check-in data, finding that enriching land use functions contributed to enhancing urban vitality. Zhang et al. [21] identified convenient transportation, compact city block forms, and diverse building esthetics as the main characteristics influencing the vitality of a city. However, these studies, which were based on the city as a whole, lacked specificity.

In micro-scale studies, different types of activity spaces in cities should be considered. The main types of activity space found in urban business circles relate to commerce, leisure, and entertainment. Commerce is one of the most important functions of a city. A reasonable commercial layout and diversified commercial types are necessary for the occurrence of urban vitality. However, scholars have paid little attention to the urban business circle, which is a representative urban public space. Gao et al. [33] explored business district vitality in Guangzhou using mobile phone data and found that factors such as metro accessibility, POI density, road density, and the land use mixture were positively correlated with business district vitality. However, this study focused on a single city. In the current study, we subdivided various types of urban amenities, explored the impact of amenities on business circle vitality in multiple cities, and analyzed their heterogeneity.

3. Materials and Methods

3.1. Study Area and Data Sources

We selected the built-up areas of 32 representative metropolises in China as the study areas: Beijing, Shanghai, Shijiazhuang, Taiyuan, Xi'an, Jinan, Zhengzhou, Shenyang, Changchun, Harbin, Nanjing, Hangzhou, Hefei, Nanchang, Fuzhou, Wuhan, Changsha, Chengdu, Guiyang, Kunming, Guangzhou, Lanzhou, Xining, Hohhot, Urumqi, Nanning, Yinchuan, Dalian, Ningbo, Haikou, Shenzhen, and Qingdao. Most of these cities are provincial capitals from a wide range of geographic regions, spanning the major economic, political, and cultural regions of China. Furthermore, the data from these cities are relatively complete, reliable, and easy to collect.

The data included the following categories: (1) The first category was street network data. City blocks were generated based on street network data obtained from Gaode Map, which is a popular electronic navigation map in China. The original data contained many details that could lead to mistakes when converting road polylines into block polygons. Therefore, we performed preprocessing to simplify the original data. (2) The second category was LBS data. The density of pedestrian flow in urban business circles was obtained from "Tencent Location Big Data". Data were collected on 23 October 2019 (weekday), and 26 October 2019 (weekend), at 10:00, 19:00, and 22:00 local time. We averaged the density of the pedestrian flow for these periods. (3) The third category was POI data. We collected POI data from multiple sources. Gaode Map mainly includes location information for POIs, such as transportation, education, healthcare, dining, shopping, finance, attractions, housing, sports, government, and companies, with a total of 13 classifications. Data on the number of residences and office buildings were obtained from "Home Link", which is a real estate brokerage firm. The location information for various indicators of retail and catering diversity was derived from "Dianping", which is a popular app used to categorize and rate local businesses. Data were collected from 18 to 20 October 2019. Table 1 lists the sources of each type of data. All data contained longitudinal and latitudinal information. We integrated all data based on latitude and longitude coordinates. There are potential problems in using different data types, such as inconsistent data locations and invalid data. Therefore, we bias-corrected and processed the acquired data.

Table 1. Data sources.

Data	Type	Data source
Street network data		Gaode Map (https://www.amap.com) (accessed on 20 October 2019)
LBS data		Tencent Location Big Data (https://heat.qq.com/bigdata/index.html) (accessed on 23, 26 October 2019)
POI data	Transportation, education and healthcare, etc., 13 classifications	Gaode Map (https://www.amap.com) (accessed on 20 October 2019)
	Residences and office buildings	Home Link (https://www.lianjia.com) (accessed on 18 October 2019)
	Various indicators of retail and catering diversity	Dianping (https://www.dianping.com) (accessed on 18 to 20 October 2019)

3.2. Methodology

3.2.1. Measuring Urban Business Circle Vitality

The explained variable in this study is urban business circle vitality, which can be measured based on the density of pedestrian flow in a business circle. We synthesized road network, POI, and LBS data to measure the density of pedestrian flow in business circles in the selected cities. First, street blocks were identified. The street block network schematic is an abstract representation of physical blocks [26]. Compared to an equidistant grid division, it retains the original spatial structure of the city and does not mechanically divide the urban amenities and environment, resulting in a schematic that better aligns with people's cognitive habits. We considered a closed polygon (mesh) as the basic unit and used the independence and adjacency of the mesh to form the polygon-growing algorithm. The core ideas of the polygon-growing algorithm are to classify the road network into closed polygons and non-closed line segments and to complete the identification of street blocks through the closed polygon extraction, mapping, and optimization of the main line and the non-closed line segment combination of the auxiliary [24].

Following street block identification, we organized the collected POI data into street blocks for statistical analyses. We classified the blocks according to the most dominant POI category and merged adjacent blocks of the same type. Based on the LBS data, we screened the blocks classified as businesses and selected those with a pedestrian flow density of more than 1000 people/hm^2 as potential urban business circles.

As a business circle is defined as a radiation area where stores expand in a certain direction and distance to attract customers [4], we identified the central gathering points for these business blocks based on the data showing the highest pedestrian flow density. We calculated the areas within a 5 min, 10 min, and 15 min walk from the central gathering points. A schematic of this process is shown in Figure 1a. On this basis, we matched the areas within a 15 min walk from the central point to city blocks. We chose 15 min because it is a relatively comfortable and acceptable walking time threshold for most people. We merged the boundaries of the blocks that intersected with those of reachable areas to form the boundaries of the urban business circle (Figure 1b). After delimiting the boundaries of the business circle, we calculated the density of the pedestrian flow in the business circle.

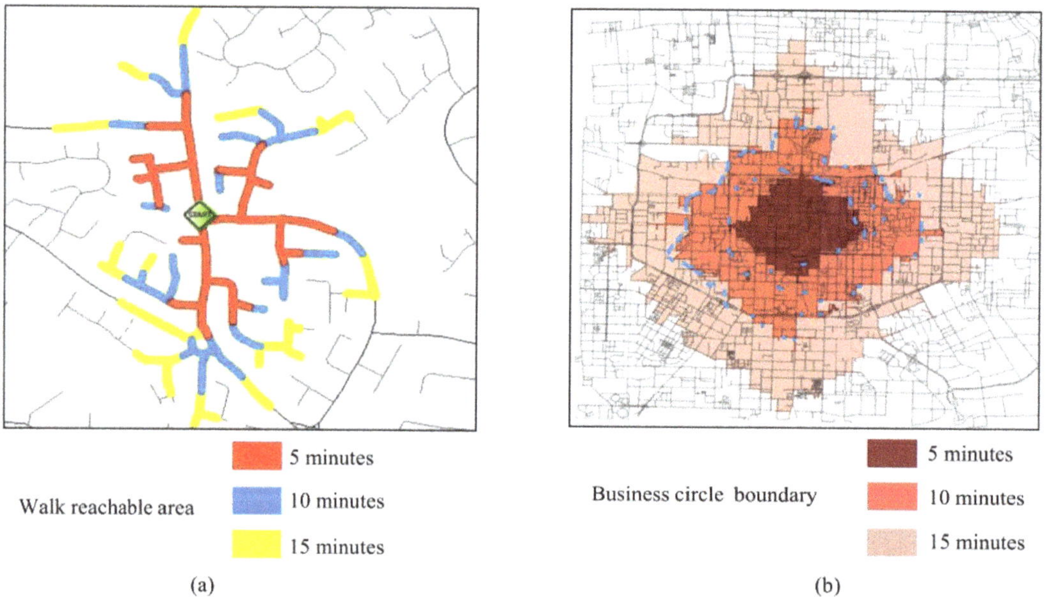

Figure 1. (**a**) Walk reachable area; (**b**) business circle boundary (walking time).

3.2.2. Measuring the Number and Diversity of Urban Amenities

In this study, we explored the impact of urban amenities on business circle vitality at the block level. The vitality of an urban business circle depends on its market potential, that is, how many potential consumers can conveniently reach the area [26]. In turn, market potential is influenced by the population living, studying, and working in the area, as well as the transportation accessibility of the area. Therefore, residential, office, educational, and transportation amenities may affect urban business circle vitality. Furthermore, the diversity of retail amenities enables consumers to achieve multiple consumption objectives in a single trip, while different types of catering amenities allow consumers to seek consumption opportunities aligned with their preferences. Therefore, the diversity of retail and catering amenities is an important factor impacting urban business circle vitality. Moreover, consumer demands are becoming increasingly complex as living standards improve. Experience amenities, such as cultural activities and recreation, are increasingly critical for attracting consumers, and their importance in contributing to business circle vitality is becoming more apparent.

In summary, the explanatory variables in this study included the following categories: (1) The first was residential amenities. We measured this using the number of residential households in the business circle. (2) The second was office amenities. We used the number of office buildings in the business circle as a proxy. (3) The third was educational amenities. This included the number of higher learning institutions in the business circle. (4) The fourth was transportation amenities. We measured this based on the number of subway stations in the business circle. (5) The first included cultural and recreational amenities. This included the number of cinemas and theaters in the business circle. (6) The sixth was the catering diversity index. This was calculated using Equation (1). (7) The seventh was the retail diversity index. This was calculated using Equation (2).

Catering amenities in this study included six categories: Chinese restaurants, Western restaurants, dessert stores, fast food restaurants, cafes, and teahouses. Descriptive statistics for the various indicators are shown in Table 2. We utilized the Shannon index, which is commonly used to calculate species diversity in biology, to construct a catering diversity index as follows:

$$Catering_i = -\sum_{j=1}^{5} \frac{R_{ij}}{R_i} \times \ln(\frac{R_{ij}}{R_i}) \quad (1)$$

where $Catering_i$ represents the diversity of catering in the ith business circle; R_i denotes the total number of catering amenities in the ith business circle; and R_{ij} denotes the number of catering amenities in category j in the ith business circle. The greater the variety of catering amenities and the more evenly the different types of catering amenities are distributed, the greater the catering diversity index.

Table 2. Descriptive statistics of catering and retail amenities.

Variable	Mean	Std.	Max.	Min.
Chinese restaurant	528.39	354.45	2142	0
Western restaurant	29.96	37.81	352	0
Dessert store	10.19	12.99	110	0
Fast food restaurant	97.12	69.44	442	0
Cafe	16.80	21.33	214	0
Teahouse	17.77	30.09	320	0
Clothing store	289.93	301.91	2100	0
Pet store	5.82	5.85	37	0
Sports store	0.73	1.51	19	0
Baby supply store	45.12	40.97	335	0
Appliance and electronics store	33.39	36.62	339	0
Supermarket	7.87	8.08	53	0

Retail amenities included six categories: clothing stores, baby supply stores, sports stores, appliance and electronics stores, pet stores, and supermarkets. Descriptive statistics for the various indicators are shown in Table 2. Using the Shannon index as a reference, we constructed a retail diversity index as follows:

$$Retail_i = -\sum_{j=1}^{5} \frac{S_{ij}}{S_i} \times \ln(\frac{S_{ij}}{S_i}) \quad (2)$$

where $Retail_i$ denotes the retail diversity of the ith business circle; S_i denotes the total number of retail amenities in the ith business circle; and S_{ij} denotes the number of retail amenities in category j in the ith business circle.

3.2.3. Control Variables

We controlled for other factors affecting business circle vitality, such as proximity to nearby business circles [33], the economic development level of the city, and the average temperature of the city [34]. The control variables in this study included the distance to the nearest neighboring business circle (DN), urban GDP, and annual average urban temperature. The data for the control variables at the city level were obtained from the China Urban Statistical Yearbook 2020.

3.2.4. Analysis Methods

To understand the effect of various urban amenities on business circle vitality, we constructed an econometric model, as shown in Equation (3). We used ordinary least squares as the regression method with which to estimate the impact of various explanatory variables on the explained variable.

$$ln(Vitality_i) = \beta_0 + \beta_1 ln(X_i) + \beta_2 lnZ_i + \varepsilon_i \quad (3)$$

where the explained variable $Vitality_i$ is the density of pedestrian flow in the business circle; the explanatory variable X_i refers to the indicators of various amenities in the business circle; Z_i is the control variable; β_0, β_1, and β_2 are the parameters to be estimated; and ε_i represents the disturbance term. Descriptive statistics of the relevant variables and indicators in this study are presented in Table 3.

Table 3. Descriptive statistics of relevant variables.

Variable	Mean	Std.	Max.	Min.
Vitality	3562.32	821.85	1100	7218
Residential	23,922.64	30,166.92	186,321	0
Office	15.94	35.72	230	0
Educational	6.57	8.32	48	0
Transportation	5.94	9.00	69	0
Cultural	4.50	5.17	13	0
Catering	0.79	0.21	1.40	0
Retail	0.72	0.18	1.09	0
DN	2.40	5.50	8.65	0
GDP	114,801.20	32,228.12	189,568	54,439
Temperature	15.59	4.62	24.40	5.10

4. Results

4.1. Analysis of Urban Business Circle Vitality

Following the methodology described above, we measure the vitality of 1471 business circles in 32 major cities across China. Among these, Beijing ranked highest with 152 business circles, Guangzhou ranked second with 109 business circles, and Shanghai ranked third with 105 business circles. A boxplot of the vitality of each city's business circles is shown in Figure 2. The x shapes in Figure 2 represent the average vitality of all business districts in the city. Overall, the vitality of business circles was higher in southern cities than in northern cities and higher in eastern cities than in midwestern cities. Business circles in Shenzhen, Guangzhou, and Shanghai had the greatest vitality. The business circle vitality of Shanghai, Changsha, and Shenzhen fluctuated markedly, whereas that of Harbin, Shenyang, and Dalian was more stable. Among all the business circles, the most vital were the Renmin South and Dong Men business circles in Shenzhen, followed by the Tianhe and Gang Ding business circles in Guangzhou.

We selected Beijing and Shanghai as examples in order to refine the urban block distribution and business circle vitality. We chose these two cities as representatives because Beijing and Shanghai are the political and economic centers of China, respectively. Figure 3a,b show the distribution and vitality measurements of street blocks with pedestrian flow densities of over 1000 people/hm^2 in Beijing and Shanghai, respectively. Based on this, we identified street blocks that could be classified as business blocks in each city.

Figure 2. Boxplot of business circle vitality in 32 major cities.

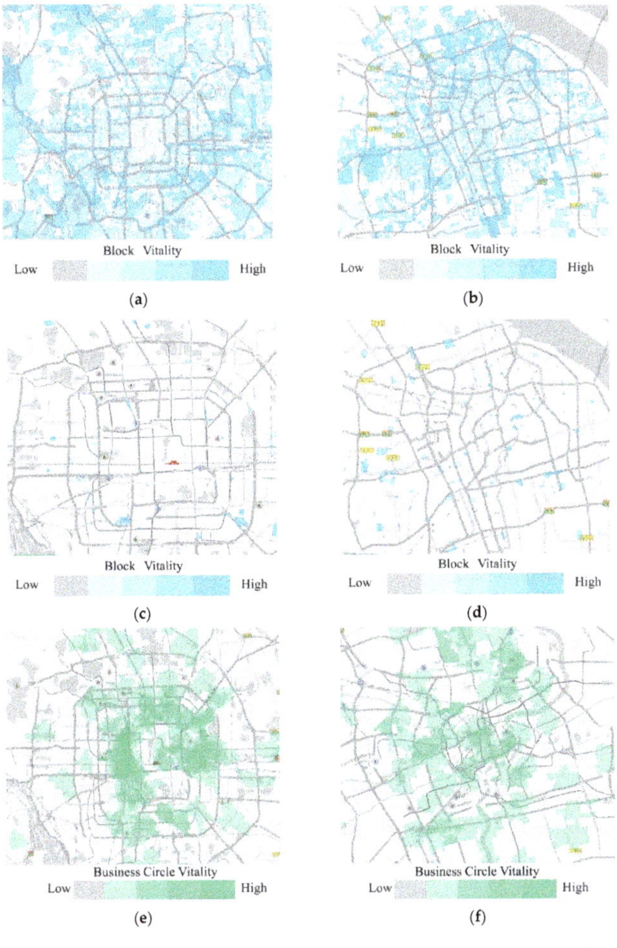

Figure 3. (**a**) Vitality of street blocks in Beijing; (**b**) vitality of street blocks in Shanghai; (**c**) vitality of business blocks in Beijing; (**d**) vitality of business blocks in Shanghai; (**e**) vitality of business circles in Beijing; (**f**) vitality of business circles in Shanghai.

Figure 3c,d show the measurement results for the vitality of street blocks classified as businesses in Beijing and Shanghai, respectively. Finally, based on the delineated boundaries of the business circles, Figure 3e,f show the vitality measurement results for the major business circles in Beijing and Shanghai, respectively. Figure 3e shows that Beijing's business circles were mainly distributed along the ring roads, showing a form of outward expansion centered on the Inner Ring. Business circles with high vitality included Guo Mao, Dong Zhimen, and San Litun. Figure 3f shows that the business circles in Shanghai with the highest vitality mainly included Wu Jiaochang, Bund, and Lu Jiazui.

4.2. Benchmark Regression Analysis

The regression results for the impact of urban amenities on business circle vitality are presented in Table 4. Model 1 includes the explained variable and all explanatory variables. Model 2 further introduces control variables at the block level in addition to the variables used for Model 1, and Model 3 introduces city-level control variables in addition to these variables. The regression results from each model show that most urban amenities had significant positive effects on the vitality of business circles. First, the positive impact of the number of residential amenities on business circle vitality was significant at the 1% level, but the influence coefficient was small. In China's urban planning, large business complexes are mostly located at sites that are independent of residential areas. This plan expects the functional zoning of large-scale commercial and residential areas to minimize the impact of commercial activities on the residential environment. Therefore, the coefficient of influence of residential amenities was small.

Table 4. Estimation results of the benchmark regression.

Variables	Model 1	Model 2	Model 3
Residential	0.010 ***	0.009 ***	0.011 ***
	(0.003)	(0.003)	(0.003)
Office	0.011	0.010	0.009
	(0.011)	(0.011)	(0.010)
Educational	0.068 ***	0.062 ***	0.059 ***
	(0.024)	(0.010)	(0.011)
Transportation	0.080 ***	0.079 ***	0.077 ***
	(0.015)	(0.015)	(0.015)
Cultural	0.150 ***	0.152 ***	0.159 ***
	(0.024)	(0.024)	(0.023)
Catering	0.499 ***	0.504 ***	0.488 ***
	(0.096)	(0.096)	(0.116)
Retail	0.977 ***	0.982 ***	0.921 ***
	(0.096)	(0.095)	(0.127)
DN		−0.026 *	−0.026 *
		(0.015)	(0.015)
GDP			0.048 **
			(0.020)
Temperature			0.007
			(0.013)
Constant	12.974 ***	12.989 ***	12.837 ***
	(0.105)	(0.105)	(0.167)
Obs	1471	1471	1471
R^2	0.143	0.148	0.176

Note: robust standard errors in parentheses. * $p < 0.1$. ** $p < 0.05$. *** $p < 0.01$.

Second, the number of office amenities had an insignificant positive impact on business circle vitality. The vacancy rate in office buildings in China has been increasing annually. Most office buildings are planned and constructed ahead of time, ignoring the current economic development of the city and failing to attract enough enterprises and people, resulting in the planning effect falling short of expectations.

Third, the number of educational amenities had a significant positive impact on business circle vitality. Students—particularly college students—are among the most dynamic consumer groups. They typically have more time and opportunities to spend than other demographic groups and are prone to impulsive spending.

Fourth, the regression results show that transportation amenities had a significant positive effect on business circle vitality, with a large coefficient of influence. Urban subways ease travel difficulties caused by congestion, reducing the time cost for consumers to travel to business circles and increasing the accessibility and market potential of business circles.

Fifth, cultural and recreational amenities had a significant positive impact on business circle vitality, with a large coefficient of 0.159. With improvements in people's living standards and the development of e-commerce, consumers' offline consumption demand has gradually tended toward entertainment and cultural experiences. Business circles are no longer just shopping spaces, but have developed into integrated service spaces.

Sixth, the diversity of retail amenities had a significant positive impact on business circle vitality, with the largest impact coefficient. Diversified retail amenities meet consumers' diverse needs. Despite the impact of e-commerce, offline retail remains irreplaceable in terms of merchandise, services, and environmental experience, which remain important factors in attracting people to business circles.

Seventh, the diversity of catering amenities had a significant positive effect on business circle vitality, with a large impact coefficient of 0.488. A wide variety of catering amenities can meet the preferences of different groups of people and attract consumers of different ages and incomes and consumers from different regions. Moreover, the public's consumption demand for business circles has evolved, shifting from a single shopping mode to deeper modes involving experience, emotion, and socialization. Catering amenities provide people with convenient places for communication and gathering.

4.3. Heterogeneity Analysis

We explored the heterogeneity of the impact of urban amenities on business circle vitality among different cities and business circles. To this end, we first divided China's southern and northern cities according to the geographical boundary of the Qinling–Huaihe Line. Columns 1 and 2 in Table 5 present the regression results. The impact coefficients of transportation amenities, cultural and recreational amenities, and catering diversity were significantly higher in southern cities than in northern cities. This may be due to the higher number of migrants and more frequent international trade and commerce in southern cities, which are typically more diverse in terms of cultural exchanges and types of cuisine.

Table 5. Estimation results for regional heterogeneity.

Variables	(1)	(2)	(3)	(4)	(5)
Residential	0.001	0.015 ***	0.010 **	−0.006	0.019 *
	(0.006)	(0.004)	(0.004)	(0.007)	(0.010)
Office	0.013	0.014	0.013	−0.038	−0.039
	(0.015)	(0.012)	(0.012)	(0.018)	(0.027)
Educational	0.070 ***	0.052 ***	0.059 ***	0.070 ***	0.096 **
	(0.017)	(0.014)	(0.013)	(0.021)	(0.039)
Transportation	0.085 ***	0.053 **	0.109 ***	0.087 ***	0.011
	(0.022)	(0.021)	(0.021)	(0.025)	(0.033)
Cultural	0.175 ***	0.148 ***	0.365 ***	0.159 ***	0.109 ***
	(0.035)	(0.030)	(0.071)	(0.034)	(0.032)
Catering	0.640 ***	0.165	0.689 ***	0.438 *	0.701
	(0.152)	(0.187)	(0.144)	(0.245)	(0.557)
Retail	0.849 ***	0.838 ***	1.020 ***	0.633 ***	1.011
	(0.152)	(0.254)	(0.141)	(0.147)	(0.659)
Control	YES	YES	YES	YES	YES
constant	12.554 ***	12.973 ***	12.972 ***	12.903 ***	12.536 ***
	(0.304)	(0.173)	(0.141)	(0.176)	(0.925)
Obs	636	835	878	353	240
R^2	0.204	0.170	0.186	0.224	0.262

Note: robust standard errors in parentheses. * $p < 0.1$. ** $p < 0.05$. *** $p < 0.01$.

Second, we categorized the 32 cities into eastern, central, and western regions according to their geographical locations. The regression results are shown in Columns 3, 4, and 5 of Table 5. Moving from east to west, the impact coefficients or significance of transportation amenities, cultural and recreational amenities, catering diversity, and retail diversity regarding business circle vitality diminished. One reason for this difference is that the traffic in eastern cities is relatively congested. Improvements in transportation amenities can support more convenient travel, thus increasing the vitality of business circles. Another factor contributing to this variation is that the populations of western cities are relatively sparse, and the consumption demand of the cultural markets is relatively small. The construction of large-scale cultural amenities can be costly and trigger local debt problems. Additionally, the eastern coastal cities opened earlier, leading to better-developed logistics and trade compared to that in western cities. As a result, catering and retail amenities have become more diverse.

Third, we divided the 32 cities into first-tier cities and others. The first-tier cities included Beijing, Shanghai, Guangzhou, and Shenzhen. Columns 1 and 2 in Table 6 present the regression results. The impact coefficients of cultural and recreational amenities, catering diversity, and retail diversity were greater for first-tier cities than for other cities. Residents of first-tier cities have relatively higher incomes and education levels and tend to participate more in cultural activities. The theaters in first-tier cities also have more frequent performances and a wider variety of programs than those in other cities. Moreover, first-tier cities are home to relatively more business enterprises and populations from different regions and countries. The greater diversity of catering and retail amenities in first-tier cities can cater to a wider range of consumer demands.

Table 6. Estimation results for heterogeneity.

Variables	(1)	(2)	(3)	(4)
Residential	0.068 *	0.009 **	0.003	0.019 **
	(0.036)	(0.004)	(0.003)	(0.008)
Office	0.021	−0.019 *	0.017 *	−0.073
	(0.019)	(0.011)	(0.010)	(0.042)
Educational	0.011	0.081 ***	0.042 ***	0.110 ***
	(0.023)	(0.013)	(0.012)	(0.027)
Transportation	0.134 ***	0.057 ***	0.080 **	0.072 ***
	(0.047)	(0.014)	(0.035)	(0.012)
Cultural	0.221 ***	0.106 ***	0.187 ***	0.071
	(0.046)	(0.025)	(0.024)	(0.060)
Catering	1.004 ***	0.335 **	0.621 ***	0.419
	(0.223)	(0.139)	(0.124)	(0.271)
Retail	0.978 ***	0.900 ***	0.992 ***	0.756 ***
	(0.223)	(0.141)	(0.233)	(0.126)
Control	YES	YES	YES	YES
constant	12.908 ***	12.737 ***	13.258 ***	12.484 ***
	(0.338)	(0.212)	(0.165)	(0.309)
Obs	406	1065	1132	339
R^2	0.195	0.201	0.116	0.280

Note: robust standard errors in parentheses. * $p < 0.1$. ** $p < 0.05$. *** $p < 0.01$.

Finally, according to the presence or absence of shopping centers in business circles, we classified the 1471 business circles into two types: business complexes and neighborhood businesses. Columns 3 and 4 of Table 6 show the regression results. For large business circles dominated by shopping centers or complexes, the coefficients of all urban amenity variables, except residential amenities, were significantly positive. This type of business circle was mostly located in the center of the city, with a large number of supporting office buildings, and was separated from the urban residential area. Therefore, the positive effect was significant for office amenities but not for residential amenities. For business circles dominated by neighborhood businesses, the coefficients were significantly positive for residential amenities and insignificant for cultural and recreational amenities. This is because most consumers in this type of business circle are nearby residents. Moreover, cultural and recreational amenities such as theaters or cinemas need to be supported by a larger consumer market and have higher market entry thresholds. They are seldom located in neighborhood business circles.

5. Discussion

The existing literature mainly focuses on the vitality of entire cities [12–14,35]. These large-scale studies have failed to differentiate between the functional areas within cities and thus lack pertinence. In the current study, we focused on targeted urban business circles. In contrast to the previous delineation of business circles based only on POI data [4,5,32], we utilized street network, POI, and LBS data to delimit urban business circles by temporal distance, which is more consistent with the characteristics of residents' actual daily economic activities. On this basis, we visualized urban business districts and business circle vitality at the street block level, thus presenting more detailed and representative measurement results.

This study focused on the impact of urban amenities on business circle vitality at the block level. The results show that urban amenities had a significant positive impact on business circle vitality. Unlike previous studies that combined all amenity data [1,16,17,19,21,23], we subdivided different types of urban amenities. The results demonstrate the important role of retail diversity, catering diversity, transportation amenities, and cultural and recreational amenities in enhancing business circle vitality, and highlight the insignificant role of office amenities. This information is important for optimizing urban amenities and enhancing the vitality of business circles. Optimizing the urban business circle requires that attention be paid to the diversification and differentiation of catering and retail amenities. Cultural and recreational amenities need to provide richer immersive experiences and diversified services. The construction of urban transportation amenities needs to be coordinated with the development of other urban amenities to create synergy. The promotion of urban business circle vitality requires the spontaneous formation of economic and social networks among various market players. The government can assist with this process, but it is not appropriate to intervene excessively by, for example, overplanning a large number of office buildings. For a long time, China's land-centered urbanization sprawl has been out of balance with actual market demand, resulting in many skyscrapers being unused or abandoned [15]. Urban planners need to be people-centered in order to adapt the layout and design of amenities to local demographic and economic conditions [36]. Governments need integrated strategic land use planning that meets actual market demands to obviate opportunistic and speculative land markets. These insights could inform strategic planning and regeneration practices for urban sustainability in China, as well as other economies facing transformation and demands for improved urban quality of life.

To our knowledge, this is the first study to extend the examination of urban business circle vitality from a single city to 1471 business circles in 32 cities. A larger sample size allowed us to achieve greater heterogeneity compared to that in previous studies [37,38]. Additionally, we explored the variation in the impact of urban amenities on business circle vitality across different cities and business circle types. We believe that these heterogeneity analyses have practical applications. Large geographical differences exist in China, such as those between the north and south and those between the east and west. Therefore, policy implementation should consider the geographical environment in which urban business circles are located and optimize the quantity, quality, and diversity of various urban amenities according to local conditions. The enhancement of urban business circle vitality cannot be achieved by blindly investing in the construction of high-end urban amenities that lack a consumption base among local residents. Urban managers should emphasize the importance of context-specific planning and design when formulating spatial strategies for urban business development, considering the specific vitality characteristics of business circles in different areas.

6. Conclusions

In this study, we explored the impact of urban amenities on business circle vitality at the block level using multi-source big data. Overall, this study serves as a valuable reference for sustainable urban development measures. The results are as follows:

(1) Most urban amenities, such as residential, transportation, educational, cultural, and recreational amenities, and the diversity of catering and retail amenities had a significant positive effect on business circle vitality, but the effect of the number of office amenities was not significant. The diversity of catering and retail amenities was one of the most critical factors.

(2) The effect of urban amenities on business circle vitality was characterized by large regional heterogeneity. The positive impacts of transportation amenities, cultural and recreational amenities, and catering and retail diversity were greater in the southern and eastern cities than in the northern and midwestern cities.

(3) There were large city-level differences in the impact of urban amenities on business circle vitality. The positive effects of cultural, recreational, catering, and retail amenities were greater in first-tier cities than in other cities.

(4) The impacts of urban amenities on different types of business circles were heterogeneous. For business circles dominated by shopping centers, the positive impacts of all urban amenities, except residential amenities, were significant. For business circles dominated by neighborhood businesses, the positive impact of residential amenities was significant, while the impact of cultural and recreational amenities was not.

This study had some limitations. For example, we could not control for all the factors affecting urban business circle vitality, such as policy systems and cultural practices, which are challenging to quantify. Although we analyzed urban business circle vitality based on multiple types of spatiotemporal big data, future studies can include additional types of data such as social media comment data, traffic trajectory data, and credit card transaction data. Finally, future researchers should consider using spatial econometric or geographically weighted regression models to better explore the impact of urban amenities on urban business circle vitality.

Author Contributions: Conceptualization, Y.J.; data curation, Y.J.; formal analysis, Y.J.; funding acquisition, Z.W.; investigation, D.Z.; methodology, Y.J.; project administration, D.Z.; software, Y.J.; validation, Y.J.; visualization, Y.J.; writing—original draft, Y.J.; writing—review and editing, Z.W. All authors have read and agreed to the published version of the manuscript.

Funding: This research was funded by Major Program of National Fund of Philosophy and Social Science of China (Grant number: 20ZDA092).

Data Availability Statement: The raw data supporting the conclusions of this article will be made available by the authors on request.

Acknowledgments: We would like to thank the reviewers for their valuable comments and suggestions, which played a positive role in improving the content of our paper.

Conflicts of Interest: The authors declare no conflicts of interest.

References

1. Lan, F.; Gong, X.Y.; Da, H.L.; Wen, H.Z. How Do Population Inflow and Social Infrastructure Affect Urban Vitality? Evidence from 35 Large- and Medium-Sized Cities in China. *Cities* **2020**, *100*, 102454. [CrossRef]
2. Shi, Y.S.; Tao, T.H.; Cao, X.Y.; Pei, X.W. The Association between Spatial Attributes and Neighborhood Characteristics Based on Meituan Take-out Data: Evidence from Shanghai Business Circles. *J. Retail. Consum. Serv.* **2021**, *58*, 102302. [CrossRef]
3. Zhang, M.; Partridge, M.D.; Song, H.S. Amenities and the Geography of Innovation: Evidence from Chinese Cities. *Ann. Reg. Sci.* **2020**, *65*, 105–145. [CrossRef]

4. Zhou, L.L.; Shi, Y.S.; Zheng, J.W. Business Circle Identification and Spatiotemporal Characteristics in the Main Urban Area of Yiwu City Based on Poi and Night-Time Light Data. *Remote Sens.* **2021**, *13*, 5153. [CrossRef]
5. Shi, Y.S.; Wu, J.; Wang, S.Y. Spatio-Temporal Features and the Dynamic Mechanism of Shopping Center Expansion in Shanghai. *Appl. Geogr.* **2015**, *65*, 93–108. [CrossRef]
6. Sung, H.; Lee, S. Residential Built Environment and Walking Activity: Empirical Evidence of Jane Jacobs' Urban Vitality. *Transport. Res. Part D-Transport. Environ.* **2015**, *41*, 318–329. [CrossRef]
7. Lee, S.H.; Kang, J.E. Impact of Particulate Matter and Urban Spatial Characteristics on Urban Vitality Using Spatiotemporal Big Data. *Cities* **2022**, *131*, 104030. [CrossRef]
8. Ming, Y.J.; Liu, Y.; Li, Y.P.; Yue, W.Z. Core-periphery disparity in community vitality in Chongqing, China: Nonlinear explanation based on mobile phone data and multi-scale factors. *Appl. Geogr.* **2024**, *164*, 103222. [CrossRef]
9. Zhang, Z.R.; Liu, J.P.; Wang, C.Y.; Zhao, Y.Y.; Zhao, X.Z.; Li, P.P.; Sha, D.X. A Spatial Projection Pursuit Model for Identifying Comprehensive Urban Vitality on Blocks Using Multisource Geospatial Data. *Sustain. Cities Soc.* **2024**, *100*, 104998. [CrossRef]
10. Chen, Y.; Yu, B.J.; Shu, B.; Yang, L.C.; Wang, R.Y. Exploring the Spatiotemporal Patterns and Correlates of Urban Vitality: Temporal and Spatial Heterogeneity. *Sustain. Cities Soc.* **2023**, *91*, 104440. [CrossRef]
11. Jiang, Y.X.; Huang, Z.; Zhou, X.; Chen, X.J. Evaluating the Impact of Urban Morphology on Urban Vitality: An Exploratory Study Using Big Geo-Data. *Int. J. Digit. Earth.* **2024**, *17*, 2327571. [CrossRef]
12. Ye, Y.; Li, D.; Liu, X.J. How block density and typology affect urban vitality: An exploratory analysis in Shenzhen, China. *Urban Geogr.* **2018**, *39*, 631–652. [CrossRef]
13. Tu, W.; Zhu, T.; Xia, J.; Zhou, Y.; Lai, Y.; Jiang, J.; Li, Q. Portraying the spatial dynamics of urban Vitality using multisource urban big data. *Comput. Environ. Urban Syst.* **2020**, *80*, 101428. [CrossRef]
14. Yue, W.; Chen, Y.; Thy, P.T.M.; Fan, P.; Liu, Y.; Zhang, W. Identifying urban vitality in metropolitan areas of developing countries from a comparative perspective: Ho Chi Minh City versus Shanghai. *Sustain. Cities Soc.* **2021**, *65*, 102609. [CrossRef]
15. Xia, C.; Yeh, A.G.O.; Zhang, A.Q. Analyzing Spatial Relationships between Urban Land Use Intensity and Urban Vitality at Street Block Level: A Case Study of Five Chinese Megacities. *Landsc. Urban Plan.* **2020**, *193*, 103669. [CrossRef]
16. Wang, X.; Zhang, Y.; Yu, D.; Qi, J.; Li, S. Investigating the spatiotemporal pattern of urban vitality and its determinants: Spatial big data analyses in Beijing, China. *Land Use Pol.* **2022**, *119*, 106162. [CrossRef]
17. Wu, W.S.; Niu, X.Y. Influence of Built Environment on Urban Vitality: Case Study of Shanghai Using Mobile Phone Location Data. *J. Urban Plan. Dev.* **2019**, *145*, 04019007. [CrossRef]
18. Li, X.; Li, Y.; Jia, T.; Zhou, L.; Hijazi, I.H. The Six Dimensions of Built Environment on Urban Vitality: Fusion Evidence from Multi-Source Data. *Cities* **2022**, *121*, 103482. [CrossRef]
19. Chen, L.; Zhao, L.; Xiao, Y.; Lu, Y. Investigating the spatiotemporal pattern between the built environment and urban Vitality using big data in Shenzhen, China. *Comput. Environ. Urban Syst.* **2022**, *95*, 101827. [CrossRef]
20. Jin, A.; Ge, Y.; Zhang, S. Spatial Characteristics of Multidimensional Urban Vitality and Its Impact Mechanisms by the Built Environment. *Land* **2024**, *13*, 991. [CrossRef]
21. Zhang, A.Q.; Li, W.F.; Wu, J.Y.; Lin, J.; Chu, J.Q.; Xia, C. How Can the Urban Landscape Affect Urban Vitality at the Street Block Level? A Case Study of 15 Metropolises in China. *Environ. Plan. B-Urban Anal. City Sci.* **2021**, *48*, 1245–1262. [CrossRef]
22. Ma, Z.P. Deep Exploration of Street View Features for Identifying Urban Vitality: A Case Study of Qingdao City. *Int. J. Appl. Earth Obs. Geoinf.* **2023**, *123*, 103476. [CrossRef]
23. Long, Y.; Huang, C.C. Does Block Size Matter? The Impact of Urban Design on Economic Vitality for Chinese Cities. *Env. Plan. B-Urban Anal. City Sci.* **2019**, *46*, 406–422. [CrossRef]
24. Gan, X.Y.; Huang, L.; Wang, H.Y.; Mou, Y.C.; Wang, D.; Hu, A. Optimal Block Size for Improving Urban Vitality: An Exploratory Analysis with Multiple Vitality Indicators. *J. Urban Plan. Dev.* **2021**, *147*, 04021027. [CrossRef]
25. Droj, G.; Droj, L.; Badea, A.C.; Dragomir, P.I. GIS-Based Urban Traffic Assessment in a Historical European City under the Influence of Infrastructure Works and COVID-19. *Appl. Sci.* **2023**, *13*, 1355. [CrossRef]
26. Cui, C.; Wang, J.C.; Pu, Y.X.; Ma, J.S.; Chen, G. Gis-Based Method of Delimitating Trade Area for Retail Chains. *Int. J. Geogr. Inf. Sci.* **2012**, *26*, 1863–1879. [CrossRef]
27. Okabe, A.; Satoh, T.; Sugihara, K. A kernel density estimation method for networks, its computational method and a GIS-based tool. *Int. J. Geogr. Inf. Sci.* **2009**, *23*, 7–32. [CrossRef]
28. Yang, J.; Zhu, J.; Sun, Y.; Zhao, J. Delimitating Urban Commercial Central Districts by Combining Kernel Density Estimation and Road Intersections: A Case Study in Nanjing City, China. *ISPRS Int. J. Geo-Inf.* **2019**, *8*, 93. [CrossRef]
29. Gao, C.; Li, S.; Sun, M.; Zhao, X.; Liu, D. Exploring the Relationship between Urban Vibrancy and Built Environment Using Multi-Source Data: Case Study in Munich. *Remote Sens.* **2024**, *16*, 1107. [CrossRef]
30. Mouratidis, K.; Poortinga, W. Built environment, urban vitality and social cohesion: Do vibrant neighborhoods foster strong communities? *Landsc. Urban Plan.* **2020**, *204*, 103951. [CrossRef]
31. Huang, B.; Zhou, Y.L.; Li, Z.G.; Song, Y.M.; Cai, J.X.; Tu, W. Evaluating and characterizing urban vibrancy using spatial big data: Shanghai as a case study. *Env. Plan. B-Urban Anal. City Sci.* **2020**, *47*, 1543–1559. [CrossRef]
32. Wu, J.Y.; Wang, B.H.; Wang, R.; Ta, N.; Chai, Y.W. Active travel and the built environment: A theoretical model and multidimensional evidence. *Transport. Res. Part D-Transport. Environ.* **2021**, *100*, 103029. [CrossRef]

33. Gao, F.; Deng, X.; Liao, S.; Liu, Y.; Li, H.; Li, G.; Chen, W. Portraying Business District Vitality with Mobile Phone Data and Optimal Parameters-Based Geographical Detector Model. *Sustain. Cities Soc.* **2023**, *96*, 104635. [CrossRef]
34. Yoo, J.; Eom, J.; Zhou, Y. Thermal Comfort and Retail Sales: A Big Data Analysis of Extreme Temperature's Impact on Brick-and-Mortar Stores. *J. Retail. Consum. Serv.* **2024**, *77*, 103699. [CrossRef]
35. Li, X.; Qian, Y.; Zeng, J.; Wei, X.; Guang, X. The Influence of Strip-City Street Network Structure on Spatial Vitality: Case Studies in Lanzhou, China. *Land* **2021**, *10*, 1107. [CrossRef]
36. Wang, Y.; Lin, S.; Tang, S. Individual preferences, jobs, or amenities? Understanding the destination choices of highly and less-skilled migrants in urban China. *Appl. Geogr.* **2024**, *162*, 103169. [CrossRef]
37. Kang, C.G.; Fan, D.W.; Jiao, H.Z. Validating Activity, Time, and Space Diversity as Essential Components of Urban Vitality. *Environ. Plan. B-Urban Anal. City Sci.* **2021**, *48*, 1180–1197. [CrossRef]
38. Dong, Q.; Cai, J.; Chen, S.; He, P.; Chen, X. Spatiotemporal Analysis of Urban Green Spatial Vitality and the Corresponding Influencing Factors: A Case Study of Chengdu, China. *Land* **2022**, *11*, 1820. [CrossRef]

Disclaimer/Publisher's Note: The statements, opinions and data contained in all publications are solely those of the individual author(s) and contributor(s) and not of MDPI and/or the editor(s). MDPI and/or the editor(s) disclaim responsibility for any injury to people or property resulting from any ideas, methods, instructions or products referred to in the content.

land

MDPI

Article

Understanding the Transformations of San Lorenzo, Rome: An Attempt at Conceptual Order between Gentrification and Urban Policy

Antonia Astore [1,*] and Luca Tricarico [2,*]

1 Gran Sasso Science Institute, 67100 L'Aquila, Italy
2 National Research Council, 00185 Rome, Italy
* Correspondence: astoreantonia@gmail.com (A.A.); luca.tricarico@cnr.it (L.T.)

Abstract: This paper is the result of a preliminary research phase aimed at identifying the phenomenon of gentrification in Rome's San Lorenzo neighborhood, examining the perception of the phenomenon and the urban planning tools that have been adopted until today to contain it. The experimental framework proposed, based on qualitative research methods, aims to discuss if some urban transformations can be related only to market phenomenon inherent to gentrification or to the lack of planning tools able to manage—through the construction of a shared view on neighborhoods development—the negative externalities resulting from the real estate investments induced by the tourism sector and linked to the presence of universities in the reference territory. The research combines an international theoretical framework; the analysis of the policies and of the urban transformation during the contemporary history of the neighborhood; and interviews to key informants such as representatives of local institutions and associations, researchers, inhabitants, and others. The research approach intends to deepen the general perception and the comprehension of the phenomenon of gentrification from the local community, in relation to his own involvement in decision-making processes about the strategic regulatory–prescriptive development of the neighborhood. The primary objective of the paper is to advocate for an analytical framework that facilitates a better understanding of gentrification within the national context and explores potential institutional responses. It seeks to clarify the concept of gentrification, which is often misunderstood when interpreted solely through its Anglo-Saxon definition. By doing so, the paper aims to contribute to informed discussions and effective policy-making regarding urban development in San Lorenzo and beyond.

Keywords: gentrification; studentification; airbnbification; spatial planning

1. Introduction

This article aims to discuss the sociological concept of gentrification, a phenomenon of spatial transformation and socio-cultural composition of neighborhood populations following processes (or more or less intentional political strategies) that are sometimes defined as urban regeneration or redevelopment [1,2]. As noted [3], the concept is often used indiscriminately in the media debate, often without a coherent empirical analysis capable of verifying its complete validity as primarily defined in Anglo-Saxon contexts [4].

This phenomenon, soon becoming a central research theme for social and urban sciences, has captured the attention of geographers, sociologists, anthropologists, economists, political scientists, and urban planners, giving rise to a rich and diverse international literature [5]. The phenomenon, in its empirical sense, therefore records the progressive replacement of the resident population with a higher social class, following their acquisition of real estate and its revaluation, facilitated by the absence of urban policies capable of opposing measures of social protection and housing inclusion [5,6]. The same phenomenon, observed in its entirety in various North American and European cities, can originate as a joint strategy of major real estate and financial players, where institutional coordination in governance and regulation is lacking or absent [7,8].

In an attempt to bridge the knowledge gap that guided the construction of the preliminary hypothesis underlying this contribution, the study aims to examine the problematic use of the concept of gentrification for two main reasons: first, in some cases, it is partially able to describe a complex set of social and inter-institutional phenomena from which negative externalities arise; and second, because it may not provide sufficient guidance to formulate urban policy recommendations aimed at managing the phenomenon.

Among these contexts, the neighborhood of San Lorenzo in Rome (Figure 1) represents a particularly significant case study [9], on which the exploratory analysis discussed in the article is based [10]. The reason for choosing this neighborhood lies also in the desire to investigate the peculiar features of the process of "collective problem construction" [11] with the aim of discussing the use of the term in relation to the institutional characteristics of the context [12]. The research question that the paper aims to address is: to what extent can the transformation of the San Lorenzo neighborhood in Rome be attributed to gentrification or to the lack of effective institutional coordination? Starting from an analysis of the literature on the phenomenon of gentrification from a sociological perspective, and an in-depth study of some key aspects of the concept itself, such as night-time gentrification, studentification [13], and airbnbification [14], the hypothesis put forward here is that the process of transformation of the San Lorenzo neighborhood cannot be defined as gentrification as it lacks the essential elements to allow the use of the term and simultaneously other factors are present that we intend to investigate. With these premises, this article is structured into five sections. After introducing the concept of gentrification and the research question, the second section reviews the methodology used to construct an investigation plan capable of adequately addressing the research objectives. The third section is dedicated to the precursor phenomena of gentrification and the literature on the process itself, with a focus on the role of consumption and culture in the transformation processes of cities in the global north [15] and on some key elements that have emerged in recent decades: night-time gentrification, studentification, and airbnbification. In the fourth section, after reviewing the history of the San Lorenzo neighborhood, the main facts related to the urban transformation of the neighborhood since the approval of the New General Regulatory Plan (NPRG) in 2008 are addressed. The main dynamics and issues that emerged during the interviews, which render the concept of gentrification only partially able to describe the urban and socio-cultural transformations of the specific context, thus inadequate, are then discussed. The concluding part is dedicated to reflections on the main emerging results, related to the lack of communication between the different levels of planning (from the bottom through citizens and residents of the neighborhood, from the top through institutions and administrations), and some future research directions.

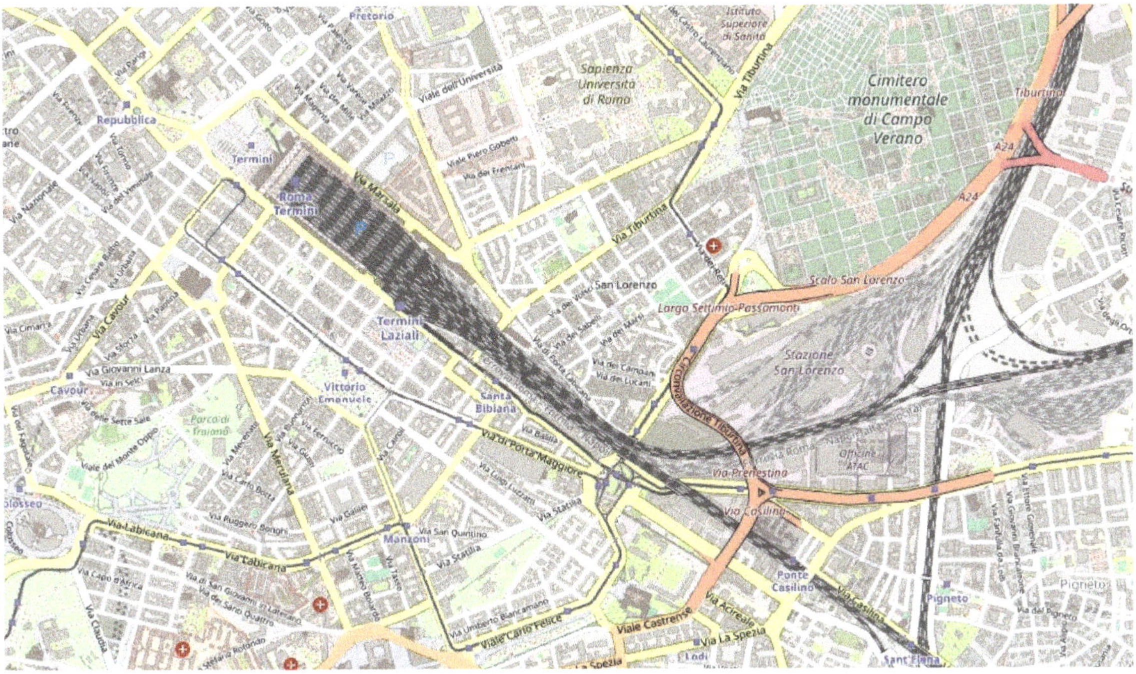

Figure 1. The neighborhood of San Lorenzo in Rome. Source: Turismo Roma.

2. Methodology

From a methodological standpoint, the research plan was structured through a literature review concerning the examined phenomena, the analysis of urban planning documents, and the utilization of qualitative research techniques.

The reference theoretical framework is based on precursor phenomena of gentrification and urban development theory (Table 1).

Table 1. Summary of Literature on Gentrification, Studentification, and Airbnbification. Source: Authors.

Key References	Perspective	Summary	Key Concepts	Main Urban Impacts	Areas Studied
[6]	Early Gentrification	Replacement of working-class residents by middle-class, leading to increased property values and social status.	Residential transformation, bourgeoisification	Displacement of workers, rapid neighborhood change	London (1950s–1960s)
[16–18]	Consumption-based approach to Gentrification; Post-Industrial Gentrification	Cities become promoters of economic growth, leading to changes in urban space to meet new middle-class needs.	New middle-class, post-industrial society, cultural class	Decline of manufacturing, rise of tertiary sector, displacement of poorer families	Canada, U.S.
[19]	Production-based approach to Gentrification	Physical change of neighborhoods by capital investment is the causal factor of gentrification; gentrification results from relatively unhampered operation of housing markets	Housing markets, capitalist development, capital reinvestment	Back to the city movement by capital	U.S., U.K., Global urban centers

Table 1. *Cont.*

Key References	Perspective	Summary	Key Concepts	Main Urban Impacts	Areas Studied
[20]	Global and Place-based Gentrification	Throughout the world there are multiple gentrifications in a pluralistic sense; Anglo-Saxon lens as a limitation for understanding the phenomenon	Place-based transformations; key elements for gentrification	Western democratic conception; "social cleansing" as a similar factor to gentrification	Global South and East; London
[21]	Night-time Gentrification	Development of nightlife economies leading to urban transformations that cater to middle-class consumption patterns.	Night-time economy, entertainment venues	Introduction of bars, nightclubs, and music venues, attracting new residents and changing urban space	Rome, Milan, Turin
[13,22,23]	Studentification	Impact of student populations on urban areas, including increased rental prices, social and cultural changes, and physical property transformations	Short-term rentals, touristification, housing market distortion, commercial transformation	Increased rental demand, reduction in family housing, socio-cultural clustering of student populations, displacement of long-term residents	University neighborhoods, particularly in major academic cities
[14,24–27]	Airbnbification	Influence of short-term rentals on urban areas, leading to increased property values, displacement of residents, and changes in local identity.	Short-term rentals, touristification	Hyper-touristicization, displacement of traditional residents, transformation of commercial offerings	Global urban centers, particularly tourist areas, Rome

The analysis of urban planning instruments proved vital in understanding the principal dynamics characterizing the ongoing transformation in the neighborhood. Of particular significance were the New General Regulatory Plan (NGRP) approved in 2008, outlining the development strategies for the city of Rome and regulating construction activity and urban transformation, and the Housing Plan approved in 2009, aimed at expanding residential units. Lastly, interviews served as a qualitative research tool to gather insights from various stakeholders involved in the neighborhood's transformation, representing the institutional context in which this process unfolded. Thirteen interviews were conducted with key informants, individuals knowledgeable about the topic or directly engaged in planning processes, or directly impacted by the neighborhood's transformation effects. The conducted interviews were of three types (the table of the interviewees is provided in Appendix A): semi-structured, focused and non-standardized [28].

Focused interviews, during which a standardized set of questions and themes were explored with public actors (individuals who, due to their position or knowledge of the issue (experts) have a direct and profound understanding of the phenomenon) such as the Former Urban Planning Assessor of Rome Capital and an urbanist; the objective was to collect precise information about the major issues investigated about the urban transformation of the neighborhood and the San Lorenzo Urban Projects.

Semi-structured interviews, guided by a set of basic questions and issues to explore, with neither the exact wording nor the order of questions predetermined, were conducted with shopkeepers such as the owners of local bars, bookstores and real estate agencies; the objective was to implement an open and informal interview and, at the same time, collect answers that would be comparable for understanding the effects of urban transformation on local stores and bars.

Non-standardized exploratory interviews, during which the conversations with the interviewees were guided by a minimum list of points to cover, have been conducted with two local activists from social centers, an independent researcher, and founders of two local

Committee; the objective was to explore general themes by changing follow-up questions depending on the interviewee's responses and the interests of the research.

Thanks to these interviews, it was possible to gain a deeper understanding of the main dynamics at play during the urban transformation process, the key actors, and how it is perceived by residents and traders.

3. Literature Review on Gentrification

3.1. Anglo-Saxon Context

The analysis of the evolution of the concept of gentrification has been developed by examining various scholarships reflecting on the relationship between this phenomenon and the evolution of consumption culture in urban space, as well as in the collective representation of the same after the assertion and diffusion of the phenomenon (Table 1).

Although the first example of planned city transformation is traced back to the Haussmannization of Paris in the late 19th century, the term "gentrification" was first introduced in academic circles in 1964 by sociologist Ruth Glass in London, and the concept has been reworked several times. Literally, the term gentrification referred to the process of replacing the existing population due to the settlement of the gentry, or the bourgeoisie, resulting in the bourgeoisification of poorer neighborhoods. Describing a transformation that is primarily residential, Glass believes that the social value and status of these homes are often inversely proportional to their condition, and in any case greatly inflated compared to their previous levels in the neighborhood. When this process of gentrification begins in a district, it proceeds rapidly until all or most of the workers occupying a home are displaced, and the social nature of the neighborhood changes [6]. The author observes how the use of urban space in London in the 1950s and 1960s became increasingly diversified, factories began to move out of the city while free spaces were filled with commercial and tertiary activities. Meanwhile, laissez-faire policies in real estate and construction began to destabilize the rental market, giving landlords carte blanche in determining rental prices. Until the mid-1970s, gentrification seemed to be limited to Anglo-Saxon reality, but soon the growth of the phenomenon was experienced on a global scale, as cities began to reemerge as "entrepreneurial" promoters of economic growth [16].

In the late 1970s, Neil Smith described gentrification as an expected product of the relatively unhampered operation of the land and housing markets [19]. According to the author, the production of urban space takes place through a sort of capital seesaw: continuous and progressive shifts of capital investments that, in turn, generate massive disinvestments in the areas from which the capital moves away. In this model of capitalist development, gentrification is seen as a phase of capital reinvestment within cities that have previously experienced forms of disinvestment, as a "back to the city movement by capital" [19]. According to his theory, the *rent gap* is the disparity between the potential ground rent level and the actual ground rent capitalized under the present land use. Hence, gentrification occurs when the gap is wide enough that developers can purchase shells cheaply, can pay the builders' costs, and interest on mortgage and construction loans, and can then sell the house for a sale price that leaves a satisfactory return to the developer. According to Smith, gentrification is a back-to-the-city movement of capital rather than people [19].

The 1980s through the 2000s can be considered the renaissance of gentrification research, as numerous scholars began examining the phenomenon through their particular research ideologies [29]. In this period, with the return of economically liberal doctrines, cities became active promoters of economic growth, giving rise to the thesis of the advent of the post-industrial society: the industrial transformations of the late 20th century lead to the decline of the manufacturing sector, growth of the tertiary sector, and with the rise of new middle classes, there is a change in values [17,18]. According to this line of thought, gentrification would consist of adapting the city space to the needs of the new cultural class [29], which began to appear on the urban and real estate scene with its tastes and lifestyles [30,31]. The improvement of working-class neighborhoods in the city center to pursue livability and well-being, thanks to the increase in creative classes and tourists, can contribute to an increase in the quality of the built space [32]. However, in the absence of forms of protection by the public sector, it can also contribute to the displacement of poorer families.

Towards the end of the 20th century, the arrival of new actors directly related to gentrification, for example through the purchase of newly renovated properties or through the opening of new venues not aligned with the socio-cultural traditions of a territory, contributed to modifying the collective representation of the territory in question, determining a transformation in terms of cultural symbols, the emergence of new lifestyles in the neighborhood, and new types of consumption habits [33].

In the 1980s, 1990s and, indeed, into the 2000s, the internationally comparative work that was being undertaken on gentrification predominantly compared Western cities [20], and studies of the phenomenon have been viewed through an Anglo-American lens [29]. Lees [20] tried to address the limitations of this gentrification lens, assessing the meaning and significance of gentrification beyond the usual suspects. In so doing, she discussed the extent to which "gentrification" is both a global and strongly place-based process. Looking at Global South and East, it comes out that there are multiple gentrifications in a pluralistic sense. For instance, there are places that do not follow the Western democratic conception of a democratic public space; places where a welfare state has been absent or poorly developed; and places where dwellings and land are subject to intense speculation [20]. According to them, "social cleansing" is exactly what makes a case of urban regeneration a process of gentrification [20], but each case is singular and strictly correlated to the social, economic, cultural and political characteristics of the place. In this context, displacement should be conceived more as a complex set of place-based processes that are spatially and temporally variable [34] rather than a brief moment in time where a particular resident is forced out of their home or neighborhood.

3.2. Italian Context

For several years, some Italian cities have been experiencing a phenomenon of emerging neighborhoods, a process of transformation concentrated in areas historically and socially characterized as working-class districts, located adjacent to historic places [35]. At the same time, the concept of gentrification has become a label for interpreting the changes occurring in cities, which materialize at the neighborhood scale. In fact, in Italy, there is a lack of marked class conflict, forced displacement of residents, and the homogeneity and rapidity with which the phenomenon has manifested in other geographical contexts [35].

The term gentrification is used as a descriptive category to explain the complex mutations of space related to the symbolic sphere and consumption [36,37]. Nevertheless, the concept has never found a clear definition and contours in Italy, not even when the debate on the phenomenon was highly active in Anglo-Saxon countries [35]. Studies on the phenomenon of gentrification in Italy are relatively recent (emerging over the past 20 years), and are often linked to the opening of new and alternative commercial activities at the neighborhood level [35]. Ticinese in Milan is described as a district of the imaginary [38], the Quadrilatero in Turin as a distinguished neighborhood [39], and Pigneto as an urban village [37]. In Italy, there is no single trajectory of neighborhood change; instead, various manifestations of urban space transformation emerge.

Gentrification policies are also shown through national goals of increased housing production and rising homeownership [40], albeit at the cost of social redistribution. Generally, the individuals most affected by the dynamics stemming from the strong control exerted by neoliberalism and the market over housing public policies are those belonging to lower-middle social classes. Considering that these individuals lack adequate resources to cope with the effects of urban changes [41], it emerges why gentrification and housing policies are closely correlated.

However, privatization in the housing sector and the promotion of homeownership in the Italian context exhibit substantial differences when compared to the Anglo-Saxon context, where the concept of gentrification was born. These differences are largely tied to the nature of home ownership and housing markets in Italy. In fact, as many people own their homes in Italy, displacement is less about economic pressure forcing residents to leave. Instead, homeowners voluntarily leave, choosing to rent their properties to tourists, or students for example, because it is more profitable [42] This creates a form of "softer displacement" that is economically driven rather than forced by unaffordability. Unlike the traditional gentrification pattern, where wealthier, higher-class newcomers replace the original working-class or low-income residents, the Italian context sees a different dynamic.

From the 1990s onwards, attention began to focus on the role of entertainment venues and nightlife in relation to the spatial and social logic of the city [21], as the night begins to represent a node of space transformation linked to non-traditional forms of consumption and entertainment; hence, nightclubs become central places for the analysis of the relationship between changing values and urban transformation. The "beauty" understood as the form and function of the venue begins to be necessary for its attractiveness: disco bars, nightclubs, new venues, spaces for dancing and live music emerge, and historical neighborhood venues are restyled—a series of changes capable of determining a clear influence of new everyday lifestyles on urban space. Following this trend, it must be considered that bars maintain a mutually reinforcing relationship with gentrification: new activities of this type meet the needs of middle-class residents, simultaneously attracting others; they act as both signals and catalysts for the gentrification process of a neighborhood [12].

Looking at this type of urban transformation recorded in some neighborhoods of cities like Rome, Milan, or Turin, it is observed how the attractiveness generated by such transformation processes is particularly directed towards a specific category of urban space consumers: students. In fact, the presence of a dynamic culture and nightlife often becomes one of the criteria for choosing the place of study [43]. The increasing number of student populations accessing higher levels of education would have urban effects on four levels: economic, social, cultural, and physical [13]. The economic dimension, i.e., the revaluation of properties in favor of student demand, leads to a decrease in owner-occupied homes and an increase in rental properties in the neighborhood chosen as a preferred destination for students. The social dimension involves a younger demographic composition and often a higher social class than the average of the residents, which inevitably generates urban transformation over the years, resulting in the displacement of existing groups and permanent residents in favor/cause of transitioning groups [44]. In fact, property owners begin to prefer renting to students rather than leasing to a single family [13,22], as it is economically more advantageous [42]. The cultural impact generated by this process is evident in the progressive clustering of young people who share a common cultural background and lifestyles [45], leading to the emergence of single-reference services (specifically for students) and spatially influencing the commercial offer of the place [46]. Lastly, the physical dimension concerns the properties hosting these populations, both in terms of private rentals and actual housing policies for students by the universities themselves [13]. Furthermore, the importance of this factor lies in the fact that studentification [23] can also be associated with a general deterioration in the physical condition of the neighborhood and its real estate assets [47], which in turn leads to further displacement of the existing population to other neighborhoods [44], which is why this process can be identified as a variant of the more general gentrification process that involves the replacement of original residents with students [13]. In this sense, social and cultural spaces, new lifestyles, and consumption habits typical of studentified areas can emerge as fertile ground for future gentrifiers. Concurrently, the territory in which residents are the same consumers of their own produced space distinguishes an area safe for commercial investments, determining a change in service offerings, attracting new investors capable of mobilizing from the neighborhood the old artisans and shopkeepers whose commercial offerings no longer meet the demand from the new local consumers, seeking commercial contexts characterized by food and beverage [42]. This phase seems to correlate with a phenomenon of social and economic polarization of cities [48], causing a portion of the resident population to be progressively "put in mobility" towards less demanding peripheral areas in terms of housing expenses per family. A series of issues may arise mainly concerning the increase in conflicts between old and new residents: taking nightlife as an example, conflict is generated by the synchronous use of urban spaces but by the presence of different needs [42].

Among the other phenomena that can play a fundamental role in facilitating the departure of old residents from their traditionally inhabited neighborhoods, there is the phenomenon known as Airbnbification [14]: the spread of the use of platforms such as Airbnb and the increase in the number of accommodations available through them are closely linked to processes of touristification and gentrification of urban areas. Although these are distinct processes, gentrification and airbnbification both contribute to the socio-cultural transformation of neighborhoods and the displacement of residents [26]. The platform's significant growth, founded in 2007, has been observed in recent years, during which studies have shown that the rise in short-term rentals leads to both the hyper-touristicization of already touristic city centers and the gentrification of yet exploited neighborhoods [25,27]. In this context, the tourist becomes the primary "temporary global inhabitant" [49], the main consumer of daily urban space, regardless of their social class. Additionally, the platform is often accused of distorting the real estate and rental markets by exponentially accelerating the process of price increases and property values, leading to a turnover of the population (of residents of traditional neighborhoods), resulting in the direct and indirect expulsion of residents. It is also accused of transforming cities into places of consumption; and distorting the idea of the sharing economy [50], considering that in 2017, only 40% of Airbnb listings in Rome were for rooms or beds in apartments occupied by others, while the rest consisted of entire apartments and could not be attributed to a sharing economy [24]. Consequently, in recent years, there has been a growing trend among administrations to attempt public control over short-term rentals, aiming to ensure fair competition and market transparency, informational, statistical, and computerized coordination of data from state, regional, and local administrations, territorial security, and the fight against irregular forms of hospitality. In Italy, for example, with the conversion into law n. 191/2023 of the "Anticipi" decree (D.L. n. 145/2023), a national identification code (CIN) is assigned to residential units intended for tourist rental contracts and short-term leases. It includes the obligation of certified notification of the start of activity (Scia) and a series of obligations aimed at ensuring compliance with any urban and landscape constraints. However, the measures adopted are suitable for containing the negative effects of hyper-touristicization of the city and airbnbification of neighborhoods. Understanding the role that night-time gentrification, studentification, and airbnbification can play in the urban and socio-cultural transformations of a neighborhood is crucial for comprehending those affecting the case study of the San Lorenzo neighborhood in Rome and associating each of these phenomena with the actual role played in the process that many residents and researchers have perceived and defined as gentrification.

4. San Lorenzo and NGRP: Origins, Characteristics and Perception of the Gentrification Phenomenon

4.1. Historical Origins of the Neighborhood

The contemporary history of San Lorenzo begins between 1884 and 1888 when the first buildings were constructed through private initiatives beyond the Aurelian Walls. At the beginning of the 20th century, the first phase of construction in the neighborhood was completed, and San Lorenzo became an extremely poor working-class neighborhood [51]. The strong housing demand from the poorest classes and new workers continued to rise thanks to the establishment of new factories, such as the Paszkowski Brewery, the Pastificio Cerere, and the former Fonderie Bastianelli. San Lorenzo also became a fertile ground for artisan workshops, especially those related to the activities of the Verano cemetery. In the wake of the first political movements that preceded the war, the neighborhood saw intense political ferment, which manifested in anti-war protests and, in the early post-war period, the opening of a Socialist Club and the development of a core group of the Arditi del Popolo, which aimed to serve as a proletarian defense organ against fascist violence [51].

In the 1930s, fascism left its mark on the area through public works, but there were few initiatives that genuinely benefited the neighborhood: the northern boundary of the San Lorenzo area was further modified by the construction of the Ministry of Aeronautics (1929–1931), the new University City (1932–1935), and the National Research Center (1939). The construction of Città Universitaria La Sapienza contributed to the neighborhood's transformation through the purchase or the use of some structures in its territory, leading to the arrival of new people in the neighborhood, namely students. Indeed, in the 1970s, real estate companies bought and renovated buildings, increasing their value due to the influx of new potential tenants (students), and residents began to leave San Lorenzo, unable to afford the rise in rental prices. During this period, some symbolic places in the neighborhood's history, like the Pastificio Cerere, were transformed into cultural and artistic centers, aiming to attract people from other neighborhoods.

4.2. The NGRP and Attempts to Build an Urban Project

At the beginning of the 2000s, hotels in the city of Rome could provide accommodation to 94,000 [52]; of this total, about 41,500 places were located in the historic center. The flow of visitors concentrated their visits in the historic center, where in 2001 fewer than 100,000 residents lived. The number of visitors daily exceeds the number of residents, creating an imbalance that impacts the commercial structure, which automatically shifts towards satisfying the new type of consumers: restaurants, pizzerias, ice cream shops, and souvenir stores. Therefore, during this period, the historic center of Rome appears as an immense tourist consumption area, and the entire city is characterized by a further increase in the weight of the tertiary sector within it [53].

In 2008, the New General Regulatory Plan for the city of Rome was approved, designed to outline the development strategies of the city in general and its neighborhoods in particular, determining their layouts and regulating building activities and thus the urban transformation of the city. Meanwhile, in March 2009, the "Piano Casa" (Housing Plan) was launched, aimed at the expansion of villas and other single-family homes, allowing the alteration of the configuration of long-established urban areas [52].

Particularly, it is important to highlight the complexity and length of the process that led to the new urban planning tool [54]. The New General Regulatory Plan of Rome was adopted in 2003 (City Council Resolution No. 33 of 20 March 2003), after nearly ten years of work [54], and approved in 2008. It is presented as an innovative tool, with the aim of defining multifunctional growth for the city and establishing new rules and guidelines to guide the development of Rome. The main points are: qualitative transformation (decentralization and polycentrism); protection of the environment and historical–cultural heritage according to the principle of sustainability; more services and urban functions for the suburbs, breaking the primacy of the disadvantaged city; new mobility system with an emphasis on rail (reduction in cars on the road); metropolitan horizon, corresponding to an articulated vision of the territory in autonomous municipalities capable of interacting [55].

The new Roman urban planning experience would focus on two fundamental objectives: modernization and redevelopment aimed at protecting the environmental and historical value of the city's remaining free spaces. Three other foundational aspects of the new Regulatory Plan are: planning by doing, the Program Agreement, and urban compensation. Planning by doing is an approach aimed at simplification by increasing synergy between public and private sectors, through the coincidence of planning and management. The Program Agreement is an administrative instrument that allows operations outside ordinary planning: public entities (such as regions, provinces, and municipalities) can coordinate their future activities for approving public works and infrastructure projects. These projects are part of the Administration's programs and have immediate funding available. Once formalized, this instrument can also serve as a variant of the urban plans for the affected areas. In such cases, the relative approval would declare the projects as being of public utility, necessity, and urgency, thereby prioritizing the community's interests. In recent years, however, the Program Agreement has mainly been used to facilitate projects not originally included in the urban plans.

Finally, the third aspect of the plan's operation involves the introduction of the so-called compensatory acquisition of land for the realization of urban standards, with the aim of avoiding the need for municipal expropriation. The combination of these tools ends up reducing the importance of public control over planning [55].

Regarding qualitative transformation, environmental sustainability, and historical–cultural heritage, they do not present a strategic and organized definition of objectives from which an intervention program can be deduced. Thus, the articulation of action programs is lacking.

Considering that the implementation of the transformations envisaged by the NGRP (New General Regulatory Plan) can occur through the adoption of executive urban planning tools, the case study represented by the San Lorenzo neighborhood is representative because an Urban Project has never been approved and implemented. The case study was chosen to try to understand the impact of the lack of urban planning capable of outlining development strategies for an area and regulating its building activity. Meanwhile, it is a representative case in order to contest the use of the term gentrification for the interpretation of urban and socio-cultural transformations of the area.

In the NGRP (2008), four "enhancement areas" had already been identified, which are "areas that fall within the historic city, but in which the original spatial characteristics have been lost and often there are disused buildings": the first is the area that includes the space between Scalo San Lorenzo, Via dei Lucani, and Via di Porta Labicana (Figure 2); the second includes part of Scalo S. Lorenzo, the Tangenziale Est, and Porta Maggiore (Figure 3), "characterized by the presence of temporary constructions used as warehouses, storage, and commercial and artisanal activities"; the third coincides with the area between Piazzale del Verano, Via Cesare de Lollis, and Via Tiburtina, and "is characterized by the presence of functions mainly related to the university" (Figure 4); finally, the area between Piazzale del Verano, Via dei Reti, and Largo Passamonti "characterized by the presence of a severely degraded and incomplete urban fabric in the definition of the road network" (Figure 5). Regarding the implementation of the planned regulations, D'Albergo and Moini identified a dual approach in the NGRP: a super-regulation, meaning the expectation of a long and hyper-structured process in the transition from the planning phase to the implementation of the forecasts, which often translates into substantial deregulation at the implementation stage [56].

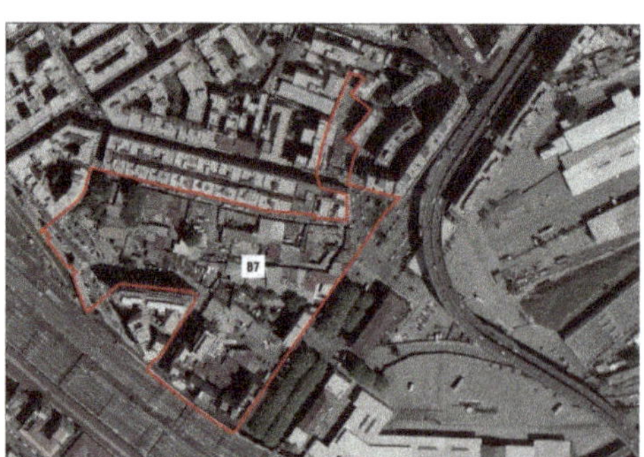

Figure 2. Enhancement Area B7: Scalo San Lorenzo, Via dei Lucani, Via di Porta Labicana. Source: Urbanistica Comune Roma Capitale.

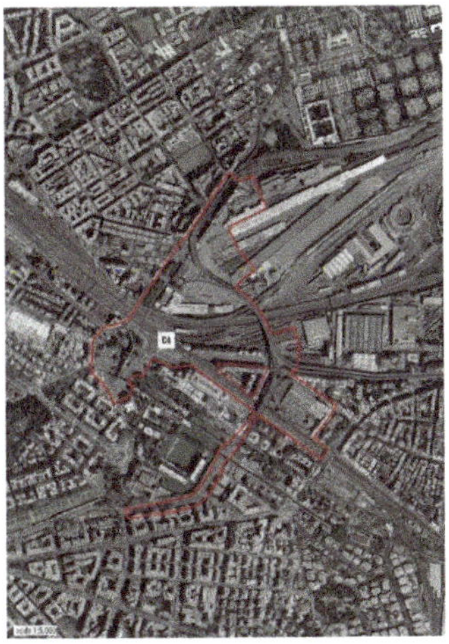

Figure 3. Enhancement Area C4: Scalo San Lorenzo, Tangenziale Est, Porta Maggiore. Source: Urbanistica Comune Roma Capitale.

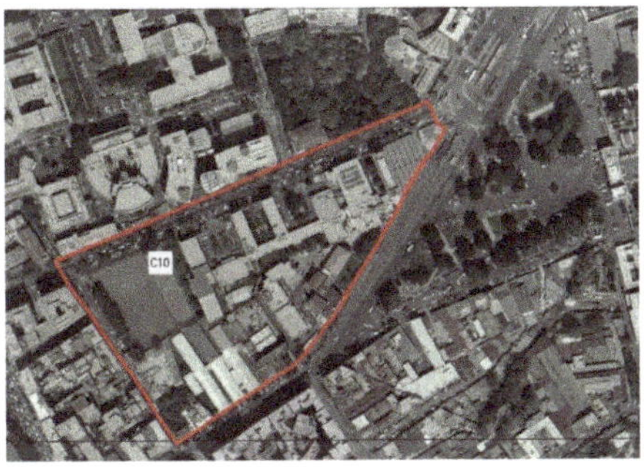

Figure 4. Enhancement Area C10: Piazzale del Verano, Via Cesare de Lollis, Via Tiburtina. Source: Urbanistica Comune Roma Capitale.

Figure 5. Enhancement Area C11: Piazzale del Verano, Via dei Reti, Largo Passamonti. Source: Urbanistica Comune Roma Capitale.

The concepts of super-regulation and de-regulation refer primarily to general attitudes and approaches to public action, specifically in urban planning. These concepts involve precise and defined forms of intervention at the urban planning level, often regulated by law. This dual mechanism has frequently allowed for the relaxation, alteration, or contradiction of the rules established by the plan itself. Consequently, it has hindered the identification of structural choices or action strategies within various interventions [52]. This phenomenon is particularly important to consider when analyzing the urban transformations characterizing the neighborhood in question.

In San Lorenzo, a process of citizen participation was planned for the development of an Urban Project requiring administrative approval. In this period, citizens report needs and requirements. Among the first to emerge is the need to avoid increasing construction in the neighborhood, as it is already densely built compared to the number of residents living there. Additionally, there is an urgent need for intervention in degraded and abandoned areas to prevent the proliferation of drug dealing zones. This process was intended to be based on constant communication between various stakeholders with the objective of collaboratively creating a project capable of predicting the neighborhood's development while mediating between different interests. The preliminary consultation phase, conducted following analysis and guidelines for drafting the Preliminary Asset Scheme, concluded in 2010. In 2013, after approximately three years of stagnation, the new Administration initiated a new citizen participation process to revive the urban project aimed at preserving the neighborhood's unique characteristics within a coherent and integrated redevelopment framework. Under the direction of Giovanni Caudo, the Urban Planning Counselor, activities resumed in 2013 to prepare a Preliminary Asset Scheme for the San Lorenzo Urban Project, and a new participation phase was launched to define the project. During this citizen participation process, involving local associations like the Libera Repubblica di San Lorenzo and residents, a series of thematic meetings were organized by the Urban Planning Department. These meetings covered topics such as enhancement areas, the relationship between the university and the neighborhood, public spaces, mobility, and area C4. Additionally, meetings were held with schools, senior centers, merchants, and artisans, and neighborhood workshops were conducted. This phase was marked by significant citizen participation, with residents beginning to voice concerns about the degradation caused by «unregulated nightlife» (Interview n.13, 1 February 2024). The new participation phase concluded on 12 January 2015, after two years of intensive meetings and analysis. The final Urban Project, incorporating all proposals and work from the two-year period, was presented at a public meeting held at the Saffi School Institute. This project was the result of mutual growth and extensive dialogue and collaboration among the various stakeholders involved. However, in 2015, a change in administration led to the project «falls into the void once again» (Interview n.13, 1 February 2024).

Meanwhile, with the approval of the urban regeneration law in July 2017 (Regional Law of 18 July 2017, n.7), the "Piano Casa" became permanent. Initially introduced in 2009 as a temporary measure intended to last 18 months to stimulate the economy through construction activities during a crisis period, it became permanent legislation, leading to the creation of what were described as «scattered volumes everywhere, without order or criterion, in San Lorenzo as in other neighborhoods» [46].

A new citizen participation process was initiated in December 2020 (Interview n.10, 25 January 2023), following the first conviction of those responsible for the death of Desirée Mariottini in an abandoned building on Via dei Lucani. This area was well-known to residents who had long requested the recovery and subsequent redevelopment of the surrounding public space. Regarding private artisanal spaces, after attempting to involve owners in a redevelopment process led by the Municipality without success, an expression of interest was launched in October 2020. This is an informative tool that the public entity can use as part of a procedure for awarding a contract. Specifically, property owners, neighborhood residents, collectives, associations, or anyone else could propose a project, which would only be approved by the Council after receiving the consent of at least 75% of the property and landowners in the area affected by the project (Interview n.10, 25 January 2024). During the participation process with various stakeholders, the primary need identified was not to intervene in the area by constructing new buildings and residences, as «San Lorenzo is saturated in terms of residences, and what is needed are services that are completely absent here» (Interview n.11, 29 January 2024). The project had to be approved first by the Municipal Council, composed of the Mayor and Councilors, and then by the Capitoline Assembly, composed of the Mayor and 48 councilors. However, the Council decided not to proceed. In 2021, during an administrative change, the project was reconsidered, and in the absence of a concrete plan, Cerberus, an American investment company, is now carrying out a project that does not meet the demands of the initially involved neighborhood residents (Interview n.8, 18 January 2023). To understand the effects of not approving an Urban Project resulting from effective communication between different stakeholders, public–private mediation, or efficient institutional coordination, it is essential to review some of the construction interventions, commercial changes, and social transformations made possible by the lack of neighborhood development planning.

4.3. Potential Effects of the Lack of an Urban Project on the Perception of Gentrification

Towards the end of the 20th century, the arrival of a new cultural class (mainly students) brought the emergence of new lifestyles and consumption patterns in the neighborhood: it was the birth of the night-time economy in the reference territory [57]. Students influence urban space and its changes, contributing to its socio-cultural transformation [13]. Meanwhile, studentification [23] can also be associated with a general deterioration in the physical condition of the neighborhood and its real estate assets [47], which in turn leads to further displacement of the existing population to other neighborhoods [44].

In the case of San Lorenzo, we cannot talk about the traditional displacement due to the gentrification process (see Section 3.2). According to Mappe Roma (https://www.mapparoma.info/zone-urbanistiche/s-lorenzo/, accessed on 5 February 2024), the number of residents decreased from 14,746 in 1981 to 8866 in 2019, but it was a voluntary choice rather than forced displacement. In fact, former residents leave not because they cannot afford the rising cost of living or property taxes (Interview n.1, 13 December 2023) but because it is more economically advantageous to rent out their homes [42]. Some original residents chose to leave the neighborhood to achieve higher rental income by leasing their homes at increasingly higher prices [46]. This "softer displacement" is also driven by factors like noise and changes in the neighborhood environment, such as the rise of nightlife and the decrease in everyday services, which make the area less appealing for families or older residents (Interview n.1, 13 December 2023).

According to the Rome Chamber of Commerce (https://www.rm.camcom.it/pagina436_dati-esercizi-commerciali.html, accessed on 28 January 2024), in 2020 San Lorenzo had a total of 206 food sector businesses, including 61 establishments with seating in the food and beverage sector: 32 bars, 29 pubs, and 41 takeaway food outlets [46]. At the time of the census, there were 34 retail food businesses, making beverage service the most common commercial activity, with around 80 establishments. This quantitative and qualitative dimension of activities depicts a neighborhood overwhelmed by food service offerings, catering to a population much larger than the neighborhood itself, absorbing external population flows drawn to the area's evening and nightlife, night users and temporary visitors [46].

Simultaneously, because short-term rentals are more economically advantageous [50], the reduction in apartments available for long-term rental by residents is likely the first direct consequence of this new type of tourism facilitated by short-term rental platforms [58]. In fact, among the other phenomena that can play a fundamental role in facilitating the departure of old residents from their traditionally inhabited neighborhoods, there is the phenomenon known as Airbnbification [14]: the spread of the use of platforms such as Airbnb and the increase in the number of accommodations available through them are closely linked to processes of touristification and gentrification of urban areas. In this context, the tourist becomes the primary "temporary global inhabitant" [49], the main consumer of daily urban space, regardless of their social class. Additionally, the platform is often accused of distorting the real estate and rental markets by exponentially accelerating the process of price increases and property values, leading to a turnover of the population (of residents of traditional neighborhoods), resulting in the direct and indirect expulsion of residents, replaced by temporary inhabitants [25,27,59]. A demographic category particularly affected by rising rental prices and decreasing housing availability is students (Interview n.9, 23 January 2024); San Lorenzo, once a student neighborhood, now tends to expel students due to these pressures.

Regarding changes in the real estate market in San Lorenzo, the constant increase in property values is noteworthy. «Value creation now no longer happens solely by building houses, as during the city's expansion phase, but by investing in and reselling existing houses, thus regenerating the built environment (Interview n.13, 1 February 2024)». This mechanism was recognized by the financial sector before urban planning. Indeed, «it captured a significant amount of value without innovation or production of goods, instead circulating capital through real estate leverage» (https://www.carteinregola.it/index.php/dieci-parole-casa-9-ottobre/, accessed on 4 February 2024). Over the past fifteen years, the number of real estate agencies in San Lorenzo has increased from two to thirteen [46]. The transformation of the neighborhood's economy reflects the social transformation following studentification, responding to market demands for rooms and apartments for out-of-town students. Recently, however, these agencies have had fewer rooms and apartments for rent, because owners, especially in the past year, prefer converting their properties into vacation rentals via platforms like Airbnb or Roomless (Interview n.9, 23 January 2024). The profits from buying apartments to be used as vacation rentals have led some private individuals to seek agencies to «rent a house to turn into a vacation home; rent to then run a B&B» (Interview n. 9, 23 January 2024), contributing to the systematic reduction in long-term rental housing and fueling the short-term tourist rental market.

The first effect of increased temporary housing is the undermining of the right to the city [60]. For example, the scarcity of affordable rentals is a major obstacle to young people's mobility and independence [46]. Other emerging effects include the transformation of a neighborhood's commercial fabric and the disappearance of the diverse cultural, commercial, and recreational venues necessary for residential life, leading to substantial changes in the local social and cultural ecology.

The lack of an Urban Project capable of anticipating the neighborhood's urban development, combined with rising property values of existing buildings undergoing "regeneration" interventions, has led to widespread construction within the neighborhood. According to R. M., an architect and urban planner in San Lorenzo, «There have been a myriad of building permits using the Housing Plan to build in the most unimaginable places, inside courtyards, on small plots of land, by renovating existing structures. The effect on the neighborhood has been tremendous because many of these developments have been used as Airbnbs, vacation homes, etc. Thus, there has been not only a physical transformation but especially a rapid change in use. This has accelerated significantly recently, utilizing the Housing Plan and any available instrument, with the complicity of the administration (...)» (Interview n.12, 31 January 2024).

Indeed, in this specific context, from the analysis of urban interventions carried out in the area and the dynamics emerging from various conducted interviews, it emerges that the lack of urban planning and regulation has led to the sudden opening of construction sites with activities demolishing artisan workshops and old disused factories. This has allowed real estate speculators to build new buildings in every available corner of the neighborhood, without safeguarding buildings considered to have high historical value (by citizens and the NPRG), threatening the existence of social centers that have historically characterized the neighborhood's social culture, and dismantling public green spaces. All these interventions have contributed to generating widespread discontent and a general lack of trust among residents towards the administration's inaction and the municipality's failure to protect the historical and socio-cultural characteristics of the area (Interview n.3, 17 December 2023). The consequences of the lack of regulation in the neighborhood's urban development emerge when examining some emblematic cases. For example, the project initiated in 2013 to dismantle the old building of the former Fonderie Bastianelli and replace it with apartment buildings with an underground garage first led to the occupation of the site by the Communia collective for a few months, followed by opposition from residents (Interview n.8, 18 January 2024). In this case, the residents aimed to protect a building of high historical value, avoid compromising a delicate balance (as an aquifer flows underground), and refrain from intervening on the part of the foundry roof constructed with asbestos-cement sheets (Interview n.3, 17 December 2023). Speaking with a resident of the neighborhood and one of the founders of the San Lorenzo Urban Project Committee (established in 2013), it emerged that «In the case of Bastianelli, they submitted the building permit based on a false declaration. From the outside, you can see the two buildings connected by a central shed that was the old foundry, which then became a garage. The story of Fonderie Bastianelli began because the lady who lives there one day saw the asbestos being removed, as it had an asbestos roof, she got alarmed, called (...), this fact made us suspicious and we discovered that there was a building permit» (Interview n.1, 13 December 2023). After the eviction of the building, citizens filed a lawsuit against the Municipality of Rome at the Regional Administrative Court (TAR) to request that the construction be halted, arguing that the project's approval violated the objectives of conserving and enhancing existing qualities and characteristics, as well as preserving the artisanal fabric of the San Lorenzo neighborhood [51]. TAR's response came on 9 April 2015, when it accepted the citizens' annulment request and condemned Rome Capital and Sabelli Trading Srl to pay court costs. Among the reasons cited was that the complex is included in Annex G1 of the NGRP Quality Charter, requiring approval from the Municipal Superintendence for Cultural Heritage. TAR thus recognized the Capitol administration's failure to examine the impact of the demolition of the former Fonderie on the surrounding urban fabric, also in terms of respecting the public interest in maintaining and preserving the historical–cultural characteristics of Rome. Subsequently, the fate of the former Fonderie Bastianelli passed into the hands of the Municipality of Rome and the Superintendence [51]. However, in 2016, the State Council overturned TAR's judgment and accepted the builders' appeal, and construction continued (Judgment 05002/2016 Reg. prov. Coll; no. 04478/2015 Reg. Ric.). Nevertheless, years later, it remains incomplete, with many apartments unsold and work appearing to have stopped on 5 April 2023 (Interview n.1, 13 December 2023).

There are some representative cases of the ongoing demolition and reconstruction activities in the neighborhood, such as Via dei Sabelli 100 (Figure 6) and the construction site at the corner of Via dei Sabelli and Via degli Ausoni (Figure 7). In the first case, a six-story building was constructed in the middle of a private enclosed courtyard which, according to the residents, did not have the proper building permit (currently at the center of a judicial process to determine its legality) (Interview n.6, 11 January 2024). The second case concerns a corner left vacant to commemorate the bombings of July 1943, as remembered by a mural that will soon be erased by the new building. In this space, the residents had expressly requested to keep it vacant to allow citizens to remember the WWII bombings. Moreover, construction should not have been permitted due to compensation requests in the post-war period by people who lived in the bombed building and chose not to rebuild their apartments in exchange for monetary compensation. However, this seems to have been disregarded, as construction activities are currently underway (Interview n.6, 11 January 2024).

Figure 6. The construction of a building in the middle of a private enclosed courtyard in Via dei Sabelli, 100. Source: authors.

Figure 7. Construction activities at the corner of Via dei Sabelli and Via degli Ausoni, a site that was intentionally left vacant to commemorate the historical memory of the bombing. Source: authors.

Among other cases of real estate speculation generating dissatisfaction among residents is the Former Dogana, acquired by The Social Hub (https://www.thesocialhub.co/it/roma/, accessed on 10 January 2024) with the goal of constructing a 21,000 sqm structure intended for tourists, students, and young professionals, offering accommodations and leisure through the extended stay formula, perfectly embodying the concept of a short-term temporal dimension [46]. The project includes the construction of 444 rooms, halfway between a luxury hotel and a student residence, with spaces for coworking and co-living. However, residents and neighborhood associations oppose its construction residents and neighborhood associations oppose its construction, arguing that San Lorenzo is already densely built, suffers from a lack of green spaces, and does not need an elite space meant for a select few (Interview n.3, 17 December 2023). A similar space is presented by the Soho House (https://www.sohohouse.com/it/houses/soho-house-rome, accessed on 15 January 2024), a private members club built on Via Cesare de Lollis, adjacent to the university. It features a luxury hotel, gym, cinema, bars, restaurants, and coworking spaces, all accessible only to members. Despite being a space for creativity and sharing, residents believe it should meet local needs rather than those of external visitors and temporary residents (Interview n.6, 11 January 2024). The opening of luxurious student accommodations and venues that do not reflect the neighborhood's socio-cultural characteristics may further drive residents away, as they feel their needs and requests are being ignored. According to one of the founders of the San Lorenzo Urban Project Committee, «Today this has become a neighborhood exploited by commercial businesses and elite spaces. It is a popular neighborhood, but it is also a central one. Then the Soho Hotel arrived, and now all these constructions aimed at short-term living are coming in, driving away the real residents of the neighborhood» (Interview n.1, 13 December 2023).

Despite San Lorenzo's severe shortage of public green spaces, one intervention that has particularly dissatisfied residents is the dismantling of public green areas at the Knights of Columbus field to build padel courts (Figure 8). This decision also clashed with the grassroots, inclusive sports model promoted by Atletico San Lorenzo. In this case, residents' discontent is particularly directed at the administration because «There are tools the administration could have used to try even to mediate, but it never did. It could have done so with the Knights of Columbus and the padel courts. Of course, there is an attitude from the administration of allowing administrative procedures to proceed, but if everything goes ahead with tacit approval, this already represents the administration's responsibility, meaning everyone has an excuse to let anything happen» (Interview n.1, 13 December 2023).

Finally, among the most exemplary cases is the fate of Via dei Lucani (Figure 9), where the construction of three buildings, a parking lot, a sports field, an area for "craft activities", and a small area for "equipped green space" is currently planned. In this case, the failure to approve a project resulting from citizen participation, hence the lack of effective communication between different stakeholders, has led to the initiation of a project that does not meet the residents' requests and needs. Moreover, the existence of one of the few remaining social centers in the area is threatened by the project because «in the place where Communia (Figure 10) is currently located, the only social stronghold in the area, part of Building C is planned. Although some small green spaces are included in the project, which the neighborhood desperately needs, the project is mainly intended for the construction of one- and two-room apartments» (Interview n.8, 18 January 2024). According to a Communia activist, «All public management aspects have been privatized. Even here [regarding the Via dei Lucani project], one of the questions we asked [Lady Moon, a vehicle company of Cerberus] was: what size are these apartments? Are they for families? No, only studios. It seems they want a neighborhood of people working in finance who spend a few days here and then leave, with no intention of creating a community neighborhood» (Interview n.8, 18 January 2024).

Figure 8. The new padel courts in the public green area of Knights of Columbus. Source: authors.

Figure 9. Construction activities in Via dei Lucani. Source: authors.

Figure 10. Communia collective in Viale dello Scalo San Lorenzo. Source: authors.

5. Discussion

A review of the literature on gentrification reveals that the phenomenon has evolved over time, transforming from a "local anomaly" to a "global urban strategy" [13]. Examining various case studies worldwide highlights the differences in the gentrification process rather than the similarities. The broadening of the concept has weakened its semantic clarity, making it increasingly less effective as a descriptive category [35].

According to Lees, necessary conditions for gentrification include class polarization, significant investment in urban regeneration, and various forms of displacement (direct, indirect, exclusionary, etc.). The two primary effects of gentrification are rehabilitation and displacement, seen as positive and negative aspects, respectively. Rehabilitation refers to improvements in residential areas and increased property values—an economic benefit of the process—while displacement emphasizes the social and psychological costs it imposes [35].

The analysis suggests that the concept of gentrification only partially applies to the complex urban and socio-cultural transformation occurring in San Lorenzo. What differs from traditional models of gentrification is the absence of a pronounced class conflict, forced displacement, or the rapid, homogeneous manifestation of the phenomenon observed in other geographic contexts, where it often serves as a form of social protest [35].

Initially, the phase of studentification could be associated with the onset of a more complex gentrification phenomenon linked to the displacement of residents due to rising property values, the emergence of single-purpose services aimed at students [46], and the regeneration and requalification of the area, capable of becoming a residence for a creative and trendy class capable of building an open and dynamic urban environment that attracts businesses and capital [43]. However, the analysis of the phase that can be identified as post-studentification, along with its conditions and characteristics, has proven crucial in understanding the inadequacy of the term in describing the urban dynamics and socio-cultural transformations of the case study. Even when considering gentrification more flexibly as the manifestation of a return wave of capital to the city that had gradually lost its value [44], this process is still identified in the struggle for the social division of urban space, where stronger actors replace weaker ones. This is a primary condition for using such a category of analysis for urban, territorial, and socio-cultural dynamics. In this context, it is important not to allow every significant dynamic that determines transformations or reproductions in the social division of urban space and the building heritage to be assimilated into gentrification [3]. In fact, in the considered neighborhood, after the studentification that began in the late 20th century and consolidated in the early 21st century, a new phase of transformation has started, during which the "newcomers" increasingly appear to be temporary residents, frequenting San Lorenzo's nightlife and renting accommodation via Airbnb. From the analysis of the interviews conducted, it emerges that the area has low attractiveness for young people due to reduced availability of long-term rental housing or the high cost of purchasing property [46], as well as for families who choose to leave the neighborhood for multiple reasons: it is economically more advantageous to offer their homes on short-term rental platforms to tourists or temporary residents, the systematic decrease in neighborhood shops alternative to the monoculture of food services, the nighttime chaos, and the lack of public green spaces (Interview n.6, 11 January 2024). These dynamics often result in, quoting an interviewee, «pain, living in a neighborhood where there was an identity-rich fabric and seeing it emptied» (Interview n.1, 13 December 2023), where the number of residents continues to decline, and where there is a «continuous opening and closing of shops» (Interview n.12, 31 January 2024). Generally, the qualitative analysis reveals that the remaining residents express dissatisfaction with various building projects aimed at exploiting the concept of real estate rent [2] and the increasing influx of tourists seeking short-term rentals, fueling a phenomenon of *Airbnbification* [14].

The investigation highlights how the transformation process that has affected the neighborhood in recent years is mainly attributable to the limits that come from the NGRP and the lack of effective institutional coordination capable of anticipating the urban and socio-cultural development and strategies of the neighborhood by mediating between different public–private entities. The five fundamental options of the Regulatory Plan lack a clear determination of one or more objectives from which programs and actions could follow. The Technical Report of the NGRP, by distinguishing between strategic plan and urbanistic plan, produces negative effects on both "types" of the Plan: thus, it renounces systemic and strategic logic, classifying the Plan only as urbanistic. In this way, it consolidates the idea of two different approaches to planning, whereas it would be more appropriate to integrate these two aspects to arrive at a single methodology for the NGRP [55].

Moreover, there has been a lack of efficient and productive communication between bottom-up movements and top-down institutions, which only occasionally occurred since 2008. The type of communication initially established through citizen participation pathways could have potentially met an advocacy and strategic plan capable of managing the area's development [61,62] and its regeneration. *Advocacy planning* involves the active support and promotion by individuals aiming to influence public policies and resource allocation within political, economic, and social systems and related institutions. Attempts to propose forms of advocacy for the local community of San Lorenzo have been experimented with in multiple instances in the past [63], all guided by the collective energy of neighborhood residents participating in planning and managing the future of the area, voicing their needs, being heard by institutions, and seeing their requests realized. However, there seems to be a lack of communication between different levels, bottom-up and top-down, which could determine the approval of an urban project capable of regulating building activity within San Lorenzo, delineating the neighborhood's urban and economic development strategies, and identifying a detailed transformation path of the space aimed at achieving objectives identified through the participation of multiple stakeholders: residents, shopkeepers, speculators, neighborhood visitors, and public administration.

6. Conclusions

The case study reveals a complex urban and socio-cultural transformation, and through the investigation, it was possible to challenge the use of the term "gentrification" in relation to the institutional characteristics of the reference context. While elements such as studentification and airbnbification are present, the classic indicators of gentrification, such as the replacement of lower-income residents by a higher socioeconomic class, or class conflict, are not entirely observed. This suggests that gentrification as an interpretative category must be reassessed, particularly when dealing with urban spaces where socio-economic dynamics do not follow conventional patterns of displacement and replacement. Among the others, one objective is to invite urban scholars and planners to develop new frameworks better suited to complex urban realities like the San Lorenzo neighborhood.

The paper indicates that the socio-cultural fabric of San Lorenzo is eroding, not due to displacement in the traditional sense, but through the fact that long-term residents leave the neighborhood because of the frustration caused by the un-regulated transformation. This suggests that urban change can have profound psychological and social impacts, even in the absence of forced displacement.

The investigations conducted so far focus on the consequences of "planning by doing" and the lack of effective communication between neighborhood residents and institutions. The absence of large-scale public–private agreements and coordinated planning in San Lorenzo highlights the risks associated with fragmented urban development. The implication here is that coordinated, long-term urban planning remains crucial for sustainable neighborhood development. Policymakers and urban planners are encouraged to adopt a more holistic and integrated approach to urban planning, ensuring that the development strategies are inclusive and reflective of the diverse needs of the community. Improved communication between institutional actors and local communities could foster more effective and equitable urban development policies [64]. Indeed, the investigations conducted so far highlight the importance of policymakers initiating an effective bottom-up enabling process and institutional coordination, creating a trading zone [65] for dialogue among the various stakeholders involved [66].

The rise of short-term rentals in San Lorenzo has emerged as a significant factor in the neighborhood's transformation. The implications for urban policy are that unregulated short-term rentals can distort housing markets, and reduce the availability of long-term housing. This suggests the need for regulatory frameworks that can mitigate the negative impact of platforms like Airbnb.

Finally, the findings suggest broader implications for urban governance. The failure of the NGRP to provide clear, strategic direction has led to inconsistent development outcomes in San Lorenzo. This indicates a need for urban policies that integrate strategic and urban planning, ensuring that neighborhood development is guided by long-term goals that consider the needs of all stakeholders.

Author Contributions: Conceptualization, A.A.; Methodology, L.T.; Validation, L.T.; Formal analysis, A.A.; Investigation, A.A.; Writing—original draft, A.A.; Writing—review & editing, A.A.; Visualization, A.A. All authors have read and agreed to the published version of the manuscript.

Funding: This research received no external funding.

Data Availability Statement: The data supporting the findings of this study consist of qualitative interviews conducted with key stakeholders. Due to privacy and ethical considerations, the interview data are not publicly available. Further information about the research or specific inquiries can be directed to the corresponding author.

Conflicts of Interest: The authors declare no conflict of interest.

Appendix A. Table of Interviewees

N.	Interviewees' Names and Roles	Association/Activities	Date and Type of Interview
1	G., one of the Committee's founders	San Lorenzo Urban Project Committee	13 December 23, non-standardized
2	S. G., expert of San Lorenzo	Independent researcher, writer and journalist	14 December 23, non-standardized
3	G. C., one of the Committee's founders	Free Republic of San Lorenzo Committee	17 December 23, non-standardized
4	C. M., owner of the bar	Bar Celestino	9 January 24, semi-structured
5	T., owner of the bookstore	Ultrapopolare Library of San Lorenzo (bookstore)	10 January 24, semi-structured
6	S. e D., owner of the bar and expert of San Lorenzo	Bar Marani and a local journalist	11 January 24, semi-structured
7	F., owner of the bookstore	Antigone (bookstore)	11 January 24, semi-structured
8	A., expert of Communia's history	Communia Activist	18 January 24, non standardized
9	I., worker in the agency	Tecnorete real estate agency in San Lorenzo	23 January 24, semi-structured
10	L. M., expert of San Lorenzo Urban Project	Former Urban Planning Assessor of Rome Capital	25 January 24, focused
11	S., expert of Esc's history	Esc Activist	29 January 24, non-standardized
12	R. M., expert of San Lorenzo	Urbanist and architect	31 January 24, focused
13	G. C., expert of San Lorenzo Urban Project	Former Urban Planning Assessor of Rome Capital	1 February 24, focused

References

1. Briata, P.G.; Bricocoli, M.; Tedesco, C. *Città in Periferia*; Politiche urbane e progetti locali in Francia, Gran Bretagna e Italia; Carocci: Rome, Italy, 2009.
2. Tocci, W. *Roma Come se: Alla Ricerca del Futuro per la Capitale*; Donzelli: Rome, Italy, 2020.
3. Laino, G. *Se Tutto È Gentrification, Comprendiamo Poco, Commento al Libro di Giovanni Semi Gentrificazione Tutte le Città Come Disneyland?* Casa della Cultura: Milan, Italy, 2016.
4. Atkinson, R. Introduction: Misunderstood Saviour or Vengeful Wrecker? The many meanings and problems of gentrification. In *Urban Studies*; Sage Journals: Thousand Oaks, CA, USA, 2003; Volume 40, pp. 2343–2350.
5. Lees, L.; Slater, T.; Wyly, E. *Gentrification*; Routledge: London, UK, 2008.
6. Glass, R. *Introduction to London: Aspects of Change*; Centre for Urban Studies: London, UK, 1964.
7. Lees, L. Super-gentrification: The case of Brooklyn heights, New York city. *Urban Stud.* **2003**, *40*, 2487–2509. [CrossRef]
8. Slater, T. 'A literal necessity to be re-placed': A rejoinder to the gentrification debate. *Int. J. Urban Reg. Res.* **2008**, *l32*, 212–223. [CrossRef]
9. Annunziata, S. *Oltre la Gentrification. Letture di Urbanistica Critica tra Desiderio e Resistenze Urbane*; Editpress: Firenze, Italy, 2022.
10. Harrison, H.; Birks, M.; Franklin, R.; Mills, J. Case study research: Foundations and methodological orientations. *Forum Qual. Sozialforschung/Forum Qual. Soc. Res.* **2017**, *18*. [CrossRef]
11. Schön, D.A. Generative metaphor: A perspective on problem-setting in social policy. *Metaphor Thought* **1979**, *2*, 137–163.
12. Ostrom, E. Background on the institutional analysis and development framework. *Policy Stud. J.* **2011**, *39*, 7–27. [CrossRef]
13. Smith, D.P. "Studentification": The gentrification factory? In *Gentrification in a Global Context*; Atkinson, R., Bridge, G., Eds.; The New Urban Colonialism; Routledge: London, UK, 2005.
14. Peters, D. Density wars in Silicon Beach: The struggle to mix new spaces for toil, stay and play in Santa Monica, California. In *Protest and Resistance in the Tourist City*; Colomb, C., Novy, J., Eds.; Routledge: London, UK, 2016.
15. Valle, M.M. Globalizing the sociology of gentrification. *City Community* **2021**, *20*, 59–70. [CrossRef]
16. Harvey, D. From managerialism to entrepreneurialism: The transformation in urban governance in late capitalism. *Geogr. Ann. Ser. B Hum. Geogr.* **1989**, *71*, 3–17. [CrossRef]
17. Ley, D. Inner city revitalization in Canada: A Vancouver case study. *Can. Geogr.* **1981**, *25*, 124–148. [CrossRef]
18. Ley, D. *The New Middle Class and the Remaking of the Central City*; Oxford University Press: Oxford, UK, 1996.
19. Smith, N. Toward a Theory of Gentrification A Back to the City Movement by Capital, not People. *J. Am. Plan. Assoc.* **1979**, *45*, 538–548. [CrossRef]
20. Lees, L.; Shin, B.H.; Lòpez-Morales, E. *Introduction: 'Gentrification'—A Global Urban Process?* The University of Chicago Press: Chicago, IL, USA, 2015.
21. Crivello, S. I locali notturni tra consumo post-moderno ed economia della notte. In *Enciclopedia Sociologica dei Luoghi a Cura di Giampaolo Nuvolati*; Amazon: La Tour d'Aigues, France, 2019; pp. 161–178.
22. Hubbard, P. Regulating the social impacts of studentification. A Loughborough case study. *Environ. Plan.* **2008**, *40*, 323–341. [CrossRef]
23. Smith, D.P. Patterns and processes of "studentification" in Leeds. *Reg. Rev.* **2002**, *12*, 14–16.
24. Celata, F. *La Airbnbificazione delle Città: Gli Effetti a Roma tra Centro e Periferia*; La Sapienza, Dipartimento Memotef: Rome, Italy, 2017.
25. Cocola-Gant, A. Holiday rentals: The new gentrification battlefront. *Sociol. Res. Online* **2016**, *21*, 112–120. [CrossRef]
26. Lees, L.; Rozena, S. The everyday lived experiences of Airbnbification in London. *Soc. Cult. Geogr.* **2023**, *24*, 253–273.
27. Wachsmuth, D.; Weisler, A. Airbnb and the rent gap: Gentrification through the sharing economy. *Environ. Plan. A* **2018**, *50*, 1147–1170. [CrossRef]
28. Gianturco, G. *L'Intervista Qualitativa. Dal Discorso al Testo Scritto*; Edizioni Angelo Guerini e Associati Srl: Milano, Italy, 2004.
29. Yeom, M.; Mikelbank, B.; Gentrification, A.N. Introduction, Overview, and Application. In *The 21st Century American City: Race, Ethnicity, and Multicultural Urban Life*; Kendall Hunt Publishing Company: Dubuque, IA, USA, 2018.
30. Bonvalet, C.; Clément, C.; Ogg, J. *Réinventer la Famille: L' Histoire des Baby-Boomers*; Presses Universitaires de France: France, Paris, 2011.
31. Bonvalet, C.; Ogg, J. *Les Baby-Boomers: Une Génération Mobile*; Amazon: La Tour d'Aigues, France, 2009.
32. Uitermark, J.; Duyvendak, J.W.; Kleinhans, R. Gentrification as a governmental strategy: Social control and social cohesion in Hoogvliet, Rotterdam. *Environ. Plan. A* **2007**, *39*, 125–141. [CrossRef]
33. Zukin, S. *The Cultures of Cities*; Blackwell: Oxford, UK, 1995.
34. Davidson, M.; Lees, L. New-Build Gentrification: Its Histories, Trajectories, and Critical Geographies. *Popul. Space Place* **2010**, *16*, 395–411. [CrossRef]
35. Annunziata, S. Se tutto fosse gentrification: Possibilità e limiti di una categoria descrittiva. In *I Territori Della Città in Trasformazione*; Balducci, A., Fedeli, V., Eds.; FrancoAngeli: Milano, Italy, 2007.
36. Gastaldi, F. *Processi di Gentrification nel Centro di Genova*; Archivio di Studi Urbani e Regionali, n. 77: Rome, Italy, 2003.
37. Scandurra, G. *Il Pigneto Gentrificato Come Testo, in Antropologia della Contemporaneità: Etnografia della Diversità e delle Convergenze Culturali*; Tesi di Dottorato Discussa all'Università Milano-Bicocca: Milano, Italy, 2005.

38. Bovone, L.; Mazzette, A.; Rovati, G. *Effervescenze Urbane*; Quartieri Creativi a Milano, Genova e Sassari; FrancoAngeli: Milano, Italy, 2005.
39. Semi, G. Il quartiere che (si) distingue. Un caso di gentrification a Torino. *Studi Cult.* **2004**, *1*, 83–107.
40. Shin, H.B. Property-based redevelopment and gentrification: The case of Seoul, South Korea. *Geoforum* **2009**, *40*, 906–917. [CrossRef]
41. Filandri, M.; Semi, G. Housing Inequality: The Need for a Shift in Public Policy Intervention. *Sociol. Res. Online* **2018**, *27*, 1–7. [CrossRef]
42. Meli, I.; Ricotta, G. *Il Conflitto Urbano nel Quartiere di San Lorenzo a Roma. Movida, Studentificazione e Hipification, in Conti U. (a Cura di), Gentrificazione. Profili e Saperi per L'analisi del Cambiamento Sociale delle città Italiane*; Franco Angeli: Milan, Italy, 2022; pp. 137–151.
43. Florida, R. *The Rise of the Creative Class, Basic Books, New York; trad. it. 2003, L'ascesa della Classe Creativa*; Mondadori: Milano, Italy, 2002.
44. Semi, G. *Gentrification. Tutte le Città Come Disneyland?* Il Mulino: Bologna, Italy, 2015.
45. Chatterton, P. The urban impossible: A eulogy for the unfinished city. *City* **2010**, *14*, 234–244. [CrossRef]
46. Barile, A.; Brollo, B.; Gainsforth, S.; Marchini, R. *Dopo la Gentrificazione. Un Quartiere Laboratorio dalla Crisi Economica All'Abitare Temporaneo*; DeriveApprodi srl: Bologna, Italy, 2023.
47. Rugg, J.; Rhodes, D.; Jones, A. Studying a Niche Market. UK students and the private rented sector. *Hous. Stud.* **2002**, *17*, 289–303. [CrossRef]
48. Geissler, J.B.; Tricarico, L.; Vecchio, G. The construction of a trading zone as political strategy: A review of London Infrastructure Plan 2050. *Eur. J. Spat. Dev.* **2017**, *15*, 1–22.
49. Sequera, J.; Nofre, J. Debates shaken, not stirred: New debates on touristification and the limits of gentrification. *City* **2018**, *22*, 843–855. [CrossRef]
50. Cocola-Gant, A.; Gago, A. Airbnb, buy-to-let investment and tourism-drive displacement. A case study in Lisbon. In *Environment and Planning A: Economy and Space*; Sage Journals: Thousand Oaks, CA, USA, 2019; pp. 1–34.
51. Fabrizi, S. *La Storia di San Lorenzo. Dalla Preistoria ai Giorni Nostri*; Cierre e Grafica Group Srl: Rome, Italy, 2021.
52. Insolera, I.; Berdini, P. *Roma Moderna. Da Napoleone I al XXI Secolo, Nuova Edizione*; ELCOGRAF S.p.A., Stabilimento di Cles (TN); Casa editrice Einaudi: Turin, Italy, 2023.
53. D'Albergo, E.; Moini, G. *Il Regime Dell'Urbe: Politica, Economia e Potere a Roma*; Carocci: Rome, Italy, 2015.
54. Casini, L. *Perequazione e Compensazioni nel Nuovo Piano Regolatore Generale di Roma*; Giornale di Diritto Amministrativo: Lucca, Italy, 2009.
55. Lupo, C. Il Nuovo Piano Regolatore Generale di Roma (2008): Dubbi, Ombre e Aspettative Tradite. 2009. Available online: http://hdl.handle.net/2067/1129 (accessed on 10 November 2023).
56. Tallon, A. *Urban Regeneration in the UK*; Routledge: London, UK, 2020.
57. Hannigan, J. *Fantasy City: Pleasure and Profit in the Postmodern Metropolis*; Routledge: London, UK; New York, NY, USA, 1998.
58. Celata, F.; Romano, A. Overtourism and online short-term rentals platforms in Italian cities. *J. Sustain. Tour.* **2008**, *30*, 1–20.
59. Yrigoy, I. Rent gap reloaded: Airbnb and the shift from residential to touristic rental housing in the Palma Old Quarter in Mallorca, Spain. In *Urban Studies*; Sage Journals: Thousand Oaks, CA, USA, 2018.
60. Lefebvre, H. *Il Diritto alla Città, tr. It*; Ombre corte: Verona, Italy, 2014.
61. Sager, T. Advocacy planning: Were expectations fulfilled? *Plan. Perspect.* **2022**, *37*, 1205–1230. [CrossRef]
62. Tricarico, L.; De Vidovich, L.; Billi, A. *Innovazione Sociale, Sviluppo Economico e Margini Territoriali: Una Riflessione per il Contesto Italiano*; LaborEst: Milano, Italy, 2021; pp. 55–63.
63. Clementi, A.; Ricci, M. *Ripensare il Progetto Urbano: Il Caso San Lorenzo a Roma*; Meltemi Editore srl: Rome, Italy, 2004; Volume 30.
64. Brignone, L.; Cellamare, C.; Gissara, M.; Montillo, F.; Olcuire, S.; Simoncini, S. Autorganizzazione e rigenerazione urbana: Ripensare le politiche a partire dalle pratiche. Tre esperienze della periferia romana. *Tracce Urbane Riv. Ital. Transdiscipl. Studi Urbani* **2022**, *8*. [CrossRef]
65. Balducci, A.; Mäntysalo, R. (Eds.) *Urban Planning as a Trading Zone*; Springer: Dordrecht, The Netherlands, 2013.
66. Billi, A.; Tricarico, L. Regional development policies in Italy: How to combine cultural approaches with social innovation. In *New Metropolitan Perspectives: Local Knowledge and Innovation Dynamics towards Territory Attractiveness through the Implementation of Horizon/E2020/Agenda2030*; Springer International Publishing: Berlin/Heidelberg, Germany, 2019; Volume 2, pp. 277–287.

Disclaimer/Publisher's Note: The statements, opinions and data contained in all publications are solely those of the individual author(s) and contributor(s) and not of MDPI and/or the editor(s). MDPI and/or the editor(s) disclaim responsibility for any injury to people or property resulting from any ideas, methods, instructions or products referred to in the content.

land

MDPI

Article

Transport Policy Pathways for Autonomous Road Vehicles to Promote Sustainable Urban Development in the European Union: A Multicriteria Analysis

Nikolaos Gavanas [1,*], Konstantina Anastasiadou [2], Eftihia Nathanail [3] and Socrates Basbas [4]

[1] Department of Planning and Regional Development, School of Engineering, University of Thessaly, 38334 Volos, Greece
[2] School of Civil Engineering, Faculty of Engineering, Aristotle University of Thessaloniki, 54124 Thessaloniki, Greece; kianasta@civil.auth.gr
[3] Department of Civil Engineering, School of Engineering, University of Thessaly, 38334 Volos, Greece; enath@uth.gr
[4] School of Rural and Surveying Engineering, Faculty of Engineering, Aristotle University of Thessaloniki, 54124 Thessaloniki, Greece; smpasmpa@topo.auth.gr
* Correspondence: ngavanas@uth.gr; Tel.: +30-24210-74451

Abstract: The European Union's policy aims for the wide-scale deployment of automated mobility by 2030, i.e., within the next programming period (2028–2034), with the deployment of autonomous road vehicles (AVs) in cities playing a key role. Researchers suggest that AV deployment will have complex impacts on urban development, which are difficult to quantify due to scarce real-life data. The present research aims to evaluate different policy pathways of AV deployment for sustainable urban development in the next EU programming period. A multicriteria analysis is conducted, combining AHP and VIKOR, with the participation of experts across Europe. Initially, the potential impacts on sustainable urban development are weighted as evaluation criteria. Then, different pathways are evaluated against these criteria, i.e., AV deployment as collective and/or private transport in specific areas and periods or in the whole Functional Urban Area (FUA) on a 24 h basis. An interesting finding is that the effect on the city's spatial development, not thoroughly examined by literature, is highly ranked by experts. Regarding policy pathways, autonomous collective transport with 24 h service of the FUA emerged as the optimum alternative. The proposed methodology provides a tool for planners, researchers, and policy makers and a framework for an open debate with society.

Keywords: autonomous road vehicles; multicriteria analysis; European Union; transport policy; sustainable mobility; urban development

1. Introduction

Urbanization has been rapidly increasing all over the world since the Industrial Revolution [1]. Nowadays, about 70% of people in the European Union (EU) live in urban areas, while the urban population is expected to further increase in the next years, with transport being a critical factor in the urbanization and urban growth rate [2]. Thus, the analysis of the interaction between transport and urban development can significantly contribute to formulating effective urban development policies within the framework of sustainability [3]. Towards this direction and identifying the potential of new technologies (such as autonomous mobility) in terms of sustainable urban development, it is important to assess the most appropriate policy pathways for the implementation of autonomous road vehicles.

Autonomous road vehicles (AVs), such as driverless cars, pods, and buses, can be defined as self-driving vehicles that are equipped with technologies that enable driving operations without requiring the driver's intervention [4–6]. In the last decade, numerous autonomous driving trials have taken place in many cities worldwide, mainly along designated routes and in controlled areas, gradually progressing towards the capability of self-driving vehicles to operate under all traffic conditions and in complex urban environments. Among the projects conducted recently (in the last five years) [7–10], many pilots test autonomous public transport vehicles, such as the trials of autonomous shuttles in Gothenburg (Sweden), the SmartBus in Aalborg (Denmark), and the Driverless Shuttle Trial along a specified route in Newcastle NSW (Australia); in other projects, autonomous passenger cars are tested, as, for example, the projects: StreetWise (Leeds, UK), testing self-driving passenger cars with security drivers to provide 24 h service; Autonomoose (Waterloo, ON, Canada), with the scope of improving autonomous driving under all weather conditions; and Drive.ai (Mountain View, CA, USA), aiming to develop an on-demand self-driving car service on public roads.

According to the current estimations by policy, research, and industry stakeholders, the year 2030 is expected to be a milestone in the evolution of autonomous road vehicles. World leading industries and forerunner countries aim for the wide-scale deployment of driverless cars by 2030 [11,12]. The assumption of the availability of fully autonomous cars by the end of the decade is commonly used in relevant impact assessments [13–15].

Apart from technology readiness [16], the actual deployment of autonomous cars in cities depends on many factors, such as the following [17–25]:

- New business models and production processes for service providers, technology developers, and manufacturers.
- Integration of autonomous road vehicles to the multimodal transport system, including the interaction between vehicles of different levels of automation in traffic.
- Regulatory reforms to allow for the seamless and safe operation of driverless cars in the street.
- Impacts on social inclusiveness, mainly related to affordability and technology literacy.
- Cyber security and data governance, regarding the big data used by the autonomous system but also the data collected and generated by the autonomous system itself.
- Ethical and liability issues, mainly regarding decision making by the autonomous car especially in case of accidents and malfunctions.
- Employment implications referring to new jobs in specialized sectors, job losses in other sectors, such as professional drivers' jobs, and upskilling potential.

It should be highlighted that the above factors are underpinned by the EU policy framework [26], which sets the year 2030 as a time horizon for coping with them and achieving the large-scale deployment of automated mobility [27]. At the urban level, the deployment of autonomous road vehicles in the next years is linked to unprecedented opportunities and unforeseen impacts. Moreover, the uncertainty of the long-term effects from the deployment of autonomous road vehicles is highlighted by literature [28,29].

These impacts expand from the direct effect on mobility conditions and road safety to the contribution in shaping the digital and green transition as well as in the spatial development of cities, which is addressed to a lesser extent by literature [18,30,31]. Thus, planners and policy makers should explore the pathways to integrate self-driving road vehicles into the sustainable urban development context, considering the above fields of potential impacts [32]. Faced with the current scarcity of real-life experience and data from the implementation of autonomous driving in cities, researchers often turn to the analysis of stakeholder opinions to analyze the uncertainties of autonomous driving [33–35].

Halfway through the current EU programming period (2021–2027), the scope of the current research is the preliminary assessment of transport policy pathways for the integration of autonomous road vehicles in European cities in the first period of adoption and large-scale deployment. The purpose is the selection of the optimal pathway, in terms of sustainable urban development, in order to support policy makers in view of the next EU programming period (2028–2034). The main objectives of the research are the following: To provide an overview of the current EU transport policy for the implementation of autonomous road vehicles in European cities; discuss the potential spatial impacts and possible effects from the deployment of autonomous mobility on the sustainable urban development; perform a multicriteria analysis with the participation of research and innovation experts for the evaluation of the above-mentioned effects on the sustainable development of European cities and of the transport policy pathways for the implementation of autonomous mobility regarding each possible effect.

The research adopts the above-mentioned assumption that autonomous road vehicles will be fully available for implementation in European cities by 2030. Moreover, based on the EU transport policy targets [27], it is assumed that these vehicles will have zero tailpipe emissions. It should be noted that neither the life-cycle emissions nor the energy mix to produce the vehicle's fuel/electricity are considered in the analysis, due to the different energy transition policies in European countries and cities. Furthermore, the current research analyses passenger mobility but not last-mile freight transportation and urban logistics.

The rest of the paper is structured as follows. The next section involves the overview of the main strategic priorities of the EU transport policy for the implementation of autonomous road vehicles in urban areas. The next section refers to the synthetic analysis of possible spatial impacts from the deployment of autonomous road vehicles on sustainable urban development, also in relation to the targets of the United Nation's Sustainable Development Goal 11. Then, the methodology of the current research's multicriteria analysis is presented. For the needs of the current research, the Analytic Hierarchy Process (AHP) is combined with the VIKOR approach. The fourth section refers to the implementation of the methodology and the fifth section presents the discussion of results. Conclusive remarks and prospects for follow-up research are presented in the last section of the paper.

2. EU Transport Policy Priorities for Autonomous Road Vehicles

The main strategic document of the European Union's (EU) policy framework in the current programming period is the European Green Deal, which aims to implement the United Nations' (UN) 2030 Agenda and the corresponding Sustainable Development Goals (SDGs) [36]. The European Green Deal outlines a series of transformative policies in all sectors of the EU economy. In the transport sector, the main objective is the acceleration of the shift towards sustainable and smart mobility. For cities, "new mobility services", such as Mobility as a Service (MaaS), and connected and automated mobility are depicted as drivers for the reduction of congestion and pollution [37]. It should be noted that in the EU policy framework, the terms: Cooperative, Connected and Automated Mobility (CCAM) and Connected and Automated Driving (CAD) are also used, referring similarly to autonomous and connected (via telecommunications) vehicles, which can guide themselves without human intervention [38,39].

In order to achieve the European Green Deal's target for a 90% reduction in transport emissions by 2050, the EU Sustainable and Smart Mobility Strategy, which describes the current European transport policy priorities, sets a series of targets with a short-term (2030), mid-term (2035), and long-term (2050) time-horizon [27]. As already mentioned, one of the strategy's short-term targets for the green and digital transition of European cities is the large-scale deployment of connected and automated mobility (CAM), the promotion of active transport and micromobility, and the multimodal cooperation through Mobility as a Service (MaaS). The strategy also highlights the need to rethink the goals of urban transport planning, especially for alleviating congestion in the urban nodes of the Trans-European Transport (TEN-T) Network to ensure the effective operation of the entire TEN-T and to support the EU's mission for 100 Climate-Neutral and Smart Cities [7].

Specific reference to the priorities for the implementation of autonomous road vehicles in urban areas is made in the New EU Urban Mobility Framework [37]. The strategic framework aims to support cities towards the transition to "safe, accessible, inclusive, smart, resilient and zero emission urban mobility". It is suggested that digitalization and automation are the tools to shape an interconnected multimodal transport system with public transport being the backbone of collective mobility in urban areas by 2030. Specifically, the following goals are set:

- Automation in rail and road public transport systems is expected to improve the quality of urban mobility services while reducing operating costs.
- Combining connected and automated transport with zero emission technologies will be one of the enablers for climate neutrality in European cities.
- Autonomous transport systems will enhance connectivity within the wider urban area and between the cities and rural areas.
- Automation will improve traffic and mobility management in congested urban areas.

Moreover, the document highlights that Connected, Cooperative and Automated Mobility (CCAM) is a competitive field for the development of research and innovation in Europe. On the other hand, the impact of automation and digitalization on urban public transport's workers, and especially on drivers, is discussed, highlighting the need for appropriate policies for the reskilling and upskilling of the work force. Finally, the framework stresses the need to integrate CCAM into the cities' transport policy and, namely, the Sustainable Urban Mobility Plans (SUMPs). It is worth mentioning that a recent Practitioner Briefing on the Sustainable Urban Mobility Plan Guidelines focuses on the challenges and potential approaches for developing a transport strategy that includes CCAM, faced with the absence of empirical data and appropriate planning tools and models for the detailed assessment of CCAM impacts [40].

3. Potential Spatial Impacts from AV Deployment on Sustainable Urban Development

Sustainable urban development is the main goal of the UN SDG11: "Make cities and human settlements inclusive, safe, resilient and sustainable" [41]. SDG 11 includes a topic dedicated to sustainable transport, with reference to safety, comfort, effectiveness, efficiency, and inclusiveness, as well as to the impacts of transport systems on air quality, Green House Gas (GHG) emissions, and public health. Moreover, the need to find sustainable pathways for urbanization is highlighted by the specific SDG, targeting enhanced accessibility to public and green space and connectivity throughout the urban area.

Autonomous road transport is a mobility option, which may contribute to the above priorities. By removing human driving, a decrease in accidents caused by human error and an increase in the ability of non-driving travelers to use road vehicles are expected. The use of AVs will change mobility choices and patterns, affecting traffic conditions, with an impact on the following: i. Energy consumption and emissions; ii. Accessibility opportunities, including access to public and green spaces; and iii. Location choices, with an effect on the patterns of urban development and the connectivity between urban and non-urban areas.

Inspired by the above-mentioned links between the targets of SDG11 and the features of autonomous road transport, the current research discusses the possible effects of AVs on sustainable urban development by analyzing the potential impacts on urban space and, namely, the following: 1. Accessibility to activity locations; 2. Location choices and spatial development of cities; 3. Availability of public space, and; 4. Quality of urban environment, in terms of mobility and environmental conditions [42]. It should be highlighted that these effects depend significantly on the use of autonomous road vehicles as privately owned or collective (shared and/or public) transport.

3.1. *Accessibility to Activity Locations*

AVs can be operated by people without a driving license and, thus, can be accessed by a greater share of potential users compared to human-driven cars [43]. This way, they can enhance the accessibility of people with disabilities, younger or older travelers, and, in general, people with reduced mobility [44]. Through improved connectivity and automation, on-demand travel solutions, following the MaaS concept, may increase accessibility for all and the full coverage of the city's land use network [45]. Provided the appropriate mix of policies, MaaS can upgrade the mobility conditions of people who do not own a car [46].

Furthermore, the accessibility of underserved and remote urban areas may be improved by using shared, autonomous mobility solutions [47]. The enhanced driving capability of autonomous cars, enabling them to move and maneuver in narrower spaces than conventional cars, allows them to access roads of poor geometry, as in the case of some historical centers or residential areas in Europe.

However, the potential of AVs to improve accessibility for all depends on their acceptance, adoption, and trust by different social groups, which is linked to the disruptive nature of autonomous mobility and the high dependence on machine-made decisions [48–51]. Another significant parameter refers to the wider notion of social inclusiveness, not only in terms of inclusive services towards people with reduced mobility but also in terms of affordable services for people of lower income and less access to technology [52]. Moreover, the contribution of AVs to the service of people with reduced mobility relies on the ability of the city's transport system to ensure seamless door-to-door mobility through appropriate physical and digital infrastructure, not only while in the vehicle but also from the trip origin to the AV pick-up location and from the drop-off location to the trip's final destination.

3.2. Location Choices and Spatial Development of Cities

Self-driving cars are expected to allow all travelers to engage in other activities (social or recreational activities, work etc.), while on the move. The fact that the time spent in traffic can be "exploited" by the driver is expected to affect the value of time, especially for commuters, professional drivers, and professionals who use driving as part of their daily work (e.g., service vehicle drivers, couriers, etc.) [53]. Consequently, the availability of private and shared autonomous cars may affect travel choices leading to the increase of the duration, distance, and frequency of trips. This, in turn, is expected to affect the location choices of households and firms through the interaction of the transport and land-use systems, as described by Wegener & Fürst [54].

The impact from the changes in location choices due to autonomous mobility on spatial development is uncertain, ranging from urban sprawl to more compact city patterns [55]. In the first case, it is possible that households will be more willing to locate at a longer distance from the city center in search of lower land values and/or better living conditions [56]. The experience from the rise of the private car in the second half of the 20th century shows that the capability for more flexible door-to-door mobility favored urban sprawl [57].

On the other hand, automation would enable the development of completely new, smarter, and more user-centric MaaS solutions [58], which may shift the urban mobility paradigm and, thus, affect the form of urban development towards more sustainable patterns. Newman and Kenworthy [59] analyze the relationship between the evolution of transport networks and the development of cities from the ancient and medieval "walking city" of mixed land use and compact development to the "transit city" of the industrialization era, which developed linearly along public transport routes with sub-centers around main terminals and, finally, to the post-World War II "automobile city" with suburbanization and land use zoning as its main features. If this approach is projected to the "MaaS city" of the digital era, it can be assumed that the city's development will depend less on the obvious choice of the "dominant" mode and more on the individual travel choices of people with different backgrounds and behaviors, having access to a variety of mobility solutions and customized transport services dynamically addressing their mobility needs. Autonomous mobility may play a key role in this evolution.

3.3. Availability of Public Space

A result from the autonomous road vehicles' uniform travel behavior and enhanced driving capabilities is the need for less roadway space compared to human-driven vehicles to navigate in traffic or to conduct parking maneuvers [60–65]. Furthermore, AVs can search for refueling/charging stations and parking spaces at a distance from the trip's destination. The spatial reorganization of parking and refueling/charging infrastructure can have a significant impact on the availability and use of space in dense city centers, recreational and touristic areas, health, or educational facilities, etc. [66]. Studies suggest that the deployment of AVs can lead to the decrease in parking demand in city centers by

67–90%, depending on the share of AVs as privately owned and shared [15]. Thus, it can be assumed that the wide-scale deployment of AVs, especially in the form of collective transport, will free up public space, which can be used for the development of sustainable mobility infrastructure or other facilities, e.g., green infrastructure [65,67]. This opportunity can be linked to the current urban planning trends for redesigning the city at the "human" scale, such as the "15 min city" in Paris, the "Superblocks" in Barcelona, the "5 min to everything" in Copenhagen, and the "20 min neighborhood" in Portland as well as in Melbourne [68,69].

In a few recent studies, the use of dedicated lanes is examined as a design scenario for the introduction of AVs in the urban road network [70–73]. The effectiveness (in terms of the use of roadway capacity and the provided level of service) from introducing a dedicated lane for AVs along a street previously used by mixed traffic, i.e., both autonomous and conventional vehicles, depends on the total traffic flow and the share and type of AVs [74]. This will consequently affect the ability to free up public space, as described above.

The mass deployment of autonomous mobility will affect the form of urban space, as physical and digital infrastructure for traffic, parking, and refueling/charging should be appropriately adjusted to ensure safe and seamless autonomous operations [75]. Moreover, the implementation of CCAM, using electric and shared AVs in MaaS systems, as suggested by the aforementioned EU policy targets, will require the integration of the transport, energy, and telecommunication networks in the context of Smart City infrastructure [76]. The integration into the telecommunication network is essential to achieve the exchange of big data between vehicles, people, infrastructure, and management systems (V2X connectivity), while the integration of the electric AVs and charging infrastructure to the city's electrical network will enable demand-responsive charging and the provision of ancillary services to the grid when the vehicle is idle [77]. However, it should be noted that the capacity for smart and integrated networks depends on the specific features of technology readiness as well as on the local needs and development objectives of each city. Thus, this aspect is not included in the analysis of Sections 4 and 5.

The challenges from the possible effects of AV deployment regarding the availability of public space, the access to facilities, and the location choices, as described in the above sections, may be further discussed in the context of urban planning theories. The deployment of AVs in cities is expected to add to the existing complexity and interconnectedness of the urban system, while it may contribute to flexible and citizen-oriented planning approaches and to the availability of data for planning in the context of smart cities. For example, the compact development and urban intensification suggested by the smart growth theory to address urban sprawl and to improve local and global sustainability advocates pedestrian and bicycle friendly neighborhoods and transit-oriented development [78]. The potential use of AVs should be closely considered in the context of such planning theories. As already mentioned, private AVs may create the conditions for further urban sprawl, while shared AVs, especially in the context of MaaS, can support similar goals if appropriately integrated into the urban planning strategy.

3.4. Mobility Conditions: Congestion and Road Safety

A main benefit that is expected from connected and autonomous mobility is the normalization of traffic flows with positive impacts on the reduction of congestion, due to more stable driving and speeding profiles [79]. Moreover, as approximately 30% of traffic in city centers is due to drivers searching for parking spaces [80,81], the aforementioned ability of AVs to park and refuel/charge away from the trip's destination may reduce parking-related congestion [82]. According to Meyer et al. [53], the implementation of fully autonomous road vehicles can increase the capacity of the road network by up to 40%. In addition, by removing human errors from driving, which correspond to 90% of road accidents in Europe [83], one of the primary goals of autonomous road vehicles is to radically reduce road fatalities [84] as well as accident-related congestion.

The potential of autonomous mobility to reduce congestion and improve road safety depends on the penetration rate of AVs and the duration of the transition period until the complete shift to road automation. During the period of transition, vehicles of different levels of automation will coexist, while they will share the road network with active transport and micromobility modes. Researchers argue that in this period the benefits for comfort, safety, and congestion may be limited or even negative because of the different requirements and capabilities of different modes and technologies [85]. For example, as already mentioned, autonomous cars need less lane width and headway distance than conventional cars. This may have an adverse effect on the mobility conditions of both human-driven cars and active transport and micromobility modes [86]. Thus, the share of public, shared, and private AVs and the duration of the transition period are two important factors to be considered for the estimation of impacts on congestion and road safety. In addition, the integration of AVs into the existing mechanized and active transport networks will require specific traffic regulations to ensure the safety of all road users and especially the more vulnerable ones [87].

The innovative features of autonomous driving in comparison to conventional driving in the city, as described in the above sections, can be summarized as follows: i. Accessibility by a wider range of travelers; ii. Ability to engage in other activities, instead of driving; iii. Safer and more comfortable traveling with smoother acceleration/deceleration; and iv. Driverless parking and/or refueling/charging. The improved accessibility, safety, and comfort of AVs may lead to the increase in the number, frequency, and distance of daily trips [87]. Shared AVs providing demand-responsive services will probably move for longer periods and cover larger areas. AVs that will park and refuel/charge at a distance from the trip destination may alleviate parking-related congestion but lead to new trips with the parking and refueling/charging stations as their new destination. Such considerations should be reflected in the discussion of the impact of autonomous road vehicles on mobility conditions.

3.5. Environmental Impacts

The potential impacts from the deployment of AVs on mobility conditions are strongly connected with the environmental impacts in terms of emissions and energy consumption. In Europe, the "100 Climate-Neutral and Smart Cities" mission, i.e., one of the five ongoing EU missions, which combines the appropriate research, policy, and legislative actions to deliver 100 carbon-neutral cities by 2030, has allocated considerable funding for research and innovation in "co-designed smart systems and services for user-centered, shared, zero-emission mobility of people and freight in urban areas", involving the support to the Connected, Cooperative and Automated Mobility (CCAM) and the 2ZERO (Towards zero emission road transport) partnerships [88]. The normalization of traffic flows and the decrease in congestion due to autonomous mobility is expected to decrease emissions and energy consumption [89]. Researchers also argue that Shared Autonomous Electric Vehicles (SAEVs) have the potential to boost the shift to electric mobility with significant environmental benefits [90]. However, the total energy efficiency and the environmental benefits of electric AVs, compared to vehicles with internal combustion engine (ICE), depend, on the one hand, on the penetration rate of AVs affecting their impact on traffic flows and, on the other hand, on the electric AV lifecycle emissions, the energy mix to produce electricity (Well-to-Wheel) and the availability and grid integration of charging infrastructure. For example, according to the simulation model of Patella et al. [91], which was set up and run for the city of Rome, the construction, maintenance, and end-of-life activities for battery electric autonomous road vehicles lead to GHG emissions, which are higher than for ICE vehicles. The same study estimates that a 100% electric AV penetration rate in the future would lead to a reduction of up to 60% in the total life cycle GHG emissions compared to the current scenario, where electric vehicles correspond to less than 1% of the city's fleet composition. The above model and other studies, such as [92,93], use different methodologies and produce different results, confirming the complexity of assessing the environmental impact of AVs. However, they all depend on assumptions on the penetration rate of AVs, with higher penetration rates linked to more environmental benefits. Thus, it is important to consider the aforementioned transition period towards the higher penetration of AVs, when vehicles of various automation levels will coexist in the city's road network. As already discussed, this coexistence may lead to local congestion, affecting energy consumption and emissions. In addition, if the wide-scale deployment of AVs leads to increased traffic volumes due to the above-described improved mobility conditions, i.e., accessibility, safety and comfort, the total energy consumption from urban mobility may be negatively affected. The increase in vehicle-kilometers is also linked to the shorter life span of each vehicle, especially in the case of shared AVs that will probably be used more intensively than private ones. Considering the effect of manufacturing, maintenance, and end-of-life processes on the life cycle impact of AVs on climate change, the life span of the vehicles may play a significant role in the overall environmental performance of the city's vehicle fleet.

Table 1 summarizes the above findings from the literature review according to the field and type (positive or negative) of possible impact. The multicriteria analysis, as presented in the following sections, considers these findings, and turns to research and innovation experts to evaluate their significance for sustainable urban development and then to assess different policy pathways for the promotion of collective and/or private AV transport in European cities.

Table 1. Potential impacts per field from the deployment of AVs in cities.

Field of Impact	Short Description of Potential Impact	Type of Potential Impact	
		Positive	Negative
Accessibility to activity locations	• Accessibility by a wider range of social groups (depending on inclusiveness, affordability, and access to technology).	X	
	• On-demand access to underserved and remote areas.	X	
	• Ability to access areas of poor road geometry.	X	
Location choices and spatial development	• Enhanced MaaS solutions for planning against private car-oriented development.	X	
	• Urban sprawl due to the reduction in the value of time for AV users.		X
Availability of public space	• Freeing up public space in congested or sensitive urban areas (depending on the increase in traffic due to induced travel demand).	X	
Quality of urban environment	• Improved mobility, road safety, and environmental conditions (depending on the share of AVs and the increase in traffic due to induced travel demand).	X	
	• Local increase in congestion, accidents, emissions, and energy consumption due to mixed (autonomous and non-autonomous) traffic in the transition period.		X
	• Increased impact on climate change due to manufacturing, maintenance, and end-of-life processes (depending on technology, policy, and business model development in the related sectors).		X

4. Materials and Methods

The scope of the current research is the analysis of different policy pathways for autonomous mobility from the perspective of sustainable urban development in Europe. According to the specific research approach, the evaluation of these pathways is linked to their potential impacts on different fields of sustainable urban development. The ability to assess these impacts by the current forecasting models face three main obstacles, i.e., the scarcity of data from the wide-scale implementation of AVs, the consequent uncertainty of impacts, and the complexity of impacts, which refers to the interactions between them and the dependence on whether the AVs are implemented as collective and/or private transport modes. Moreover, the comprehensive description of these impacts requires the analysis of both quantitative and qualitative parameters. In order to overcome these obstacles, the current forecasting models adopt various assumptions on potential changes in travel choices or on different AV penetration rates, which seldom derive from specific policies [42]. On the contrary, the present research is based on the principle that the use and penetration of autonomous road vehicles in cities will depend on wider policy objectives, expressed by strategic planning decisions, for promoting sustainable urban development. Given that the wide-scale deployment of AVs will take place by 2030, as outlined by the corresponding EU policy target, the proposed methodology aims to appropriately structure the decision problem of how to realize this target with the highest-possible contribution to the sustainable development of European cities.

Within this scope, the Multicriteria Analysis (MCA) is selected as the appropriate methodology to structure and solve the above decision problem by including all the necessary quantitative and qualitative parameters to address the complexity and uncertainty of the subject and by capitalizing on the knowledge and insight of a group of independent experts in the field of research and innovation. Another characteristic of MCA is the transparency in the appraisal of the examined alternatives, as the expert preferences are linked to the overall research objective [94].

Different MCA methods are usually combined in the literature according to the scope, approach, and resources of each case [20,95–97]. In the current research, the Analytic Hierarchy Process (AHP) [98] is applied to structure the hierarchy of the decision problem by analyzing its fundamental parts and reflecting the significance of each one of them. More specifically, AHP is used for extracting the weights of the criteria in relation to the goal, and the performance of the alternatives in relation to each criterion. A benefit of AHP is that it allows group decision making, with the consistency of the group's answers being checked through the consistency ratio [72,99–101]. VIKOR (VIseKriterijumska Optimizacija I Kompromisno Resenje) is then applied for the overall ranking of the alternatives [102]. The main advantage of VIKOR over AHP in terms of ranking is that the former requires the meeting of the condition of "acceptable advantage", referring to the highest group utility, as well as the condition of "acceptable stability", referring to the lowest individual regret. Another advantage is that the difference between two successively ranked alternatives is large enough for one to be characterized as undoubtedly better than the other. The combination of AHP and VIKOR in consecutive stages of the process allows for the exploitation of the advantages of both methodologies.

The steps of the methodology are briefly presented in Figure 1.

The implementation of the methodology leads to the calculation of the S_i, R_i, and Q_i values for each alternative (Step 12). These values are used to rank the alternatives in terms of preference. Namely, the most preferable alternative is characterized by the lowest S_i, R_i, and Q_i values, while the least preferable one is characterized by the highest S_i, R_i, and Q_i values. Moreover, the condition of acceptable advantage, i.e., the best alternative based on the above-mentioned values of maximum group utility and minimum individual regret, is checked along with the condition of acceptable stability, i.e., a large enough difference between two successively ranked alternatives in order for one to be characterized as undoubtedly better than the other.

The list of equations implementation of the AHP and VIKOR, respectively, are presented below.

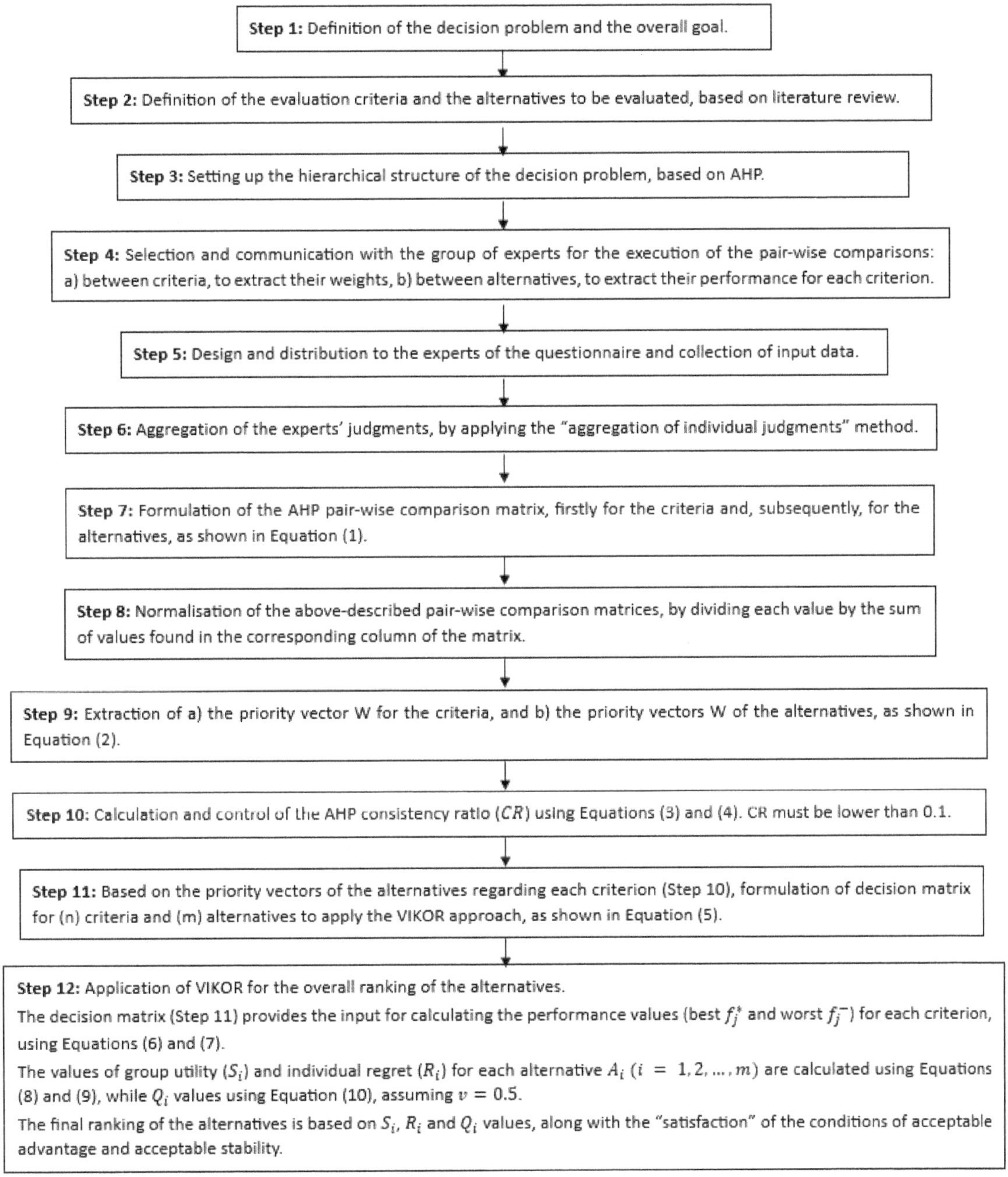

Figure 1. Methodological steps.

4.1. Equations Used for the Application of AHP

$$A = \begin{bmatrix} a_{11} & a_{12} & \dots & a_{1n} \\ a_{21} & a_{22} & \dots & a_{2n} \\ \dots & \dots & \dots & \dots \\ a_{n1} & a_{n2} & \dots & a_{nn} \end{bmatrix}, \tag{1}$$

where

- $a_{ij} = w_i/w_j$ = element of matrix A, i.e., the relative importance of the criterion (i) over the criterion (j), or of the alternative (i) over the alternative (j), with i = 1, 2, . . ., n and j = 1, 2, . . ., n; ($\alpha_{ij} = 1/\alpha_{ji}$ and $\alpha_{ii} = w_i/w_i = 1$).
- w_i, w_j = weight coefficients of the criteria or of the alternatives (i) and (j), respectively.

$$(A - \lambda_{max}) \cdot W = 0, \tag{2}$$

where

- A = pairwise comparison matrix.
- $W = (w_1, w_2, \dots, w_n)^T$ = priority vector for each hierarchy level.
- λ_{max} = principal eigenvalue of matrix A.

$$CR = CI/RI < 0.10, \tag{3}$$

- CI = consistency index, calculated by the Equation:

$$CI = \frac{\lambda_{max} - n}{n - 1}, \tag{4}$$

- RI = random consistency index, as shown in Table 2.

Table 2. Random consistency index values for n elements.

n	1	2	3	4	5	6	7	8
RI	0.00	0.00	0.58	0.90	1.12	1.24	1.32	1.41

In order to ensure the consistency of the pairwise comparison matrix, CR must be lower than 0.1. In case it is calculated higher than 0.1, the comparison must be repeated for the corresponding pairs that generate the inconsistency.

4.2. Equations Used for the Application of VIKOR

$$D = \begin{array}{c} \\ A_1 \\ A_2 \\ \dots \\ A_m \end{array} \begin{array}{c} \begin{array}{cccc} C_1 & C_2 & \dots & C_n \end{array} \\ \begin{bmatrix} x_{11} & x_{12} & \dots & x_{1n} \\ x_{21} & x_{22} & \dots & x_{2n} \\ \dots & \dots & \dots & \dots \\ x_{m1} & x_{m2} & \dots & x_{mn} \end{bmatrix} \end{array}, \tag{5}$$

where x_{ij} = performance of the alternative A_i regarding criterion C_j, with $i = 1, 2, \dots, m$ and $j = 1, 2, \dots, n$.

Performance values for benefit functions:

$$f_j^* = \max_i(x_{ij}) \text{ and } f_j^- = \min_i(x_{ij}), \tag{6}$$

Performance values for cost functions:

$$f_j^* = \min_i(x_{ij}) \text{ and } f_j^- = \max_i(x_{ij}), \tag{7}$$

$$S_i = \sum_{j=1}^{n} \left[\left(f_j^* - x_{ij} \right) / \left(f_j^* - f_j^- \right) \right], \tag{8}$$

$$R_i = \max_j \left\{ w_j \cdot \left[\left(f_j^* - x_{ij} \right) / \left(f_j^* - f_j^- \right) \right] \right\} \tag{9}$$

where w_j = weight of each criterion, with $j = 1, 2, \ldots, n$.

$$Q_i = v \cdot \frac{S_i - S^*}{S^- - S^*} + (1 - v) \cdot \frac{R_i - R^*}{R^- - R^*}, \tag{10}$$

where $S^* = min_j S_i$, $S^- = max_j S_i$, $R^* = min_j R_i$ and $R^- = max_j R_i$.

5. Analysis and Results

5.1. Definition of the Decision Problem and the Overall Goal

Assuming the wide-scale availability of AVs by 2030, according to the above literature review, the reference period (2028–2034), i.e., the next EU programming period, is considered by the specific research as the period of wide-scale integration of AVs into the transport networks of European cities. Within this time horizon, the objectives of the analysis are:

a. The evaluation of the significance of possible spatial effects regarding their contribution to the overall goal of sustainable urban development for the next EU programming period (2028–2034).
b. The assessment of the positive contribution of specific alternative transport policy pathways regarding each possible spatial effect for the period 2028–2034.
c. The assessment of the alternative transport policy pathways in relation to the overall contribution to sustainable urban development for the same period.

5.2. Definition of Criteria and Alternatives

The criteria of the analysis describe the possible spatial effects due to AV deployment in a city, which are summarized in Table 1. These effects can be either positive or negative. In this context, the selected criteria include the following:

- Effect on physical access to activities by vulnerable groups, e.g., people with impairments or reduced mobility of lower income households, etc.—PHYS AC
- Effect on the city's spatial development, e.g., leading to urban sprawl or enhanced accessibility of remote and/or under-serviced urban areas, etc.—SP DEV
- Effect on public space availability, e.g., due to less requirements in roadway and parking capacity or additional infrastructure for autonomous transport operations, etc.—PUBL SP
- Effect on congestion and related emissions and energy consumption, e.g., due to smoother traffic flows or induced transport demand, etc.—EMIS/EN
- Effect on road safety, e.g., due to the decrease in human errors or due to the coexistence of vehicles of different autonomy levels, etc.—R SAF

The above effects from the large-scale deployment of AVs depend significantly on the way that AVs are used, i.e., as privately owned or as part of a collective (shared and/or public) transport system. Policy makers, supported by the scientific advice of research and innovation experts, can direct the deployment to ensure a positive contribution to sustainable urban development, i.e., to enhance the positive impacts and mitigate or decrease the negative ones. This can be achieved through appropriate transport policy pathways, which will then be implemented through the corresponding policy mix (infrastructure, technology, incentivization, regulatory framework, etc.). In this context, the following alternative policy pathways are examined:

- Implementation of autonomous collective transport to service specific routes and areas (e.g., congested and/or underserved) and/or specific periods (e.g., peak and/or low demand)—COL SPEC R/P
- Implementation of autonomous collective transport to service specific routes and areas and/or specific periods and incentivization for private autonomous mobility (e.g., economic incentives and/or exclusive services)—COL SPEC R/P & PR INC
- Implementation of autonomous collective transport to service the whole Functional Urban Area, FUA, (full network coverage) at all times—COL FUA ALL T
- Implementation of autonomous collective transport to service the whole Functional Urban Area, FUA, at all times and incentivization for private autonomous mobility—COL FUA ALL T & PR INC

It should be noted that the Functional Urban Area (FUA) consists of a city and its commuting zone. Based on the definitions by OECD [103], the FUA consists of a densely inhabited city and a less densely populated commuting zone whose labor market is highly integrated with the city.

5.3. Hierarchical Structure of the Decision Problem

The hierarchy of the decision problem is presented in Figure 2. The top level of the hierarchical structure refers to the overall goal, the middle to the related criteria, and the bottom to the alternatives.

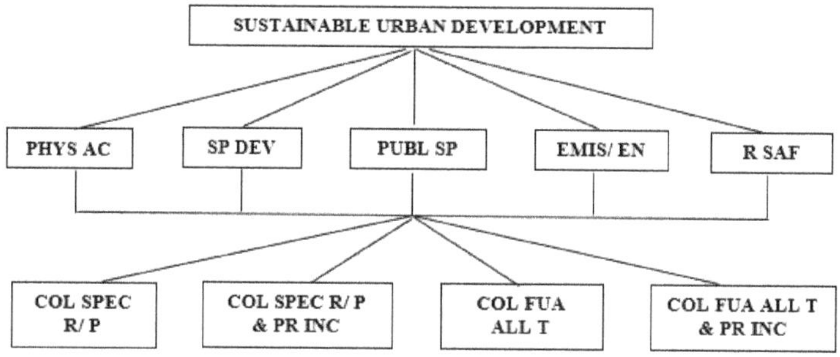

Figure 2. Hierarchy of the decision problem based on AHP.

5.4. Selection of the Group of Experts

The assessment of the pairwise comparisons of criteria and alternatives to extract the weights of criteria and the performance of the alternatives in terms of each criterion should be conducted by a group of 8–15 independent experts, as recommended in relevant literature [72,96,104]. The execution is conducted anonymously, so that the judgments are expressed independently and freely [72,104–106]. In the present research, 10 experts in the field of transport research and innovation with experience in autonomous road vehicles were selected. More specifically, the selected experts represent academic and research institutions from Germany, Greece, Hungary, Luxembourg, the Netherlands, Switzerland, and the UK, while the group also includes a member of the European Commission's Directorate General (DG) for Research and Innovation. The survey complied with the General Data Protection Regulation (GDPR).

5.5. Weighting of Criteria

The above-mentioned experts participated in the survey through an appropriately designed questionnaire, supported by bilateral communication with the research team. In particular, the experts initially evaluated the significance of the given criteria, i.e., possible effects, regarding their contribution to the overall goal of sustainable urban development by executing a total of 10 criteria pairwise comparisons based on Saaty's scale (Table 3). An indicative part of the pairwise comparisons is shown in Table 4.

Table 3. Relative importance scale for the pairwise comparison of the criteria.

Importance Value	Explanation
1	Equal importance
3	Moderate importance of one over another
5	Essential or strong importance of one over another
7	Very strong importance of one over another
9	Extreme importance of one over another
2, 4, 6, 8	Intermediate values between the two adjacent judgments

Table 4. Indicative part of criteria pairwise comparison regarding the overall goal, i.e., sustainable urban development.

The criterion on the **left** is **more important** than the one on the **right** (select the degree of relative importance)									The two criteria are of **equal importance**	The criterion on the **right** is **more important** than the one on the **left** (select the degree of relative importance)								
Physical access to activities	9	8	7	6	5	4	3	2	1	2	3	4	5	6	7	8	9	*City's spatial development*

The "aggregation of individual judgments" method was applied for the aggregation of the experts' answers, using the geometric mean (GM) of the values attributed to each criterion, while the consistency check of the experts' answers was executed using the AHP consistency ratio (CR) from Equation (3). The experts' answers, along with the corresponding geometric mean values, are shown in Table 4. It should be noted that, if an expert selects the criterion on the left side of Table 5, the selected value is used in the analysis, while if an expert selects the criterion on the right side of the Table, the reverse of the selected value is used.

Table 5. Expert judgment and geometric mean (GM) values for the criteria compared in pairs.

CRITERIA/EXPERTS	EX1	EX2	EX3	EX4	EX5	EX6	EX7	EX8	EX9	EX10	GM
PHYS AC vs. SP DEV	3	1/8	1/3	7	1/5	7	5	3	1/8	4	1.2483
PHYS AC vs. PUBL SP	6	1/7	1/3	5	3	5	3	5	1/4	2	1.6619
PHYS AC vs. EMIS/EN	7	1/7	1/5	9	5	1	7	3	1/3	1/6	1.2651
PHYS AC vs. R SAF	1	1/6	1/6	1/5	5	4	1	1/4	1/8	1/6	0.4745
SP DEV vs. PUBL SP	4	4	3	3	5	1/4	1/3	3	8	1/2	1.9308
SP DEV vs. EMIS/EN	5	6	1/5	5	3	1/7	1/2	1/3	8	1/7	1.0937
SP DEV vs. R SAF	1/3	8	1/7	1/7	7	1/5	1/5	1/5	1/8	1/7	0.3746
PUBL SP vs. EMIS/EN	4	3	1/6	5	1/3	1/4	1	1/4	7	1/7	0.8548
PUBL SP vs. R SAF	1/5	5	1/6	1/5	1/5	1	1/3	1/6	1/8	1/8	0.2994
EMIS/EN vs R SAF	1/6	6	1/3	1/7	1/5	5	1/5	1/3	1/8	1/3	0.4094
PHYS AC vs. SP DEV	3	1/8	1/3	7	1/5	7	5	3	1/8	4	1.2483
PHYS AC vs. PUBL SP	6	1/7	1/3	5	3	5	3	5	1/4	2	1.6619
PHYS AC vs. EMIS/EN	7	1/7	1/5	9	5	1	7	3	1/3	1/6	1.2651
PHYS AC vs. R SAF	1	1/6	1/6	1/5	5	4	1	1/4	1/8	1/6	0.4745
SP DEV vs. PUBL SP	4	4	3	3	5	1/4	1/3	3	8	1/2	1.9308

The AHP pairwise comparison matrix for the criteria (Table 6) is formulated by inserting the above presented geometric mean values in Equation (1). Table 7 presents the normalized pairwise comparison matrix for the criteria and their corresponding weights, represented by the priority vectors from Equation (2). The results from the respective consistency control are also presented.

Table 6. Pairwise comparison matrix for the five criteria.

	PHYS AC	SP DEV	PUBL SP	EMIS/EN	R SAF
PHYS AC	1	1.2483	1.6619	1.2651	0.4745
SP DEV	0.8011	1	1.9308	1.0937	0.3746
PUBL SP	0.6017	0.5179	1	0.8548	0.2994
EMIS/EN	0.7905	0.9143	1.1698	1	0.4094
R SAF	2.1074	2.6695	3.3401	2.4426	1

Table 7. Normalized pairwise comparison matrix, priority vector and consistency control for the five criteria.

	PHYS AC	SP DEV	PUBL SP	EMIS/EN	R SAF	PRIORITY VECTOR (W)
PHYS AC	0.1887	0.1966	0.1826	0.1901	0.1855	0.1887
SP DEV	0.1511	0.1575	0.2121	0.1643	0.1465	0.1663
PUBL SP	0.1135	0.0816	0.1099	0.1284	0.1170	0.1101
EMIS/EN	0.1491	0.1440	0.1285	0.1502	0.1601	0.1464
R SAF	0.3976	0.4204	0.3669	0.3670	0.3909	0.3886
	λ_{max} = 5.0264		CI = 0.0066		CR = 0.0059 < 0.10 ✓	

✓refers to the last parameter (CR).

5.6. Evaluation of Alternatives Regarding Each Criterion

After the evaluation of criteria, the experts evaluated the given alternatives, i.e., transport policy pathways, by taking into consideration the extent of their positive contribution regarding each criterion, through the respective pairwise comparisons leading to six pairwise comparisons for each one of the five criteria. The comparisons are executed based on the scale of Table 8. An indicative part of these comparisons is illustrated in Table 9.

Table 8. Relative preference scale for the alternatives regarding each criterion.

Importance Value	Explanation
1	No preference
3	Moderate preference relation
5	Essential or strong preference relation
7	Very strong preference relation
9	Absolute preference relation
2, 4, 6, 8	Intermediate values between the two adjacent judgments

Table 9. Indicative part of pairwise comparisons between the alternatives regarding each criterion.

The alternative on the **left** is **preferable to** the one on the **right** (select the degree of relative preference)									**No preference** between the two alternatives	The alternative on the **right** is **preferable to** the one on the **left** (select the degree of relative preference)								
Autonomous collective transport to service specific routes and areas and/or specific periods	9	8	7	6	5	4	3	2	1	2	3	4	5	6	7	8	9	***Autonomous collective transport to service specific routes and areas and/or specific periods and incentivization for private autonomous mobility***

Following the same methodological steps as for the criteria, the input data from the experts' pairwise comparisons of alternatives per criterion, as well as the geometric mean value for each one of the five criteria were calculated. The normalized pairwise comparison matrices for the alternatives, the priority vectors, and the consistency control for each criterion are shown in Table 10.

Table 10. Normalized pairwise comparison matrices, priority vectors, and consistency control for the alternatives in terms of each of the five criteria.

Criterion: PHYS AC	COL SPEC R/P	COL SPEC R/P & PR INC	COL FUA ALL T	COL FUA ALL T & PR INC	PRIORITY VECTOR (W)
COL SPEC R/P	0.2293	0.1739	0.2259	0.2758	0.2262
COL SPEC R/P & PR INC	0.1517	0.1150	0.1234	0.0802	0.1176
COL FUA ALL T	0.4426	0.4067	0.4362	0.4317	0.4293
COL FUA ALL T & PR INC	0.1764	0.3044	0.2145	0.2123	0.2269
	λ_{max} = 4.0619 CI = 0.0206 CR = 0.0229 < 0.10 ✓				
Criterion: SP DEV	COL SPEC R/P	COL SPEC R/P & PR INC	COL FUA ALL T	COL FUA ALL T & PR INC	PRIORITY VECTOR (W)
COL SPEC R/P	0.2451	0.2642	0.2188	0.3491	0.2693
COL SPEC R/P & PR INC	0.0936	0.1009	0.1163	0.0719	0.0957
COL FUA ALL T	0.5612	0.4348	0.5011	0.4364	0.4834
COL FUA ALL T & PR INC	0.1001	0.2002	0.1638	0.1426	0.1517
	λ_{max} = 4.0751 CI = 0.0250 CR = 0.0278 < 0.10 ✓				
Criterion: PUBL SP	COL SPEC R/P	COL SPEC R/P & PR INC	COL FUA ALL T	COL FUA ALL T & PR INC	PRIORITY VECTOR (W)
COL SPEC R/P	0.1534	0.2838	0.1288	0.2339	0.2000
COL SPEC R/P & PR INC	0.0462	0.0854	0.1250	0.0557	0.0780
COL FUA ALL T	0.7031	0.4033	0.5903	0.5620	0.5646
COL FUA ALL T & PR INC	0.0974	0.2276	0.1559	0.1484	0.1573
	λ_{max} = 4.2341 CI = 0.0780 CR = 0.0867 < 0.10 ✓				
Criterion: EMIS/EN	COL SPEC R/P	COL SPEC R/P & PR INC	COL FUA ALL T	COL FUA ALL T & PR INC	PRIORITY VECTOR (W)
COL SPEC R/P	0.1710	0.2714	0.1445	0.3118	0.2247
COL SPEC R/P & PR INC	0.0539	0.0856	0.1145	0.0534	0.0768
COL FUA ALL T	0.7081	0.4473	0.5984	0.5126	0.5666
COL FUA ALL T & PR INC	0.0670	0.1957	0.1426	0.1222	0.1319
	λ_{max} = 4.2383 CI = 0.0794 CR = 0.0883 < 0.10 ✓				
Criterion: R SAF	COL SPEC R/P	COL SPEC R/P & PR INC	COL FUA ALL T	COL FUA ALL T & PR INC	PRIORITY VECTOR (W)
COL SPEC R/P	0.1567	0.1989	0.1227	0.2387	0.1793
COL SPEC R/P & PR INC	0.1025	0.1302	0.1618	0.1021	0.1242
COL FUA ALL T	0.5853	0.3687	0.4583	0.4223	0.4587
COL FUA ALL T & PR INC	0.1555	0.3022	0.2571	0.2369	0.2379
	λ_{max} = 4.1028 CI = 0.0343 CR = 0.0381 < 0.10 ✓				

5.7. Application of VIKOR for the Overall Ranking of the Alternatives

VIKOR is applied to assess the final ranking of the alternatives. For this purpose, the priority vectors of the alternatives (Table 10), which correspond to the ranking of the alternatives for each criterion, are used as input to develop the decision matrix of Table 11, based on Equation (5). It should be noted that all the criteria in the current research are considered as benefit criteria (benefit functions), due to the way that the questionnaire was formulated, asking experts to evaluate which alternative is preferable to the other regarding each criterion.

Table 11. Decision matrix for the application of VIKOR.

	PHYS AC	SP DEV	PUBL SP	EMIS/EN	R SAF
COL SPEC R/P	0.2262	0.2693	0.2000	0.2247	0.1793
COL SPEC R/P & PR INC	0.1176	0.0957	0.0780	0.0768	0.1242
COL FUA ALL T	0.4293	0.4834	0.5646	0.5666	0.4587
CPF	0.2269	0.1517	0.1573	0.1319	0.2379

Based on Table 11, the f_j^* and f_j^- values are calculated using Equations (6) and (7), respectively:

- $f_j^* = max_i(x_{ij})$ = {0.4293 0.4834 0.5646 0.5666 0.4587}
- $f_j^- = min_i(x_{ij})$ = {0.1176 0.0957 0.0780 0.0768 0.1242}

Table 12 presents the values of S_i, R_i, and Q_i for v = 0.5 (according to Equations (8)–(10), respectively), which are calculated using the criteria weights (priority vector W) of Table 10. Table 10 also presents the final ranking of alternatives, where lower values of S_i, R_i, and Q_i correspond to a higher ranking.

Table 12. S_i, R_i, and Q_i values and alternatives ranking.

	COL SPEC R/P	COL SPEC R/P & PR INC	COL FUA ALL T	COL FUA ALL T & PR INC
S_i	0.7240	1.0000	0.0000	0.7433
R_i	0.3246	0.3886	0.0000	0.2564
Q_i	0.7796	1.0000	0.0000	0.7016
RankS_i	3	4	1	2
RankR_i	3	4	1	2
RankQ_i	3	4	1	2

According to Table 12, the alternative COL FUA ALL T (implementation of autonomous collective transport to service the whole Functional Urban Area at all times) has the minimum Q_i and, thus, it is selected as the best alternative. As regards the condition of acceptable advantage, where $0.7796 - 0.000 = 0.7796 > 1/(m-1) = 1/(4-1) = 1/3 = 0.3333$, and of acceptable stability, the specific alternative is also first in rank by S_i and R_i. Consequently, the alternative COL FUA ALL T is considered the optimum alternative.

6. Discussion

Regarding the results from the implementation of the combined AHP–VIKOR methodology on the weighting of criteria and the corresponding priority vector (Table 7), the criterion of road safety (R SAF) was undoubtedly first with a weight of 38.86%. This result is aligned with the current literature, which suggests that the elimination of human error due to autonomous driving will decrease road accidents. The physical access of vulnerable groups to activity locations (PHYS AC) scored a weight of 18.87% and was ranked second, in line with another widely reported competitive advantage of autonomous road vehicles referring to their ability to provide mobility for travelers who are unable to access or drive conventional cars. An interesting finding was that the effect on the city's spatial development (SP DEV), ranked third with the significant weight of 16.63%, even though this aspect is not widely covered by the existing literature. The weight related to congestion, emissions, and energy consumption (EMIS/EN) was found to be lower, i.e., 14.64%. This result possibly reflects the uncertainties regarding the effects of autonomous road vehicles on mobility conditions, as discussed in Section 3. The effect on public space availability (PUBL SP) ranked last with a weight of 11.01%, indicating, however, the potential of autonomous road vehicles to contribute as one of the planning tools for shaping urban space.

Regarding the evaluation of alternatives, the group of independent experts indicated that the implementation of autonomous collective transport always servicing the whole of the Functional Urban Area (COL FUA ALL T) is the optimum transport policy pathway in terms of promoting sustainable urban development in Europe for the next programming period (2028–2034) (Table 12). The implementation of the above alternative combined with the incentivization of private autonomous mobility (e.g., through economic incentives and/or exclusive services) (COL FUA ALL T & PR INC) was ranked second, closely followed by the implementation of autonomous collective transport to service specific routes and areas (e.g., congested and/or underserved) and/or specific periods (e.g., peak and/or low demand) (COL SPEC R/P). The alternative of implementing autonomous collective transport for specific routes and areas and/or specific periods while incentivizing private autonomous mobility (COL SPEC R/P & PR INC) was ranked last.

According to the priority vectors of the alternatives (Table 10), the overall optimum alternative, i.e., COL FUA ALL T, is also characterized by the best performance in terms of each criterion with considerable difference from the second-best alternative. On the other hand, the alternative COL SPEC R/P & PR I is ranked last in terms of all criteria. The alternatives COL FUA ALL T & PR INC and COL SPEC R/P are interchangeably ranked second and third depending on the examined criterion. Specifically, the alternative COL SPEC R/P is the second-best alternative in terms of the positive contribution to the city's spatial development, public space availability, and congestion. The alternative COL FUA

ALL T & PR INC is the second-best alternative in terms of the accessibility of vulnerable groups to activity locations, closely followed by COL SPEC R/P, and in terms of road safety. However, the considerably higher weight attributed by the experts to the criterion of road safety leads to the ranking of COL FUA ALL T & PR INC as second in the overall ranking.

The present assessment provides a preliminary evaluation of pathways for the deployment of autonomous road vehicles in cities, prioritizing collective (public and shared) transport services based on AVs. From a strictly cost-based approach, there are cost factors related to the current business models that may affect the viability of this pathway, which involve the cost-efficient supervision of passengers, vehicle repairing processes, and cleaning protocols [75]. The present research also confirms that the uncertainties regarding the impacts of AVs on mobility conditions and urban development patterns have not yet been fully addressed by the scientific community. Similar conclusions are drawn by a recent participatory visioning and multicriteria analysis conducted for Manchester (UK) and Melbourne (Australia) [107]. Furthermore, the research results highlight that the deployment of AVs in the context of either collective or private transportation will significantly affect urban development patterns (towards higher availability of public space or more urban sprawl, respectively), as also illustrated by the work of Legênea et al. [108]. From a reverse perspective, Wadud and Mattioli [109] assess the total cost of ownership and use of AVs in different scenarios incorporating private or on-demand vehicles to conclude that the rural location of households may increase the possibility that automated vehicle ownership will cost less compared to shared automated mobility services.

Regarding the limitations of the present research, the main limitation is related to the constraints of the applied methodology in the maximum number of the elements (criteria and alternatives), which can be compared in pairs. This is due to the fact that the human brain is allegedly capable of efficiently comparing in pairs up to 7 ± 2 elements at a time [110]. Thus, the present research provides a preliminary assessment of pathways, without accounting for all possible combinations of autonomous transport services and city contexts. Moreover, in order to address the low maturity and disruptive potential of AV deployment, the research team engaged a sufficient number (as discussed in Section 5.4) of experts in research and innovation with significant experience in the field of autonomous mobility, representing different organizations across the EU. In the near future, as cities experience the wide-scale implementation of AVs, this preliminary analysis could be used as the basis for a survey covering the whole spectrum of stakeholders (e.g., research and innovation experts, local authorities, industry, service providers, businesses, civic society organizations, etc.) in higher numbers to analyze their different perspectives. Furthermore, it can be complemented by methods and models to assess the cost-effectiveness of the alternative pathways in the context of a comprehensive transition study [111].

7. Conclusions

The specific research focuses on the role of autonomous mobility for the sustainable development of European cities with its main innovation being the assessment of transport policy pathways for the implementation of autonomous road vehicles in view of the next EU programming period (2028–2034). By assuming the wide-scale deployment of autonomous mobility by 2030, as suggested by the literature, and by considering that the sustainable development of cities is a global goal, as indicated by the SDGs, the results of the specific analysis provide useful insights for planners and policy makers not only in Europe but also in other cities of the world. Moreover, the current research contributes towards bridging a gap in the literature by delivering a policy-oriented framework for the evaluation of alternative pathways for the implementation of autonomous road vehicles based on their potential spatial impacts on sustainable urban development.

The EU policy framework in the current programming period refers to the deployment of autonomous transport as one of the transformative policies for the green and digital transition of urban mobility. The framework also draws attention to the need to assess specific challenges for the effective implementation of autonomous mobility in terms of socio-economic and environmental sustainability. The review of international literature regarding these challenges leads to different, and sometimes controversial, results. An overall conclusion from the literature review is the uncertainty, due to the scarcity of empirical data and real-life experience, and the complexity, due to the wide range of impacts and their interactions, in the assessment of anticipated effects. A common finding from the synthetic analysis of recent research and policy documents is that the impacts of autonomous mobility on cities depend on the type of implementation, i.e., as privately owned vehicles or as collective transport modes. The present research explores different pathways that depend on the type of implementation.

The above-described uncertainty and complexity leads to the need of analyzing in an integrated way both quantitative and qualitative parameters related to both positive and negative impacts. The MCA approach, with the participation of a group of independent experts in research and innovation across Europe, was adopted by the present research to address this methodological challenge. Specifically, the combination of AHP and VIKOR is used, building on the advantages of both methodologies. The methodology provides an effective structure of the decision problem and a scientifically sound method for the evaluation of alternative transport policy pathways in relation to specific criteria for sustainable development. The main constraint of the AHP methodology refers to the number of the examined criteria and alternatives, which should be adequate for the comprehensive analysis of the problem but low enough for the experts to be able to efficiently execute the corresponding pairwise comparisons. Consequently, the present research focuses on specific types of criteria (mainly on potential spatial impacts) and the analysis of general categories of alternatives, i.e., the use of AVs as private and/or collective transport with partial or full spatial and time coverage.

The application of the methodology concludes that road safety plays a crucial role in the effective implementation of autonomous mobility in terms of sustainable urban development for the next EU programming period (2028–2034). Social inclusion, expressed in terms of the accessibility of vulnerable users, as well as congestion are also regarded as important. It should be noted that the above issues are often mentioned by literature as the main advantages of autonomous mobility. The survey, however, highlighted additional aspects, which are not yet thoroughly examined by literature, such as the effects on the spatial patterns of urban development and, less so, the availability of public space. These aspects should be further explored by the scientific community.

In terms of the transport policy pathways deriving from the analysis, the implementation of autonomous mobility as a collective transport system can contribute to the cities' spatial development, the alleviation of congestion, and the enhancement of public space if implemented in specific areas and routes and at specific time periods. To unlock the full potential of autonomous mobility by including to the above the improvement in road safety and the accessibility of vulnerable users, it is important to develop an autonomous collective transport system with 24 h service of the whole Functional Urban Area. At the same time, the benefits for road safety and the accessibility of vulnerable users are expected to be significant through the incentivization of private autonomous mobility.

The present research accepts, as a given, the EU policy target for large-scale deployment of automated mobility by 2030. However, according to the research findings, one may question whether this target can be reached in time by all European cities. More specifically, the development and implementation of a collective transport system that combines autonomous shared and public transport solutions with the ability to cover the wider urban area at all times is a great challenge by today's standards. It requires significant investments in digital and physical infrastructure, capacity building, and changes in governance and management, cooperation, and information exchange between private and public stakeholders, innovative business models, policies, and regulations, as well as other measures aiming to change current travel patterns and behaviors. It also requires the reconsideration of other ongoing policies referring to sustainable urban mobility and development, which should be adjusted accordingly. According to the authors' view, the way forward for the EU policy to address this challenge involves four interconnected dimensions. Firstly, it is needed to capitalize on the long tradition and pioneering role of European cities in the promotion of sustainable urban development. Good practices and lessons learnt regarding sustainable mobility can be used as a guide to ensure the effective deployment of autonomous road vehicles. Secondly, technological innovation should be led by policy goals, such as the vision for zero road fatalities, the global goal for decarbonization, and the right of all citizens to have access to affordable and effective mobility services. Thirdly, policy measures should consider the different contexts, needs, and goals of each region, city, and social group. Lastly, flexible planning strategies and monitoring mechanisms for evidence-based policy making should be designed to account for the uncertainty of the long-term impacts of autonomous mobility on cities. Overall, the EU transport policy for the next programming period should regard the implementation of autonomous road vehicles as a tool for transformative change in sustainable urban mobility.

In conclusion, this specific research provides a coherent evaluation of transport policy pathways for the anticipated wide-scale deployment of autonomous mobility for passenger transport in European cities. In this way, it enhances the preparedness of planners, researchers, and policy makers to exploit the potential of the specific technology towards sustainable urban development. The main findings can be communicated to the public in terms of an open debate to increase awareness and to incorporate the users' opinion in a bottom-up future research approach. To fully cover the subject, follow-up research is planned by the authors, which will address the autonomous mobility solutions for urban logistics. Moreover, there is a potential to expand the current methodology to cover the gap from the evaluation of transport policy pathways at the EU level to their adaptation at the local level, taking into consideration each city's different degree of technological readiness, as well as its specific spatial characteristics and urban development goals. Another important aspect for follow-up research is the analysis and evaluation of implementation scenarios for the proposed transport policy pathways, taking into account the official instruments, policy tools, and jurisdictions in EU transport policy, as well as the regional and local differences. Finally, the present research defines specific criteria for sustainable urban development, which can be used as categories for the design of a set of indicators that could be used by cities in parallel to the implementation of autonomous road vehicles with the purpose of monitoring and assessing their impact.

Author Contributions: Conceptualization, N.G.; methodology, N.G., K.A., E.N. and S.B.; validation, E.N. and S.B.; formal analysis, K.A. and N.G.; investigation, K.A. and N.G.; data curation, K.A.; writing—original draft preparation, N.G. and K.A.; writing—review and editing, E.N. and S.B.; visualization, K.A. and N.G.; supervision, E.N. and S.B.; project administration, N.G. and K.A. Authorship is limited to those who have contributed substantially to the work reported. All authors have read and agreed to the published version of the manuscript.

Funding: This research received no external funding.

Data Availability Statement: The data presented in this study are available on request from the corresponding author due to data privacy restrictions.

Acknowledgments: The authors would like to acknowledge the valuable contribution of experts who have participated in the multicriteria analysis.

Conflicts of Interest: The authors declare no conflicts of interest.

References

1. Ji, H.; Li, X.; Geng, Y.; Chen, X.; Wang, Y.; Cheng, J.; Chen, Z. Delineation of urban development boundary and carbon emission effects in Xuzhou city, China. *Land* **2023**, *12*, 1819. [CrossRef]
2. Nikolaou, P.; Basbas, S. Urban development and transportation: Investigating spatial performance indicators of 12 European Union coastal regions. *Land* **2023**, *12*, 1757. [CrossRef]
3. Banister, D.; Lichfield, N. The key issues in transport and urban development. In *Transport and Urban Development*; Banister, D., Ed.; E & FN Spon: London, UK, 1995; pp. 1–17.

4. Li, L.; Huang, W.L.; Liu, Y.; Zheng, N.N.; Wang, F.Y. Intelligence Testing for Autonomous Vehicles: A New Approach. *IEEE Trans. Intell. Veh.* **2016**, *1*, 158–166. [CrossRef]
5. Gordon, T.J.; Lidberg, M. Automated driving and autonomous functions on road vehicles. *Veh. Syst. Dyn.* **2015**, *53*, 958–994. [CrossRef]
6. Pillah, S. *Automated Vehicles in the EU*; European Parliamentary Research Service: Brussels, Belgium, 2016.
7. European Commission. *European Missions. 100 Climate-Neutral and Smart Cities by 2030. Implementation Plan*; European Commission: Brussels, Belgium, 2020.
8. Cooperative Research Centres Programme, Department of Industry, Sciences and Resources, Australian Government. *iMOVE Australia. Smart Mobility Projects and Trials Across the World*; CRC Programme: Canberra, Australia. Available online: https://imoveaustralia.com/smart-mobility-projects-trials-list/# (accessed on 28 October 2024).
9. Dai, J.; Cara Wang, X.; Ma, W.; Li, R. Future transport vision propensity segments: A latent class analysis of autonomous taxi market. *Transp. Res. Part A Policy Pract.* **2023**, *173*, 103699. [CrossRef]
10. Fernández-Medina, K.; Kent, J.; Holcombe, A. *StreetWise Trials: Technical Report. Published Project Report PPR966*; TRL Limited: Berkshire, UK, 2020. Available online: https://trl.co.uk/Uploads/TRL/Documents/StreetWise%20trials%20-%20Technical%20report.pdf (accessed on 28 October 2024).
11. ERTRAC Working Group Connectivity and Automated Driving. *Connected Automated Driving Roadmap*; ERTRAC: Brussels, Belgium, 2019.
12. KPMG International. *Autonomous Vehicles Readiness Index. Assessing Countries' Openness and Preparedness for Autonomous Vehicles*; KPMG International: Amstelveen, The Netherlands, 2018.
13. Akimoto, K.; Sano, F.; Oda, J. Impacts of ride and car-sharing associated with fully autonomous cars on global energy consumptions and carbon dioxide emissions. *Technol. Forecast. Soc. Change* **2022**, *174*, 121311. [CrossRef]
14. Deichmann, J.; Ebel, E.; Heineke, K.; Heuss, R.; Kellner, M.; Seiner, F. *Autonomous Driving's Future: Convenient and Connected*; McKinsey Global Pubsishing: New York, NY, USA, 2023.
15. Milakis, D.; Snelder, M.; van Arem, B.; van Wee, B.; de Almeida Correia, G.H. Development and transport implications of automated vehicles in the Netherlands: Scenarios for 2030 and 2050. *Eur. J. Transp. Infrastruct. Res.* **2017**, *17*, 63–85. [CrossRef]
16. Padmaja, B.; Moorthy, C.V.; Venkateswarulu, N.; Bala, M.M. Exploration of issues, challenges and latest developments in autonomous cars. *J. Big Data* **2023**, *10*, 61. [CrossRef]
17. Rahman, M.M.; Thill, J.C. Impacts of connected and autonomous vehicles on urban transportation and environment: A comprehensive review. *Sustain. Cities Soc.* **2023**, *96*, 104649. [CrossRef]
18. Litman, T. *Autonomous Vehicle Implementation Predictions. Implications for Transport Planning*; Victoria Transport Policy Institue: Victoria, Canada, 2023.
19. Milakis, D.; van Arem, B.; van Wee, B. Policy and society related implications of automated driving: A review of literature and directions for future research. *J. Intell. Transp. Syst.* **2017**, *21*, 324–348. [CrossRef]
20. Anastasiadou, K.; Gavanas, N.; Pitsiava-Latinopoulou, M.; Bekiaris, E. Infrastructure planning for autonomous electric vehicles, integrating safety and sustainability aspects: A multi-criteria analysis approach. *Energies* **2021**, *14*, 5269. [CrossRef]
21. Constantini, F.; Thomopoulos, N.; Steibel, F.; Curl, A.; Lugano, G.; Kováčiková, T. Data protection in a GDPR era: An international comparison of implications for autonomous vehicles. In *Policy Implications of Autonomous Vehicles*; Advances in Transport Policy and Planning Book Series; Milakis, D., Thomopoulos, N., van Wee, B., Eds.; Elsevier: Amsterdam, The Netherlands, 2020; Volume 5, pp. 191–213. [CrossRef]
22. Martens, K.; Beyazit, E.; Henenson, E.; Thomopoulos, N.; Milakis, D.; Mladenovic, M.; Negulescu, M. *A WG2 Thematic Report: Autonomous and Connected Transport as Part of an Inclusive Transport System, COST Action CA16222 WISE-ACT*; COST Association: Brussels, Belgium, 2021.
23. Sheehan, B.; Murphy, F.; Mullins, M.; Ryan, C. Connected and autonomous vehicles: A cyber-risk classification framework. *Transp. Res. Part A Policy Pract.* **2019**, *124*, 523–536. [CrossRef]
24. Zhang, Q.; Wallbridge, C.D.; Jones, D.M.; Morgan, P.L. Public perception of autonomous vehicle capability determines judgment of blame and trust in road traffic accidents. *Transp. Res. Part A Policy Pract.* **2024**, *179*, 103887. [CrossRef]
25. Tamakloe, R.; Park, D. Discovering latent topics and trends in autonomous vehicle-related research: A structural topic modelling approach. *Transp. Policy* **2023**, *139*, 1–20. [CrossRef]
26. European Commission. *Commission Staff Working Document Accompanying the Document: Sustainable and Smart Mobility Strategy—Putting European Transport on Track for the Future, SWD(2020)331 Final*; European Commission: Brussels, Belgium, 2020.
27. European Commission. *Sustainable and Smart Mobility Strategy—Putting European Transport on Track for the Future, COM(2020)789 Final*; European Commission: Brussels, Belgium, 2020.
28. Gruel, W.; Stanford, J.M. Assessing the Long-Term Effects of Autonomous Vehicles: A speculative approach. *Transp. Res. Procedia* **2016**, *13*, 18–29. [CrossRef]
29. Pimenta, A.R.; Kamruzzaman, M.; Currie, G. Long-term effects of autonomous vehi-cles on the built environment: A systematic scoping review towards conceptual frameworks. *Transp. Rev.* **2023**, *43*, 1083–1117. [CrossRef]
30. Ghehri, A.; Mouftah, H.T. Autonomous vehicles in the sustainable cities, the beginning of a green adventure. *Sustain. Cities Soc.* **2019**, *51*, 101751. [CrossRef]
31. Duarte, F.; Ratti, C. The Impact of Autonomous Vehicles on Cities: A Review. *J. Urban Technol.* **2018**, *25*, 3–18. [CrossRef]

32. Faisal, A.; Yigitcanlar, T.; Kamruzzaman, M.; Currie, G. Understanding autonomous vehicles: A systematic literature review on capability, impact, planning and policy. *J. Transp. Land Use* **2019**, *12*, 45–72. [CrossRef]
33. Strömberg, H.; Martins Silva Ramos, E.; Karlsson, M.; Johansson, M.; Ekman, F.; Bligård, L.O.; Jakobsson Bergstad, C. A future without drivers? Comparing users', urban planners' and developers', assumptions, hopes, and concerns about autonomous vehicles. *Eur. Transp. Res. Rev.* **2021**, *13*, 44. [CrossRef]
34. Hamadneh, J.; Duleba, S.; Esztergár-Kiss, D. Stakeholder viewpoints analysis of the autonomous vehicle industry by using multi-actors multi-criteria analysis. *Transp. Policy* **2022**, *126*, 65–84. [CrossRef]
35. Feys, M.; Rombaut, E.; Macharis, C.; Vanhaverbeke, L. Understanding stakeholders' evaluation of autonomous vehicle services complementing public transport in an urban context. In Proceedings of the 2020 Forum on Integrated and Sustainable Transportation Systems (FISTS), Delft, The Netherlands, 3–5 November 2022. [CrossRef]
36. European Commission. *The European Green Deal, COM(2019)640 Final*; European Commission: Brussels, Belgium, 2019.
37. European Commission. *The New EU Urban Mobility Framework, COM(2021)811 Final*; European Commission: Brussels, Belgium, 2021.
38. High Level Group on the Competitiveness and Sustainable Growth of the Automotive Industry in the European Union. *Ensuring That Europe Has the Most Competivie, Innovative and Sustainable Automotive Industry of the 2030's and Beyond. GEAR 2030-Final Report*; European Commission: Brussels, Belgium, 2017.
39. European Commission. *Connected and Automated Mobility*; European Commission: Brussels, Belgium, 2023. Available online: https://digital-strategy.ec.europa.eu/en/policies/connected-and-automated-mobility (accessed on 26 October 2024).
40. Rupprecht Consult-Forschung & Beratung GmbH. *Road Vehicle Automation in Sustainable Urban Mobility Planning, Practitioner Briefing, European Platform for Sustainable Mobility Plans*; Rupprecht Consult-Forschung & Beratung GmbH: Köln, Germany, 2019.
41. United Nations, Department of Economic and Social Affairs, Sustainable Development. *Sustainable Development Goal (SDG) 11. Make Cities and Human Settlements Inclusive, Safe, Resilient and Sustainable*; United Nations: New York, NY, USA. Available online: https://sdgs.un.org/goals/goal11 (accessed on 26 October 2024).
42. Gavanas, N. Autonomous Road Vehicles: Challenges for Urban Planning in European Cities. *Urban Sci.* **2019**, *3*, 61. [CrossRef]
43. Simpson, C.; Ataii, E.; Kemp, E.; Zhang, Y. *Mobility 2030: Transforming the Mobility Lanscape*; KPMG International: Amstelveen, The Netherlands, 2019.
44. Pandyaswargo, A.H.; Siregar, T.H.; Onoda, H. Exploring Japan's older adults mobility challenges and the potential role of autonomous vehicles. *Transp. Res. Part A Policy Pract.* **2023**, *176*, 103818. [CrossRef]
45. Williams, B. *Automated Vehicles and MaaS: Removing the Barriers*; Wiley: Hoboken, NJ, USA, 2021.
46. International Transport Forum. *Integrating Public Transport into Mobility as a Service. Summary and Conclusions*; ITF Roundtable Reports, No. 184; OECD Publishing: Paris, France, 2021.
47. Eppenberger, N.; Richter, M.A. The opportunity of shared autonomous vehicles to improve spatial equity in accessibility and socio-economic developments in European urban areas. *Eur. Transp. Res. Rev.* **2021**, *13*, 32. [CrossRef]
48. Miller, K.; Chng, S.; Cheah, L. Understanding acceptance of shared autonomous vehicles among people with different mobility and communication needs. *Travel Behav. Soc.* **2022**, *29*, 200–210. [CrossRef]
49. Dirsehan, T.; Can, C. Examination of trust and sustainability concerns in autonomous vehicle adoption. *Technol. Soc.* **2020**, *63*, 101361. [CrossRef]
50. Chalermpong, S.; Thaithatkul, P.; Ratanawaraha, A. Trust and intention to use autonomous vehicles in Bangkok, Thailand. *Case Stud. Transp. Policy* **2024**, *16*, 101185. [CrossRef]
51. Chen, Y.; Khan, S.K.; Shiwakoti, N.; Stasinopoulos, P.; Aghabayk, K. Analysis of Australian public acceptance of fully automated vehicles by extending technology acceptance model. *Case Stud. Transp. Policy* **2023**, *14*, 101072. [CrossRef]
52. Nikitas, A.; Tsouamou Njoya, E.; Dani, S. Examining the myths of connected and autonomous vehicles: Analysing the pathway to a driverless mobility paradigm. *Int. J. Automot. Technol. Manag.* **2019**, *19*, 10–30. [CrossRef]
53. Meyer, J.; Becker, H.; Bosch, P.; Axhausen, K.W. Autonomous vehicles: The next jump in accessibilities? *Res. Transp. Econ.* **2017**, *62*, 80–91. [CrossRef]
54. Wegener, M.; Fürst, F. *Land-Use Transport Interaction: State of the Art. Deliverable 2a of the project TRANSLAND, 4th RTD Framework Programme, European Commission*; Institut für Raumplanung, Fakultät Raumplanung, Universität Dortmund: Dortmund, Germany, 1999.
55. Larson, W.; Zhao, W. Self-driving cars and the city: Effects on sprawl, energy consumption, and housing affordability. *Reg. Sci. Urban Econ.* **2020**, *81*, 103484. [CrossRef]
56. Heinrichs, D. Autonomous driving and urban land use. In *Autonomous Driving*; Maurer, M., Gerdes, J., Lenz, B., Winner, H., Eds.; Springer: Berlin/Heidelberg, Germany, 2016; pp. 213–233. [CrossRef]
57. European Environment Agency. *Urban Sprawl in Europe. The Ignored Challenge*; EEA Report No 10/2006; Office for Official Publications of the European Communities: Luxembourg, 2006.
58. Coppola, P.; Silvestri, F. Autonomous vehicles and future mobility solutions. In *Autonomous Vehicles and Future Mobility*; Coppola, P., Esztergar-Kiss, D., Eds.; Elsevier: Amsterdam, The Netherlands, 2019; pp. 1–15. [CrossRef]
59. Newman, P.W.; Kenworthy, J.R. The land use-transport connection. An overview. *Land Use Policy* **1996**, *13*, 1–22. [CrossRef]
60. Ruso, A.; van Ommeren, J.; Dimitropoulos, A. *The Environmental and Welfare Implications of Parking Policies*; OECD Environment Working Paper No. 145; OECD Publishing: Paris, France, 2019. [CrossRef]

61. Nourinejad, M.; Bahram, S.; Roorda, M.J. Designing parking facilities for autonomous vehicles. *Transp. Res. Part B Methodol.* **2018**, *109*, 110–127. [CrossRef]
62. Zhu, C.; Susskind, J.; Giampieri, M.; O'Neil, H.B.; Berger, A.M. Optimizing sustainable suburban expansion with autonomous mobility through a parametric design framework. *Land* **2023**, *12*, 1786. [CrossRef]
63. Plihal, J.; Nedoma, P.; Sestak, V.; Herda, Z.; Aksjonov, A. Transport automation in urban mobility: A case study of an autonomous parking system. *Vehicles* **2022**, *4*, 326–343. [CrossRef]
64. Park, J.; Jang, S.; Ko, J. Effects of exclusive lanes for Autonomous Vehicles on urban expressways under mixed traffic of autonomous and human-driven vehicles. *Sustainability* **2024**, *16*, 26. [CrossRef]
65. Chen, T.; Sun, C. An optimization design of hybrid parking lots in an automated environment. *Sustainability* **2023**, *15*, 15475. [CrossRef]
66. Rebalski, E.; Adelfio, M.; Sprei, F.; Johansson, D.J.A. Brace for impacts: Perceived impacts and responses relating to the state of connected and autonomous vehicles in Gothenburg. *Case Stud. Transp. Policy* **2024**, *15*, 101140. [CrossRef]
67. Silva, D.; Földes, D.; Csiszár, C. Autonomous Vehicle use and urban space transformation: A scenario building and analysing method. *Sustainability* **2021**, *13*, 3008. [CrossRef]
68. Logan, T.M.; Hobbs, M.H.; Conrow, L.C.; Reid, N.L.; Young, R.A.; Anderson, M.J. The x-minute city: Measuring the 10, 15, 20-minute city and an evaluation of its use for sustainable urban design. *Cities* **2022**, *13*, 103924. [CrossRef]
69. Fayyaz, M.; González-González, E.; Nogués, S. Autonomous mobility: A potential opportunity to reclaim public spaces for people. *Sustainability* **2022**, *14*, 1568. [CrossRef]
70. Solmaz, R.R.; Haneen, F.; Taale, H.; van Arem, B.; Hoogendoorn, S.P. Design and operation of dedicated lanes for connected and automated vehicles on motorways: A conceptual framework and research agenda. *Transp. Res. Part C Emerg. Technol.* **2020**, *117*, 102664. [CrossRef]
71. Liu, Z.; Song, Z. Strategic planning of dedicated autonomous vehicle lanes and autonomous vehicle/toll lanes in transportation networks. *Transp. Res. Part C: Emerg. Technol.* **2019**, *106*, 381–403. [CrossRef]
72. Anastasiadou, K. Sustainable mobility driven prioritization of new vehicle technologies, based on a new decision-aiding methodology. *Sustainability* **2021**, *13*, 4760. [CrossRef]
73. World Economic Forum; The Boston Consulting Group. *Reshaping Urban Mobility with Autonomous Vehicles. Lessons from the City of Boston*; World Economic Forum: Geneva, Switzerland, 2018.
74. Ye, L.; Yamamoto, T. Impact of dedicated lanes for connected and autonomous vehicle on traffic flow throughput. *Phys. A Stat. Mech. Appl.* **2018**, *512*, 588–597. [CrossRef]
75. Bosch, P.M.; Becker, F.; Becker, H.; Axhausen, K.W. Cost-based analysis of autonomous mobility services. *Transp. Policy* **2018**, *64*, 76–91. [CrossRef]
76. World Economic Forum; Bain and Company. *Electric Vehicles for Smarter Cities: The Future of Energy and Mobility*; World Economic Forum: Geneva, Switzerland, 2018.
77. Iacobucci, R.; McLellan, B.; Tezuka, T. Modeling Shared Autonomous Electric Vehicles: Potential for transport and power grid integration. *Energy* **2018**, *158*, 148–163. [CrossRef]
78. Melia, S.; Parkhurst, G.; Barton, H. The paradox of intensification. *Transp. Policy* **2011**, *18*, 46–52. [CrossRef]
79. Fagnant, D.J.; Kockelman, K. Preparing a nation for autonomous vehicles: Opportunities, barriers and policy. *Transp. Res. Part A Policy Pract.* **2015**, *77*, 167–181. [CrossRef]
80. Shoup, D.C. Cruising for parking. *Transp. Policy* **2006**, *13*, 479–486. [CrossRef]
81. Transport & Environment. *Driverless Cars Increase Congestion—But Could Cut Massive Parking Times*; Transport & Environment: Brussels, Belgium, 2017. Available online: https://www.transportenvironment.org/articles/driverless-cars-increase-congestion-could-cut-massive-parking-times (accessed on 26 October 2024).
82. Tettamanti, T.; Horcher, D.; Varga, I. The impact of autonomous vehicles on urban traffic network capacity: An experimental analysis by microscopic traffic simulation. *Transp. Letters. Int. J. Transp. Res.* **2019**, *12*, 540–549. [CrossRef]
83. Michalowska, M.; Ogłoziński, M. Autonomous Vehicles and Road Safety. In *Solutions in Today's Transport. TST 2017. Communications in Computer and Information Science*; Mikulski, J., Ed.; Springer: New York, NY, USA, 2017; Volume 715, pp. 191–202. [CrossRef]
84. Soteropoulos, A.; Berger, M.; Ciari, F. Impacts of automated vehicles on travel behaviour and land use: An international review of modelling studies. *Transp. Rev.* **2019**, *39*, 29–49. [CrossRef]
85. Parkin, J.; Clark, B.; Clayton, W.; Ricci, M.; Parkhurst, G. Autonomous vehicle interactions in the urban street environment: A research agenda. *Proc. Inst. Civ. Eng. Munical Eng.* **2018**, *171*, 15–25. [CrossRef]
86. Lee, S. Opinions of active transportation users on policies to ensure their perceived safety in the era of autonomous vehicles. *Case Stud. Transp. Policy* **2023**, *12*, 101002. [CrossRef]
87. Metz, D. Developing policy for urban autonomous vehicles: Impact on congestion. *Urban Sci.* **2018**, *2*, 33. [CrossRef]
88. European Commission. *EU Mission: Climate-Neutral and Smart Cities*; European Commission: Brussels, Belgium, 2020. Available online: https://research-and-innovation.ec.europa.eu/funding/funding-opportunities/funding-programmes-and-open-calls/horizon-europe/eu-missions-horizon-europe/climate-neutral-and-smart-cities_en (accessed on 25 October 2024).
89. Taiebat, M.; Brown, A.L.; Safford, H.; Xu, M. A Review on Energy, Environmental, and Sustainability Implications of Connected and Automated Vehicles. *Environ. Sci. Technol.* **2018**, *52*, 127. [CrossRef]

90. Bauer, G.S.; Greenblatt, J.B.; Gerke, B.F. Cost, Energy, and Environmental Impact of Automated Electric Taxi Fleets in Manhattan. *Environ. Sci. Technol.* **2018**, *52*, 4920–4928. [CrossRef]
91. Patella, S.M.; Scrucca, F.; Asdrubali, F.; Carrese, S. Carbon Footprint of autonomous vehicles at the urban mobility system level: A traffic simulation-based approach. *Transp. Res. Part D Transp. Environ.* **2019**, *74*, 189–200. [CrossRef]
92. Greenblatt, J.B.; Shaheen, S. Automated Vehicles, On-Demand Mobility, and Environmental Impacts. *Curr. Sustain. /Renew. Energy Rep.* **2015**, *2*, 74–81. [CrossRef]
93. Liu, F.; Zhao, F.; Liu, Z.; Hao, H. Can autonomous vehicle reduce greenhouse gas emissions? A country-level evaluation. *Energy Policy* **2019**, *132*, 462–473. [CrossRef]
94. Thomopoulos, N.; Grant-Muller, S. Incorporating equity as part of the wider impacts in transport infrastructure assessment: An application of the SUMINI approach. *Transportation* **2013**, *40*, 315–345. [CrossRef]
95. Curiel-Esparza, J.; Mazario-Diez, J.L.; Canto-Perello, J.; Martin-Utrillas, M. Prioritization by consensus of enhancements for sustainable mobility in urban areas. *Environ. Sci. Policy* **2016**, *55*, 248–257. [CrossRef]
96. Karahalios, H. The application of the AHP-TOPSIS for evaluating ballast water treatment systems by ship operators. *Transp. Res. Part D Transp. Environ.* **2017**, *52*, 172–184. [CrossRef]
97. Menon, R.R.; Ravi, V. Using AHP-TOPSIS methodologies in the selection of sustainable suppliers in an electronics supply chain. *Clean. Mater.* **2022**, *5*, 100130. [CrossRef]
98. Saaty, T.L. *The Analytic Hierarchy Process*; McGraw-Hill: New York, NY, USA, 1980.
99. Turcksin, L.; Bernardini, A.; Macharis, C. A combined AHP-PROMETHEE approach for selecting the most appropriate policy scenario to stimulate a clean vehicle fleet. *Procedia Soc. Behav. Sci.* **2011**, *20*, 954–965. [CrossRef]
100. Basbas, S.; Makridakis, C.M. A review of the contribution of Multi-Criteria Analysis to the evaluation process of transportation projects. *Int. J. Sustain. Dev. Plan.* **2007**, *2*, 387–407. [CrossRef]
101. Tsamboulas, D.A. A tool for prioritizing multinational transport infrastructure investments. *Transp. Policy* **2007**, *14*, 11–26. [CrossRef]
102. Opricovic, S.; Tzeng, G.H. Compromise solution by MCDM methods: A comparative analysis of VIKOR and TOPSIS. *Eur. J. Oper. Res.* **2004**, *156*, 445–455. [CrossRef]
103. Dijkstra, L.; Poelman, H.; Veneri, P. *The EU-OECD Definition of a Functional Urban Area. OECD Regional Development Working Papers 2019/11*; OECD: Paris, France, 2019. [CrossRef]
104. Martin-Utrillas, M.; Reyes-Medina, M.; Curiel-Esparza, J.; Canto-Perello, J. Hybrid method for selection of the optimal process of leachate treatment in waste treatment and valorization plants or landfills. *Clean Technol. Environ. Policy* **2015**, *17*, 873–885. [CrossRef]
105. De Loë, R.C.; Melnychuk, N.; Murray, D.; Plummer, R. Advancing the State of Policy Delphi Practice: A Systematic Review Evaluating Methodological Evolution, Innovation, and Opportunities. *Technol. Forecast. Soc. Chang.* **2016**, *104*, 78–88. [CrossRef]
106. Esmaeilpoorarabi, N.; Yigitcanlar, T.; Guaralda, M.; Kamruzzaman, M. Evaluating place quality in innovation districts: A Delphic hierarchy process approach. *Land Use Policy* **2018**, *76*, 471–486. [CrossRef]
107. Acheampong, R.A.; Legacy, C.; Kingston, R.; Stone, J. Imagining urban mobility futures in the era of autonomous vehicles—Insights from participatory visioning and mul-ti-criteria appraisal in the UK and Australia. *Transp. Policy* **2023**, *136*, 193–208. [CrossRef]
108. Legênea, M.F.; Auping, W.L.; de Almeida Correia, G.H.; van Arem, B. Spatial impact of automated driving in urban areas. *J. Simul.* **2020**, *14*, 295–303. [CrossRef]
109. Wadud, Z.; Mattioli, G. Fully automated vehicles: A cost-based analysis of the share of ownership and mobility services, and its socio-economic determinants. *Transp. Res. Part A* **2021**, *151*, 228–244. [CrossRef]
110. Triantafyllou, E. *Multi–Criteria Decision Making Methods: A Comparative Study*; Kluwer Academic Publishers: Dordrecht, The Netherlands, 2000.
111. Zolfagharian, M.; Walrave, B.; Raven, R.; Romme, A.G.L. Studying transitions: Past, present, and future. *Res. Policy* **2019**, *48*, 103788. [CrossRef]

Disclaimer/Publisher's Note: The statements, opinions and data contained in all publications are solely those of the individual author(s) and contributor(s) and not of MDPI and/or the editor(s). MDPI and/or the editor(s) disclaim responsibility for any injury to people or property resulting from any ideas, methods, instructions or products referred to in the content.

land

MDPI

Review

Urban Stormwater Management Using Nature-Based Solutions: A Review and Conceptual Model of Floodable Parks

Silvia Martín Muñoz [1], Simon Elliott [1], Jonas Schoelynck [1] and Jan Staes [1,2,*]

[1] ECOSPHERE Research Group, Department of Biology, University of Antwerp, 2000 Antwerp, Belgium; silvia.martinmunoz@uantwerpen.be (S.M.M.)
[2] Institute for Environment and Sustainable Development, University of Antwerp, 2000 Antwerp, Belgium
* Correspondence: jan.staes@uantwerpen.be; Tel.: +32-3-2652385

Abstract: Climate change is causing the alteration of atmospheric dynamics, leading to extreme precipitation events and floods. On the other hand, landscape modification and increased imperviousness due to urbanization exacerbate the impacts of flooding. In order to become more permeable, cities are increasingly embracing aquatic Nature-based Solutions which, using natural processes, allow for the mitigation of water-related hazards. One of these solutions is floodable parks, where pluvial runoff is conveyed for its temporal storage into, firstly, permanent retention ponds and, eventually, the partial or totality of their surface. Floodable parks are still a novel aquatic Nature-based Solution and have not yet been investigated. In this paper, a systematic review on current floodable parks was performed in order to study (1) the conditions needed for their implementation, (2) their design, and (3) the connection between design and ecosystem services. A subsequent systematic review was performed to understand (4) the processes occurring within the park. With the obtained information, a conceptual model of floodable parks was developed. The results indicate that both the vegetation surrounding the permanent pond of the floodable park and the biodiversity within the pond enhance the performance of this solution and allow potential water reuse. The implementation of floodable parks will therefore facilitate the transformation of urban areas to create sustainable, climate-resilient, and circular cities.

Keywords: climate change; flooding; urban stormwater management; nature-based solutions; ecosystem services; floodable parks; biodiversity; water reuse; circularity; conceptual model

1. Introduction

Rapid urbanization is a common phenomenon to both developed and developing countries, leading to uncontrolled urban sprawl and unsustainable land-use changes. Due to anthropogenic impacts such as deforestation, river plain modification, and increased imperviousness, flooding has intensified in the past decades [1]. Floods are currently the most common natural disaster in cities globally [2]. They caused a loss of approximately USD 826 billion in the period between 1960 and 2019 [3]; only the European floods in July 2021 costed USD 46 billion. The frequency of flooding has risen by 48% compared to historic levels [4], with climate change being the main driver of this increase [5]. Extreme droughts are also associated with the alteration of atmospheric dynamics, which have raised the urban water demand and caused severe water use restrictions in the most vulnerable regions [6].

Nature-based Solutions (NbSs) can be used as a tool to create greener and more sustainable cities and help tackle the aforementioned challenges [7]. NbSs are measures which use natural processes to provide environmental, social, and economic benefits and to increase resilience to climate-change hazards. NbSs have been promoted across Europe, the USA, and Australia through guidelines and policies which encourage water-sensitive urban design (WSUD), low-impact development (LID) practices, and best management practices

(BMPs), all being concepts that can be used interchangeably [8]. Aquatic NbSs (aNbSs) address, among others, water-related problems such as droughts and floods. Popular aNbSs are green roofs, bioswales, and artificial constructed or floating wetlands, as well as detention and retention ponds. Remarkably, all of these aNbSs can be integrated into the urban matrix in public spaces. In this paper, wet retention ponds will be studied within the concept of a floodable urban park for pluvial flood management.

Floodable urban parks with retention consist mainly of a permanent pond area with landscaped banks and surroundings which allow the storage of local runoff water [9]. Unlike normal parks with ponds or wadis, this aNbS is designed for its complete inundation (both banks and surroundings) in order to attenuate surface runoff during extreme precipitation events. Floodable parks with retention ponds (henceforth called floodable parks) can also provide a rich and stable ecosystem for biodiversity due to their permanent waterbody [10]. An additional benefit of this type of solution is that it can potentially be connected with treatment plants for the post-rainfall treatment of the stored water, such as in the "La Marjal" floodable park (Alicante, Spain), a model example of a circular water system [11]. Therefore, floodable parks are meant not only to tolerate runoff water, but also to store, harvest, and convey it, potentially working together with existing infrastructure as multifunctional modules [2].

Although floodable urban parks can incorporate in their design multiple aNBSs (e.g., rain gardens, vertical gardens), the main element which provides runoff attenuation is the permanent pond. Taking as reference previous works on floodable parks for coastal and fluvial flooding [12–15], we define floodable urban parks as recreational spaces to which pluvial water is conveyed into a stagnant waterbody for its retention, infiltration, and drainage and which can be temporarily flooded. However, not only water flow-regulating services are delivered by floodable urban parks, but also other co-benefits, such as water quality regulation, cultural and educational services, aesthetic value, biodiversity conservation, and even potable water. Despite the multiple ecosystem services (ESs) which are created by this aNbS, it has not been systematically studied under an ES perspective, nor have these ESs been correlated to the design.

This review seeks to address the following research questions: (i) what are the topographical and climatic conditions needed for the implementation of floodable parks, (ii) what are the key attributes required in the design of floodable parks, (iii) how are these attributes connected to the ESs and benefits that these parks deliver, and (iv) what are the occurring processes within the permanent pond which contribute to ES delivery. The aim of this article is thus to establish a holistic understanding of urban floodable parks and propose a graphical definition that can be used in further studies. In order to do so, a conceptual model was developed based on existing study cases. A conceptual model helps to promote an integrative approach when planning floodable parks as a sustainable solution for water regulation in cities.

2. Materials and Methods

2.1. Case Study Selection of Floodable Parks

An initial literature search was conducted using the Scopus and Web of Science online databases to find case studies from which to derive a conceptual model of a floodable park. Different combinations of search strings were used to search within the title, abstract, and keywords fields, tackling (1) infrastructure/solution, (2) location, (3) function/benefits, and (4) period of publication. The criteria of selection were: (i) it must be available in English or Spanish, (ii) the case study should be placed in urban or peri-urban areas, and (iii) it should have been constructed to prevent pluvial (not fluvial) flooding only. The initial search was (1) "floodable parks" OR "floodable urban landscape" OR "sustainable urban drainage systems" OR "water nature-based solution" OR "aqua nature-based solution" OR "aqua NBS", (2) urban OR city, (3) "water management" OR "water reuse" OR "flood resilience" OR runoff, and (4) Jan 2014–Dec 2023. After this first search, 241 papers were found in Scopus and 152 in Web of Science. Since the search was too broad, a second search string was used: (1) "floodable parks" OR "floodable urban landscape" OR "sustainable urban drainage systems" OR "water nature-based solution" OR "aqua nature-based solution" OR "aqua NBS", (2) urban OR city, (3) "ecosystem services" OR "ecological benefits" OR "ecosystem benefits" OR "benefits", (4) January 2014–December 2023. In this case, 88 papers were found in Scopus and 58 in Web of Science. After title and abstract screening, it was observed that there were few examples in literature and therefore, that this solution has not yet been scientifically studied nor monitored. Therefore, the search was broadened to water engineering projects and municipality and architectural firm websites as well as newspapers. The following keywords were used, both in English and Spanish: "floodable parks", "sustainable urban drainage system", "flood-resilient urban park". Since the focus of this paper is on pluvial flood management, projects in which floodable parks were used to manage fluvial floods were excluded. Therefore, the case studies were selected for the presence of artificial ponds or pools within the urban park and importantly, for their ability to be temporarily inundated. Snowball techniques (e.g., taking referenced floodable parks from initially sampled respondents) were also used to look for the case studies. The search process is summarized in Figure 1. In total, 12 case studies were selected (Table 1).

Table 1. Summary of the selected case studies. The name of the park, year of construction, country, region, reference source, and its year of publication are presented.

Name	Year of Construction	Country	Region	Reference Source	Year of Publication
El Recorral Forest Park	2019	Spain	Alicante	Sánchez-Almodóvar et al. [16]	2022
Los Alcázares	Ongoing implementation	Spain	Murcia	Tragsatec Presupuestary document	2023
Via Parque	2023	Spain	Alicante	Sánchez-Almodóvar et al. [17]	2021
La Marjal	2015	Spain	Alicante	Sánchez-Almodóvar et al. [17]	2022
La Quebradora	Ongoing implementation	Mexico	Ciudad de Mexico	Perroti [18]	2022
Hans Tavsens Park	Not implemented yet	Denmark	Copenhagen	Soul of Norrebro booklet [19]	2019
Celso Peçanha	Not implemented yet	Brazil	Mesquita	Pitzer J. et al. [19,20]	2019
Maple Valley Park	2013	Taiwan	Taichung	Hsu C. & Hung C. [21]	2019
Enghaven Park	2019	Denmark	Copehnaghen	Braae, E., & Riesto, S. [22]	2018
Elsa Eschelsson's Park	2022	Sweden	Uppsala	http://lanzine.com/ (accessed on 10 April 2024) [23]	2022
Benthemplein	2013	Netherlands	Rotterdam	Peinhardt [24]	2021
Sidwell Friends School	2007	USA	Washington, D.C	Quach [25]	2021

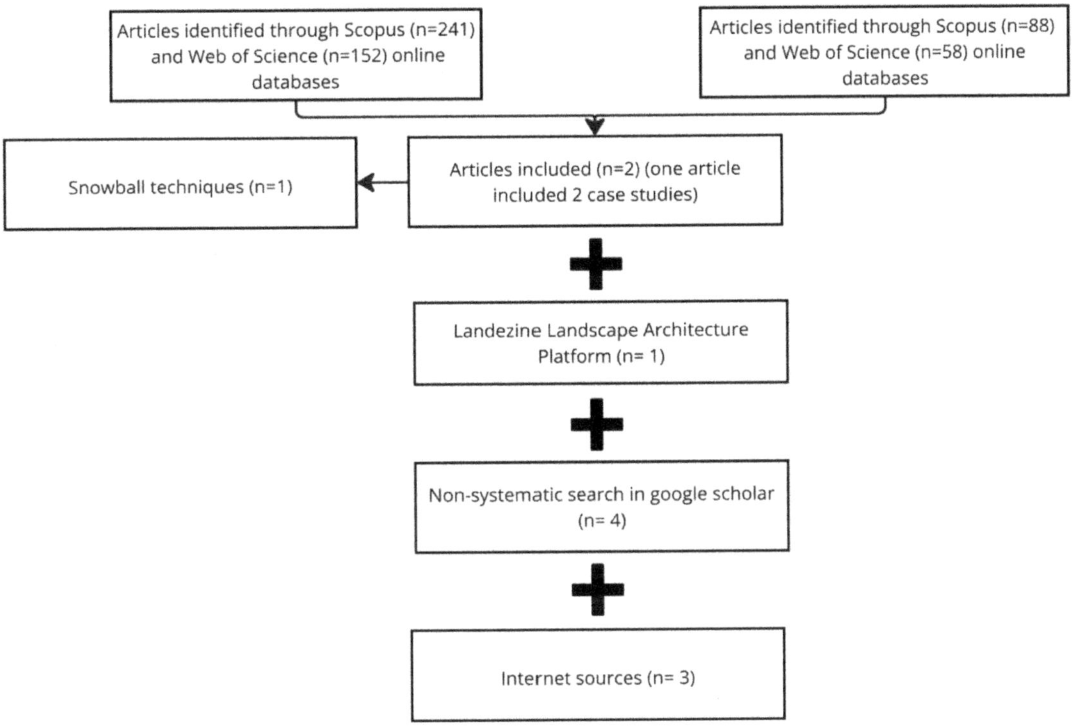

Figure 1. Scheme of the literature search for the case studies.

2.2. Identification of the Topographical, Social, and Climatic Conditions for the Implementation of Floodable Parks

The case studies were analyzed, and the topographical, social, and climatic conditions required for the implementation of floodable parks were evaluated. These were categorized in a systematic way to standardize the main drivers across the different case studies.

2.3. Identification of Key Elements for the Design of Floodable Parks

Key elements were identified from the case studies' designs and standardized across the case studies. Attributes were categorized into "water management elements", "recreational elements", and "biodiversity elements". "Water management elements" were further classified into "aNbS elements", "technical elements" and "water reuse". Within "biodiversity", a differentiation was made between "nuisance species management" and "promotion of biodiversity". These key elements complemented the processes of the conceptual model.

2.4. Identification of ESs and Benefits Delivered by Floodable Parks and the Association of These to the Identified Key Elements

According to the ESs classification of the European Environment Agency (EEA) (Classification of ecosystem services (EEA) (UNCEEA/5/7)), the benefits provided by the floodable parks were standardized and categorized into "provisioning", "regulating" and "cultural" ESs. These were further divided into the corresponding proposed classes by the EEA [26].

2.5. Identification of Processes and Influencing Factors Occurring in Floodable Parks

The studied processes are of a hydrological nature. Although a former study [27] has separated the main attributes supporting water circularity in NbSs into (1) the layer of plants and roots and (2) the body of water and its bed, processes taking place in these compartments were described jointly due to their interconnected relation. The identified processes occurring in aNbSs were pre-classified into "water detention and retention processes", "water purification processes" (later divided into "physicochemical water purification" and "biological water purification"), and "water reuse". This step resulted in a total of four processes considered for the conceptual model.

In order to identify these processes, a literature search was performed using the Scopus and Web of Science databases and filtered by title, abstract, keywords, and all fields, respectively. The following string was used for the period of 2000 to 2024: "retention pond" OR "detention pond" AND urban OR city AND "water quality" OR "water pollution" OR "water storage" OR "water reuse" OR "water purification" OR "water treatment" AND pluvial OR runoff OR rain. Through snowball techniques, using the references of the obtained papers, the literature review was completed (Figure 2). The sources of these papers can be found in Appendix A (Figure A1).

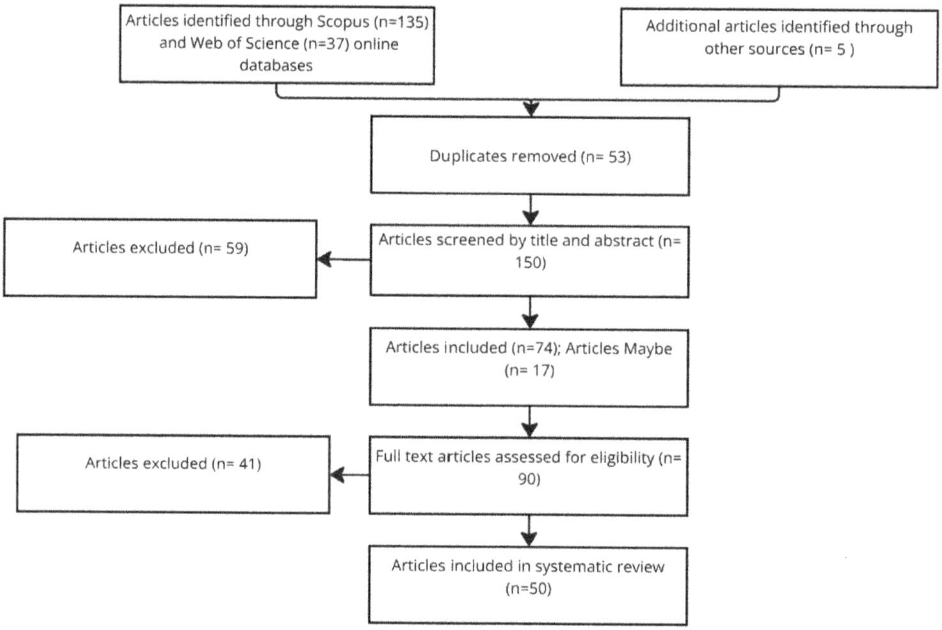

Figure 2. Scheme of the literature search for the processes occurring in floodable parks.

2.6. *Development of the Conceptual Model*

Based on the design characteristics of the floodable parks (e.g., arrangement of the modules, presence of wetland habitat, overflow possibilities, etc.), the processes and the identified ESs, a conceptual model was developed.

3. Results

3.1. *Case Studies*

Out of the 12 studies, four were found in Spain, one in Mexico, one in Norway, two in Denmark, one in the Netherlands, one in the United States, one in Taiwan, and one in Brazil. Although most of them were already built, two of them (i.e., Los Alcázares and La Quebradora) were still in construction and another two (i.e., Hans Tavsens Park, Celso Peçanha) were not implemented yet. An overview of the main design features of the selected parks is shown in Tables 2 and 3.

Table 2. Comparison of main drivers, functioning, and biodiversity incorporation of the selected case studies.

Name	Country	Drivers	Description of the System	Treated Water Reuse	Biodiversity
El Recorral Forest Park	Spain	Runoff water was reaching nature parks. There was a need for water discharge points for treated wastewater	1. Upstream tanks for first flush water storage and conveyance into treatment water 2. When tanks are filled, the park acts as an intermediate discharge point 3. Staggered ponds with design that enhances water purification 4. Water from the ponds descends to natural ravine where discharge is authorized	• Irrigation (i.e., parks, gardens, golf course) • Fire extinguishment • Street cleaning • Pond's filling	• Autochthonous vegetation • Creation of areas for nesting, feeding and refuge zones for amphibians and birds, • Insectivorous species such as bats and birds are attracted for biological pest control
Los Alcázares	Spain	Landscape modification (urbanization and agriculture) enhanced runoff water and its accumulation into unwanted discharge points (i.e., roads, sea).	1. Runoff water conveyed to the park's pond 2. When filled, water flows into channels along the park 3. When overflowed, the rest of the park is flooded gradually 4. If the maximum volume for storage is exceeded, water goes to a spillway 5. The park is designed with permeable zones to enhance infiltration 6. Additional water storage in an onsite tank	• Water reuse for internal irrigation of the park green zones after treatment	• Autochthonous xeric vegetation • Trees for shadow • Creation of areas for nesting, feeding and refuge zones for amphibians, birds, insects and bats • Insectivorous passerine birds for biological pest control • Bee hotel
Via Parque	Spain	Existing infrastructure was observed to be insufficient to hold runoff water from heavy rain events	1. Upstream tank (insufficient storage volume) collects first flush runoff water and conveys it to a wastewater treatment plant 2. Rest of runoff water/rainwater fills the pond gradually 3. The park is designed with permeable zones to enhance infiltration 4. Pond's biologically purified water is infiltrated into a below-pond tank for direct water reuse or soil water recharge 5. If it does not meet water quality standards, water is conveyed to wastewater treatment plant 6. After maximum flooding level of the pond, water goes to spillway 7. Design of park with permeable zones to enhance infiltration	• Irrigation of the park, city greenery and University Street • Street cleaning • Cleaning of onsite tank	• Permanent and temporary water zones to create different niches for biodiversity • Vegetation selection for water purification • Aeration to avoid toxic algae blooms
La Marjal	Spain	Natural ravines transformed into urban areas, resulting on small rain events causing high water accumulation. Existing tank observed to be insufficient.	1. Upstream existing tank 2. When filled, water goes to the park 3. Gradual filling of the park 4. When overflown, water goes to an underground tank 5. If water storage is exceeded, water goes to spillway 6. Water is conveyed to wastewater treatment plant	• Irrigation of the park • Aquifer recharge	• Constructed Wetlands (CWs) • Aeration to avoid toxic algae blooms • Creation of areas for nesting, feeding and refuge zones for amphibians and birds • Insectivorous species such as bats and birds are attracted for biological pest control

Table 2. *Cont.*

Name	Country	Drivers	Description of the System	Treated Water Reuse	Biodiversity
La Quebradora	Mexico	Water shortage, floods in urban areas due to landscape modification and lack of green areas.	1. Runoff water is collected in upstream tanks 2. When filled, runoff water is conveyed via existing piping network into the park 3. The spiral pond gradually fills 4. When the capacity for water retention is exceeded, water goes into a sport area with permeable surface 5. A concrete skate park with slope is gradually filled as well 6. There is an extra tank within the park in cases the storage volume is exceeded 7. Water is conveyed to the onsite wastewater treatment plant The park is designed with permeable zones to enhance infiltration	• Aquifer recharge	-
Hans Tavsens Park	Denmark	Runoff water caused damages to urban infrastructure and was discharged into a natural lake	1. Water flows from upstream urban areas to the park 2. Existing drainage system collects first flush water and conveys it to wastewater treatment plants 3. Park fills gradually 4. The park is designed with permeable zones to enhance infiltration 5. Rain water is conveyed downstream along the main discharge street, which will capture the runoff water thanks to its redesign with permeable surfaces and tanks. 6. Before it reaches the natural lake, water is conveyed into Constructed Wetlands for its further purification 7. Water from the park and the lake is continuously pumped (even in dry periods) to a purification system 8. This purification system also treats first flush runoff water before discharge into the lake 9. In case of exceeded storage volume, the water is conveyed to the lake without purification	• Private irrigation of rainwater from tanks • Irrigation of the park • Soil water recharge	• Vegetation will be selected to decrease air and noise pollution and provide habitat for birds and pollinators • Constructed Wetland (CWs) for further purification
Celso Peçanha	Brazil	Extreme flood events caused infrastructure damages and discharged polluted water to the river system	1. Runoff water is conveyed to the park 2. This park is near a river, so in order to avoid fluvial flood, a one-way floodgate will ensure only rain water is conveyed into the park 3. Gradual filling of the park 4. The park is designed with different levels which can withstand different flooding thresholds. The infrastructure will be constructed accordingly. 5. The park is designed with permeable zones to enhance infiltration	-	• Vegetation will be planted in order to increase volume storage capacity

Table 2. *Cont.*

Name	Country	Drivers	Description of the System	Treated Water Reuse	Biodiversity
Maple Valley Park	Taiwan	Water shortage due to direct discharge into the sea.	1. Runoff water from upstream is conveyed to the park, entering through a gate opened via remote control 2. The park is designed with permeable zones to enhance infiltration	• Irrigation of the park	• The vegetation planted in the park has increased the city's green coverage by 28,000 m^2
Enghaven Park	Denmark	Existing infrastructure was observed to be insufficient to manage current rain events, leading to runoff accumulation in urban areas.	1. Runoff water from nearby roofs and houses flows to the park 2. The lowest part of the park (a dry detention basin) gradually fills 3. When the capacity is exceeded, the rest of the park (including permanent ponds) is filled gradually	• Irrigation of the park • Street cleaning	• 11,000 perennial autochthonous plants have been planted, acting as a stepping stone in the green corridor of the city
Elsa Eschels-son's Park	Sweden	High economic losses due to floods in the past decade.Need to incorporate multifunc-tional areas in a growing urban population.	1. Runoff water from nearby areas is collected in the stormwater pond 2. Pond fills gradually 3. When overflown, water is conveyed into the permeable area (where the water infiltrates) and to the concrete sports space (where the water is stored) 4. When the water storage capacity is exceeded, the water flows into a spillway	• Pond's infiltration is prevented to avoid contaminated water reaching the local aquifers	• Vegetation is chosen to provide habitat for biodiversity • Stepping stone in the green corridor connecting nearby forest
Benthemplein	Holland	Lack of water-regulating practices in the neighbour-hood.	1. Rainwater flows to the stormwater management basins 2. Water is infiltrated and drained through gutters into the soil and waterway	• Soil water recharge	• No vegetation
Sidwell Friends School	USA	Integrated Water Management education	1. Rainwater falls from the roof to the park, which is filled gradually 2. The water flows through the vegetated terraces to the pond and underground cistern 3. Runoff water goes into settling tanks 4. The water from the tank and cistern is pumped to the upper terraces for its biological purification and infiltration 5. This cycle is repeated five days before the water is reused	• Toilet flushing • Pond's refilling during dry months	• Constructed Wetlands • Nature is an important element of this system, where 50 native plant species attract birds and insects.

Table 3. Comparison of main water management features of the selected case studies.

Name	Country	Return Period (Years)	Total Area (m^2)	Floodable Area (m^2)	Pond Area (m^2)	Maximum Inundation Volume (m^3)	Ratio (m^3/total m^2)	Ratio (m^3/floodable m^2)
El Recorral Forest Park	Spain	-	-	-	-	5304	-	-
Los Alcazares	Spain	100	110,000	110,000	45,000	490,000	4.45	4.45

Table 3. *Cont.*

Name	Country	Return Period (Years)	Total Area (m^2)	Floodable Area (m^2)	Pond Area (m^2)	Maximum Inundation Volume (m^3)	Ratio (m^3/total m^2)	Ratio (m^3/floodable m^2)
Via Parque	Spain	3	16,045	6000	978.27	12,500	0.78	2.08
La Marjal	Spain	50	36,700	36,700	6674	45,000	1.23	1.23
La Quebradora	Mexico	500	38,470	24,085	13,985	67,000	1.74	2.78
Hans Tavsens Park	Denmark	100	6600	4000	-	18,000	2.72	4.50
Celso Peçanha	Brazil	50	700,000	22,856	-	-	-	-
Maple Valley Park	Taiwan	-	-	-	-	200,000	-	-
Enghaven Park	Denmark	100	35,000			22,600	-	-
Elsa Eschelsson's Park	Sweden	10	-	6600	4000	1540	-	0.23
Benthemplein	Holland	-	5500	-	-	1800	0.33	-
Sidwell Friends School	USA	1	-	-	-	-	-	-

3.2. *Topographical, Social, and Climatic Conditions for the Implementation of Floodable Parks*

The topographical, social, and climatic conditions of each case studied varied, highlighting the need to adjust NbSs according to a specific local context. However, some generalizations could be withdrawn from the case studies. Particularly, all mentioned the need to locate the floodable park in a landscape depression or lower plain to avoid the outflow of the water once it had been stored in the detention pond of the park. Geotechnical investigations were performed in each case to ensure that groundwater levels were deep enough and therefore, that potential seepages could not reduce the storage capacity of the pond. These investigations are also important to assess the type of soil, geology, and land stability prior to the construction.

When analyzing the conditions that drove the implementation of the floodable park, in 10 of the cases, the existing infrastructure used to mitigate flooding (e.g., tanks, pipes, treatment plants as discharge points) was evidenced to be insufficient, even if seven reported that the natural topography enhanced runoff. Landscape modification was also frequently observed, either by urbanization (66%) or by the alteration of the natural water flow (25%), showing how the modification of the natural topography enhances water runoff in cities. The presence of urban water runoff was mentioned in 66% of the cases, caused by the accumulation of high water levels in urban zones due to the occurrence of big (66%) or small (16%) precipitation events. A total of 50% reported traffic collapse or damage to infrastructure due to pluvial water-related hazards. A lack of flood-managed areas and unwanted discharge points each accounted for 16.66% of the reasons to build the floodable park.

Regarding the social context of the localities at which floodable parks were implemented, a lack of green space, lack of social space, unused land, and lack of sports facilities were mentioned in 58%, 25%, 25%, and 16.66%, respectively. Interestingly, two cases mentioned the success of the implementation of floodable parks in the past as a driver to build another floodable park in the municipality, showing the effectiveness of this aNbS in tackling extreme precipitation events. This implementation is normally driven by the willingness of the municipalities to apply sustainable measures, since in 50% of the cases reported, the local authorities increased awareness and funding for this type of projects.

Lastly, it was noticeable how little consideration was given to biodiversity as a reason for the implementation of the aNbS. Only 25% mentioned the lack of green zone connectivity within the urban matrix and only one case study mentioned the need for conservation of the former habitat (i.e., wetlands) as a driver for the construction of the floodable park.

3.3. *Key Elements for the Design of Floodable Parks*

The design of a floodable park is fundamental for its correct functionality and appropriate runoff water conveyance, retention, storage, infiltration, and drainage. The design can also deliver water quantity and quality regulation by managing the water system's hydroperiod, choosing an adequate pond bed, integrating certain plant species, and avoiding stagnation of the water. However, the services delivered by floodable parks are not restricted to hydrological climate-related problems and therefore, they have to be designed for a broad variety of purposes. The concept of multifunctionality has been recently studied regarding the design of NbSs, since these are meant to address social, economic, and environmental issues simultaneously [28]. Key elements from the designs of the case studies were identified and categorized into "water management elements", "recreational elements", and "biodiversity elements".

3.3.1. Water Management Elements

Aquatic NbS

Within the parks, the most prominent water management elements were permeable vegetated zones and pavements (66%). In fact, three case studies mention the integration of other NbSs (e.g., rain gardens and vertical gardens) within the design of the park. For example, dry retention ponds (25%) and artificial wetlands or wetland mimicking systems (58%) were incorporated in the case studies; 33% acknowledged the enhancement of biological purification.

Technical

A great proportion (83%) of the floodable parks allowed the gradual filling of the park by favoring high percolation, controlling water through channels, or by constructing a terraced design. A total of 66% of the parks connected upstream collectors, tanks, and pipes (either already existing or built for the purpose), although only 25% mention the use of an emergency spillway for safe overflow when the storage capacity is exceeded. Interestingly, 41% of the floodable parks incorporated a tank inside of the park which could be placed below the pond or next to it as a backup in case of the overflow of the pond. Also, 33% of the ponds integrated a pump system to increase aeration to prevent water stratification, and a 16% used mechanisms to avoid the deposition of suspended particles. The flooding was mentioned to last less than 48 h only in two case studies for which the full construction proposal was obtained, but we believe this can be generalized to the conceptual model of a floodable park [29].

Technical elements such as dikes, dams, and smart systems, as well as erosion and landslide prevention elements were also considered in the design of 41% of the floodable parks, although only 16% reported using existing infrastructure (i.e., tanks or the pipe network). The pond's bed was not specified in most of the cases, but discrepancies in the material and design used were found (e.g., reticular permeable material (25%) vs. impermeable bed (8%)).

In addition, differentiation between polluted (first flush) and non-polluted water was performed in 33% of the case studies, discharging the water at different authorized locations depending on the water quality.

Water Reuse

Regarding water reuse, 91% reported the reuse of treated water, either using the treated water for internal uses (i.e., irrigation of the park (58%) or pond recharge (33%)), or using this water for external uses, either irrigation (i.e., public green spaces) (16%) or street cleaning (33%). This water was also used for aquifer recharge in 25% of the case studies. One of the case studies mentioned the reuse of water for fire extinction.

3.3.2. Biodiversity

Promotion of Biodiversity

Interestingly, 66% of the case studies chose a wetland mimicking system within their ponds, and 41% favored the growth of riparian vegetation in the pond perimetry by controlling the O_2, nutrients, and sediment and reducing contaminants. Rendering a place for biodiversity was also considered in the design of the floodable parks: 33% placed nesting zones for birds, 16% gave space for amphibians, and 16% built pollination sites.

Regarding the selected vegetation, 41% of the case studies chose autochthonous plants and 16% mentioned avoiding invasive species. In 16% of the cases, rain-fed species were favored as well as those which had ecological value and the capacity to prevent erosion. However, just one mentioned the incorporation of flood-adapted species. Interestingly, one case mentioned taking measures (i.e., extensive management and gradients) for enhancing a heterogenous hydroperiod. While promoting biodiversity, 33% of the case studies mention the functionality of the vegetation (i.e., shadow, noise reduction, and aesthetic value) and 25% the creation of stepping stones in the city's green corridor, which shows how the design, the ES delivery, and the support for biodiversity are deeply intertwined.

Nuisance Species Management

Mosquitoes and toxic algae blooms are often a trade-off of the incorporation of stagnant water elements within cities. Therefore, floodable parks should be managed to decrease the presence of nuisance species. Mentioned management strategies for decreasing mosquito populations were the introduction of insectivorous animals (e.g., mosquitofish, bats, and birds) (25%) and the selection of vegetation which attracts predators of mosquitoes and other pests (8%). Regarding the prevention of toxic algae blooms, aeration is favored through the integration of a cascade of fountains in the design (8%), as well as the use of ultrasound emissions for O_2 production (8%) and circular water pumping (25%).

3.3.3. Recreational Elements

Recreational elements in floodable parks were not found to differ much from regular urban parks. Sports facilities were incorporated in 75% of the studies, game spaces and resting places in 33%, and picnic areas and a lecture zone in 16%. However, some elements in their designs highlight the multifunctional character of these parks, such as the presence of educative posts (50%), community gardens (41%), viewing points for contemplation and elevated corridors (41%), connection with other green areas of the neighborhood (25%), and the integration of cultural elements (25%). Existing features (e.g., church, school, and rivers) were used in three of the case studies as the framework of the project. One floodable park included a commercial zone within the park and another a pavilion or auditorium for performances. Another case study mentioned the use of the park as a botanical garden for educational purposes. However, in only two were mentioned the inclusion of warning elements so citizens could leave the park before major flooding. One case study mentions the use of floodable recreation infrastructure.

3.4. ESs and Benefits Delivered by Floodable Parks and the Association of These to the Identified Key Elements

None of the analyzed case studies used the concept of ESs to describe the floodable parks, although three mention the concept of NbSs. The innovation of this paper is to undertake an ESs perspective which could be used to compare and frame floodable parks together with other well-studied NbSs (e.g., green roofs, vertical gardens, bioswales, artificial wetlands, etc.). Therefore, the depicted benefits were translated into different ESs according to the EEA [26].

The main ESs, mentioned in 91% of the case studies, was water regulation. Recreation and community activities also played an important role as a service delivered by floodable parks, accounting for another 91%. Knowledge and educational value were highly considered, as 66% of the floodable parks incorporated informative panels or activities (e.g., tours and teaching) in their design in order to raise awareness about water-related hazards. Heritage services were delivered in 41% of the cases. Despite not being a main reason for implementing the floodable park, biodiversity as a benefit was mentioned in 66% of the case studies. A total of 25% of the parks highlighted the potential improvement of mental and physical health which can be obtained in these parks. Aesthetic value, social connection, and a connection to nature were described as benefits in two of the parks respectively. Remarkably, climate regulation, air purification, and noise reduction were not core services for which these parks were constructed. Lastly, although this aNbS can be connected to treatment plants for water reuse, only two case studies mentioned water circularity as a service that these parks can offer.

3.5. Processes and Influencing Factors Occurring in Floodable Parks

Floodable parks are aNbSs which normally rely on the presence of a retention pond in order to accommodate the excess of runoff water. Meanwhile, they provide additional benefits (aesthetic, recreational amenities, groundwater recharge, sub-potable water supply, and new habitats for wildlife). After the review of the case studies, it was observed that the retention pond acts as the main water-regulating element of floodable parks. Processes occurring therein were classified into (1) water detention and retention processes, (2) water purification processes, and (3) water reuse. However, there was no detailed explanation of how these processes allowed the ecological functioning of the floodable park, and therefore, these were further investigated through a second systematic literature review. Both the biological and physicochemical components of the three hydrological processes are described below, according to their respective states of the art obtained from this second review.

3.5.1. Water Detention and Retention Processes

One of the most important features to consider when designing the detention/retention pond (hereafter called pond) within a floodable park is the water residence time [30]. This has been recommended to be between 24 to 48 h by institutions such as the EPA and water agencies [31,32]. Ponds are designed to gradually release runoff water into downstream aquatic ecosystems via horizontal flow, thus decreasing potential downstream soil erosion and flooding. However, when the water residence time is not correctly determined when designing these ponds, runoff water cannot be efficiently discharged and is held for a longer time than required. This causes the water to rise beyond the designed level, eventually producing higher pond outflows [33].

Three factors of pond design are crucial for a pond to correctly manage excess water. The first factor is the distance between the inputs and outputs, which has to be maximal in order to allow the water to flow and sediments to settle [34]. The second is the correct orifice size of these inputs and outputs. Orifice size has to consider the fluctuations of the runoff inflow and discharge, as well as the pond drainage rate [33,35–37]. The third feature of stormwater management practices is size [35,37,38]. Water retention efficiency relies on a sufficiently large area (2–5 % of the watershed area) [39] which can hold the runoff water below a certain depth (i.e., 1.5–3 m) [33,37,40]. In fact, small surface pond areas hinder mixing occurring by the wind, causing stratification and reduced water residence time, which results in ecological damage and increased greenhouse gas [36].

Another parameter which determines the water retention is the permeability of the pond's bed [35]. The higher the permeability (e.g., sandy silt, gravel and sand), the higher the groundwater–surface water interactions and the higher the water residence time [36]. It is of foremost importance that the pond's bed remains uncompacted and that proper maintenance (i.e., removal of leaf litter, revegetating bare areas, and the minimization of soil disturbance during construction) is performed [35].

The imperviousness of the surroundings also influences the performance of wet detention ponds. Urban aquatic ecosystems are usually subjected to rapid increases and decreases in water levels (i.e., flashy hydrology) due to a lack of permeable surfaces. Therefore, infiltration and evapotranspiration of the surroundings are two main processes that determine the correct performance of retention ponds [33]. Infiltration and evaporation within the pond also condition the water residence time.

Lastly, the water residence time determines not only the flow attenuating efficiency, but also other processes such as water purification (see Section 3.5.2). The longer the residence time, the more effectively coarse particles and particulate-bound pollutants are removed via gravity sedimentation [30,34–36,41] and the more time biological communities have to degrade or uptake these contaminants [30]. Therefore, longer retention times have been found to promote more adsorption, while higher groundwater seepage rates have been linked to microbe export from biofilters, decreasing removal efficiency [42].

The water residence time conditions the species living in the pond as well. Slower rates of water level decline have been positively correlated to native obligate tree and native shrub coverage [33].

3.5.2. Water Purification Processes

Urban aquatic ecosystems receive incoming runoff water from their surroundings. This water carries various sources of nitrogen and phosphorous (e.g., fertilizers, lawn and garden waste, and pet waste) and minerals (e.g., calcium and sodium bicarbonate), which respectively increase nutrient loadings as well as pH and conductivity. This consequently changes the soil and sediment's chemistry and the present biological communities [41]. Other pollutants found in the runoff water include suspended solids (e.g., erosion and litter), hydrocarbons (e.g., fuels), metals (e.g., roads), and even fecal coliforms (e.g., humans and animals) [43–45]. Wet retention ponds have been demonstrated to significantly remove contaminants apart from attenuating the runoff peak [43,46].

Physicochemical Water Purification

Pollutants of different natures are removed from the water by precipitation or binding to sediment particulates and soluble organics (i.e., heavy metals) [47,48], adsorption and ion exchange via soil filtration (i.e., ammonia and phosphorus) [35], volatilization (i.e., NH_3 and hydrocarbons) [27], and photochemical oxidation (hydrocarbons) [43]. The physicochemical water purification focuses mainly on the processes of sedimentation and filtration, by which organic matter, metals, and suspended solids are deposited and trapped in the sediment layer and in the infiltration substrate, purifying the water [34,49]. Other physicochemical processes that occur in these ecosystems are straining, mixing, dilution, and aeration [49].

Pollutants in the runoff water include (1) dissolved pollutants and (2) pollutants associated to sediment particles that accumulate in the waterbody's bed. Different processes are required for the removal of each type of pollutant. While sorption and biodegradation (see Section "Water Detention and Retention Processes") are the dominant processes for eliminating soluble pollutants like biocides and pharmaceuticals [50], sedimentation is the primary physicochemical mechanism by which pollutants are removed from the water column [37,43,44]. According to the EPA, retention ponds have a removal efficiency of 70–90% of total solids, 50–70% of total phosphorous, 30–50% of total nitrogen, and 50–80% of metals [51].

The efficiency of sediment removal depends on factors such as the type, size, shape, and density of the soil particles [50]. Vopicka (2009) [43] found that, contrary to sand, silt and clay are positively associated to metal and heavy-end hydrocarbons due to the negative charge of clay particles to which these contaminants adhere. The size of the clay particles is also relevant, since pollutants concentrate in clays with smaller particles. This should be considered in the design of the pond, since sand normally deposits on the inlet and more loamy soils (with contaminants) may accumulate further in the pond, where dredging is more difficult.

Factors such as temperature can decrease the efficiency of the removal. In the study of a novel solution consisting of chained bioretention cells and a retention pond, Pineau et al. (2021) [46] reviewed the effect of temperature on metal removal and concluded that, although with sufficient percolation capacity, the temperature should not affect total metal removal, the removal of some heavy metals (e.g., zinc (Zn), lead (Pb) decreased with a lower temperature. Also, they observed that the influx of de-icing salts alters the transformation of metals in the soil of bioretention and retention ponds, increasing their dissolution in water and discharge in the efflux. However, the negative effect of salts on the removal of metals in submerged conditions was lower compared to the unsubmerged conditions of the bioretention cells, which highlights the importance of the permanent water presence of retention ponds.

Biological Water Purification

Although sedimentation can remove coarse and heavy particles with attached pollutants, biological processes are required to remove lighter particles and dissolved fractions [44]. These fractions are usually the most nutrient-rich and contain high metal concentration [37]. In addition, already settled matter and retained nutrients might, over time, be released from the pond, due to wind mixing, mineralization of organic particles, and/or transformations into a reduced state. Therefore, biological purification should be encouraged.

The biological component of ponds entails mainly microbial and vegetative processes [44]. Three major pollutant removal mechanisms are (1) nutrient uptake, (2) microbial decomposition and breakdown of organic matter on floating mats and plant root systems, and (3) filtering of sediment and associated pollutants by root systems [52]. Biological water purification also includes processes such as oxygen transfer [53].

The microbial mineralization of organic elements (phosphorus, nitrogen, and carbon) allows the energy and matter transfer to the primary producers [54]. In fact, phosphorous removal relies mostly on uptake by vegetation. Phosphorus transformations (e.g., from unreactive to reactive forms, and from dissolved into particulate forms) determine its bioavailability and therefore the phytoplankton and plant uptake capacity [55], showing how deeply intertwined purification processes of different natures (e.g., physicochemical and biological) are. Dissolved inorganic nitrogen species, especially ammonium (NH_4^+), are considered the most available for immediate assimilation by aquatic microorganisms and plants. However, this uptake is often seasonal, and the stored nutrients are released as dissolved and particulate organic nitrogen after senescence. Long-term removal of nitrogen includes biological processes such as ammonification (mineralization of organic nitrogen into ammonia and ammonium, NH_3/NH_4^+), nitrification (oxidation of NH_3/NH_4^+ into nitrate, NO_3^-), and denitrification (aerobic reduction of NO_3 into nitric oxide (NO), nitrogen gas (N_2) and nitrous oxide (N_2O) [47]. On the other hand, hydrocarbons can be biologically degraded through processes such as fermentation and aerobic/anaerobic respiration, although due to the hydrophobic nature of the latter, sedimentation and sorption of particulates dominate as the main removal mechanism [43,44].

Benthic infauna also influence biogeochemical cycles and nutrient flux directly through feeding, respiration, secretion, and excretion, and indirectly through bioturbation, bioirrigation, and resuspension [56]. These processes alter the redox potential and oxygen content in the soil (creating oxic and anoxic zones), promote solute exchange, and contribute to the sequestration and transformation of nutrients and carbon [56]. They can also alter the biotic component of the sediment by feeding on microbes and introducing gut bacteria through excretion.

An important influencing factor in these processes is the oxygen content in the water and sediments, since aerobic and anaerobic bacteria target different pollutants and act at different biodegrading rates. In aerobic ponds, algae growth can increase the oxygen in the water column, which is later consumed by aerobic bacteria [27]. Oxygen content in the water and sediment can be altered through the modification of the water residence time, the type of aNbS (e.g., vertical versus horizontal flow constructed wetlands) and the presence of plants.

Vegetation can be used as a phytoremediation technology to reduce, degrade, and immobilize environmental toxins and thus restore the area [35]. Plant roots embody the most important treatment component of the vegetative system. In fact, it was found that nutrient removal was highest in ponds that had maximum contact with plant's rhizomes [34]. The roots not only uptake contaminants, but also provide an environment where microorganisms can perform their biodegradation processes (e.g., denitrification for N removal) [34,57]. Therefore, the roots and biofilm perform adsorption, absorption, and uptake processes in ponds, contributing to its purification.

The efficiency of plants as pollutant removals depends on their size, species, and root mass. Floating wetlands have been studied to enhance nutrient uptake through assimilation and promote nitrogen removal via denitrification; emergent aquatic plants can decrease phosphorus, and submerged aquatic plants are better for decreasing minerals and therefore the pH in the pond's water [41]. Macrophytes with voluminous root systems (e.g., *Phragmites australis*) have been recommended for water-based NbSs such as ponds and constructed wetlands [58], and therefore can be applied for floodable parks as well. Other species that can be used for this purpose are, e.g., *Carex virginata* and *Juncus edgariae*, since they have been observed to develop greater biomass than other species (e.g., *Eleocharis acuta* and *Schoenoplectus tabernaemontani*) and accumulate metals present in the runoff water. The contaminated biomass can be later removed via regular harvesting. To this end, Floating Treatment Wetlands (FTWs) show an advantage compared to regular Constructed Wetlands (CWs), since the latter requires a complete excavation of the soil which is a rather invasive operation [59].

3.5.3. Water Reuse

Water scarcity is another of the consequences of climate change. The growing human population exacerbates water stress by increasing water demand and groundwater abstraction. In the past decade, stormwater reuse has been considered in urban areas in order to more efficiently utilize the available water resources [60,61]. Different aNbSs, such as artificial wetlands, permeable surfaces, bioswales and rain gardens have demonstrated their capacity for water treatment and purification for its subsequent reuse [61]. However, further treatment is needed for the utilization of stormwater infiltrated and retained in the retention ponds. Several impediments hinder its reutilization, like the aforementioned presence of oils, litter, heavy metals, nutrients, sediments, organic matter and bacteria, which may persist despite the physicochemical and biological purification [61–63]. Pathogenic microorganisms have been identified as the main concern regarding stormwater reuse, since it directly impacts human health if users have contact with untreated urban stormwater [63].

Natural disinfection through processes such as filtration, sedimentation, dehydration, and sun irradiation can provide a certain level of runoff treatment. However, the removal of pathogens achieved by NbSs does not reach the national standards of EU countries in which floodable parks have been implemented (e.g., Spain). In addition, although natural filtration can remove fecal bacteria, this might percolate into the groundwater [45]. Further disinfection by UV, chlorination, or ozonation is a necessary final step before the reuse of reclaimed water [27]. Certain elements can be included in the design of the aNbS, such as biofilters, sand-gravel filter [38], biochar, and real-time control (RTC) [37], in order to increase the suitability of the water quality for harvesting and reuse [61]. Once treated, the water can serve multiple purposes such as irrigation, cleaning and cooling of public spaces, fire hydrants, replenishment or development of artificial lakes, and even potable water [61]. Ponds can also act as a potential reservoir of treated water [61] and collected rainwater can be used directly, since the level of pathogens is considered lower, and therefore its quality is good enough without any further treatment [64]. Nevertheless, risk management and financial assessment for stormwater reuse is needed for the upscale of this type of solution.

Furthermore, retention ponds have also been studied for energy recovery using the stored volumes after urban floodings. However, the efficiency of this system for this purpose depends on the total runoff, the timing of the runoff, and the topography and geometry of the retention pond. Thus, this type of water reuse is usually recommended for countries with abundant rainfall [65]. Low heads (small waterfalls or systems with non-negligible flows) require hydropower converters with significantly low costs of installation and maintenance, although the higher the head, the better the production of energy (for the same volume). A threshold volume is required for the system to operate correctly and design considerations (e.g., outlet diameter) can alter the efficiency [65]. Nevertheless, Ramos et al. (2013) [66] investigated a retention pond in which 210 MW h/year could be produced (for an average year hourly power of 800 W).

3.6. Conceptual Model of Floodable Parks

A conceptual model was developed after the identification of the main processes and key elements of the design and its relationship with the ESs delivered by the floodable parks of the case studies (Figure 3). Although the main attribute of the floodable parks was usually a permanent water body, the surrounding vegetation (where in fact other NbSs were incorporated, such as rain gardens, vertical gardens, or dry detention basins) and paved surfaces (i.e., permeable pavements) were fundamental for the correct functioning of the floodable parks. Wetland mimicking systems were incorporated in most of the reviewed cases, supporting the identified (and further reviewed) processes that take place within a retention pond.

It was observed from the case study review that water retention and water purification were intimately related to the presence of an aNbS, which allowed the gradual filling of the park by enhancing water infiltration. However, aNbSs were not enough, so technical elements were implemented (e.g., upstream or onsite tank, dams or erosion-prevention mechanisms, channels and pipes, smart systems for water circulation, or a reticular pond bed) to further reinforce and improve the water retention and water (physicochemical) purification processes.

The final aim of these processes was the subsequent water reuse. Thus, water from the tanks and water from the ponds (either being first forwarded to a treatment plant or not) was used for internal or external uses. Internal uses were normally the irrigation of the vegetation of the park and the recharge of the pond itself, while external uses included irrigation of public spaces, street cleaning, and fire extinguishment. Water reuse, despite not being a tangible element as the aNbS and technical elements described above, was key to the design of the reviewed floodable parks, and therefore, was included by the authors in the conceptual model as part of the water management of the parks. Water reuse was found to be linked to the process of water retention, since this determines the availability of water for other uses, and most importantly, water purification (biological and physicochemical). However, for this water to be utilized, normally a further treatment was needed to reach the quality standards for the aforementioned uses so the water was conveyed to wastewater treatment plants.

All the abovementioned elements and processes contributed to the water-regulating ecosystem service delivery of the park.

Moreover, biodiversity was a key element of the floodable park design. Biodiversity, which was supported in the different aNbSs of the park and in the pond, has been studied to favor water regulating processes, as described in Section "Biological Water Purification". Hence, it should be promoted by the selection of the correct vegetation species (i.e., autochthonous, non-invasive, climatically-adapted, and supporting water regulating processes) and the creation of refugia (i.e., nesting places and pollination sites), which increase the biological connectivity of the city. This enhancement of biodiversity was observed in most of the case studies. On the other hand, flooding might increase the presence of nuisance species (i.e., mosquitoes and toxic algae). Therefore, different attributes were used in the floodable parks to reduce them. Natural attributes included the attraction of insectivorous species for biological control, whereas technical attributes were water circularization and ultrasound emissions for increasing O_2 concentration in the water. Biodiversity was portrayed in the conceptual model both as a co-benefit and a disservice, since it promotes services but can potentially affect human health through the increases of these species.

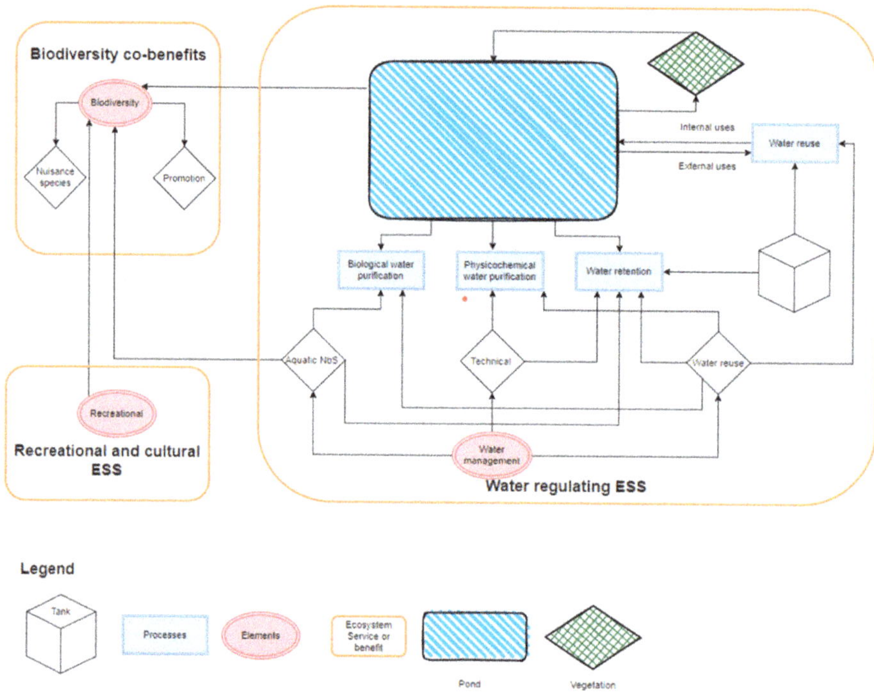

Figure 3. Conceptual model of a floodable park. The blue striped rectangle represents the pond and the green checkered diamond the permeable surfaces present in the park. The chosen processes are represented as blue rectangles, the identified key elements of the design of floodable parks are represented as red circles, and further divisions of both are represented as white diamonds. A water tank is shown as a cube. All these elements are grouped according to the main identified ecosystem services. Connections between each other are represented with arrows.

Biodiversity was interconnected with recreation elements in our model. For example, vegetation was used for several purposes (e.g., shadow, noise reduction, and air purification), which improved the conditions for relaxation, sports, and leisure. Connectivity with other green areas of the neighborhood increased the spatial transference of these services. On the other hand, biodiversity also promoted cultural and social services through the creation of educational and community gardens. Floodable parks had to incorporate recreational elements in their design, which are adapted to inundation. These were, for example, elevated corridors or floodable infrastructure. Also, warning systems warned citizens to leave the park before a major inundation.

4. Discussion

Humans have altered the natural topography of landscapes in such a way that common meteorological phenomena such as precipitation are now recurrently causing disruptions in our society. Urban areas have been designed to discharge pluvial runoff but, as it was found in the review of the case studies, existing water management infrastructure is currently not capable of withstanding the increasing volumes of longer and more frequent extreme precipitation events. Therefore, it has recently been acknowledged that, instead of building our metropolises to be impermeable, cities have to start welcoming water into their designs. This need is reinforced by the effects of the opposite climatic impact of global change, drought. Water is becoming a scarce resource and thus, any opportunity for reusing it needs to be undertaken.

Floodable parks are a novel solution which were conceived within this new conception of cities. In the introduction, these parks were defined as recreational spaces to which pluvial water is conveyed into a stagnant waterbody for its retention, infiltration, and drainage and which can be temporarily flooded. However, after the review of the case studies and the processes taking place within their floodable parks, noteworthy importance has to be given to the pervious areas that surround the permanent pond. Although they do not capture the main mass of water entering the park, these areas are fundamental for allowing the gradual filling of the pond and decreasing the peak flow volume which may otherwise be discharged downstream with fatal consequences. In our conceptual model, these pervious areas were pictured as vegetation, but they include several aNbSs such as dry basins, rain gardens, wetland mimicking systems and permeable pavements. Thus, floodable parks can be re-defined as "recreational spaces to which pluvial water is conveyed into a stagnant waterbody for its retention and drainage and which can be temporarily flooded as a result of the infiltration occurring in the surrounding aNbS".

These findings align with prior work on detention ponds, where it is recommended that detention ponds incorporate other stormwater control measures either (a) upstream of the detention pond (e.g., bioretention swales and other infiltration practices for water infiltration), (b) along the pond (e.g., forebays and vegetated buffer strips to decrease nutrient content in the water, sediment input, and water erosion), and (c) within the pond (e.g., floating wetlands to enhance nutrient uptake and nitrogen removal via denitrification) [41,54]. Hancock et al. (2010) [33] state that, amongst others (i.e., under-sizing), reduced infiltration and evapotranspiration hinder the correct functioning of retention ponds, sustaining the need for additional elements in their surroundings to improve their water regulation. They also write that the proportion of green areas around these solutions are frequently overestimated, which creates a mismatch between the calculated volume that the pond can hold and the observed one. Integrating other aNbSs in the surroundings of the pond would assure that the park can adequately manage current and future runoff by increasing the area in which the water is accommodated and thus, reducing the runoff peak flow and discharge through infiltration.

Several studies have demonstrated that retention ponds alone cannot meet the water-mitigating requirements. Hess et al. (2022) [41] studied stormwater wetlands which were either connected or not to detention ponds to see whether the discharge was slower when these two aNbSs were connected. Although the water level decline was 11% slower when both were connected, it was still 13% faster than non-urban wetlands, suggesting that detention ponds alone cannot recreate pre-development hydrological dynamics and thus lessen urban impacts on downstream ecosystems. Hancock et al. (2010)'s findings corroborate the results of similar studies in which retention ponds, regardless of their size and design, are observed to exceed the calculated outflows, again recognizing their inefficiency in reducing overall runoff volume to pre-development conditions. Thus, the surrounding of the pond by aNbSs is indispensable for the pond within the floodable park to function correctly. Further research on the processes occurring in floodable parks should be performed in order to understand, in an integrative way, the intertwined functioning of both the retention pond and the surrounding aNbSs.

Other studies have observed the inadequacy of retention ponds to remove the dissolved pollutants of the runoff water [47]. Although different elements (e.g., filters) can be incorporated into the design of the ponds in order to improve the retention of the dissolved pollutants [38], the microbial degradation and uptake of these pollutants by vegetation should be enhanced. Hence, biodiversity should be an intrinsic element within the pond's design. Many of the selected case studies incorporated a wetland mimicking system. Andradótir et al. (2014) [67] agree that ponds with wetland mimicking systems, such as those found in the studied floodable parks, are an ideal solution, since both the removal of particulate species through gravitational settling and the uptake of dissolved nutrient species by plants and algae take place. This is of foremost importance for those species and heavy metals (e.g., Zn and Cu) which have both dissolved and particulate phases and for which only physical methods are not sufficient [38,59].

Despite the advantages of incorporating natural elements in the design of aNbSs, it was observed in this study that it was normally not a factor driving the construction of the studied floodable parks. However, their incorporation provided benefits beyond water regulation (e.g., shadow, noise reduction, aesthetic value, and social interaction), increasing recreation and sense of belonging. These findings align with those of Jabbar et al. (2021) [68], who observed that UK citizens considered park trees important not only for aesthetic beauty and wildlife habitat but also for flood protection, pollution reduction and climate regulation, and 97.7% of them considered the biodiversity of trees essential in urban parks. However, the selection of the plant species has to consider both local climate adaptation and flood adaptation, which can be challenging in arid regions.

One proposed solution which may fit within the floodable park concept is Floating Treatment Wetlands (FTWs). FTWs, as their name indicates, are emergent macrophytes grown hydroponically on the surface of the pond. FTWs have several advantages over the traditional Constructed Wetlands (CWs), as McAndrew et al. (2016) describe [69]. In the first place, in CWs, the vegetation is rooted in the sediment, limiting the root exposure and therefore the nutrient capture. On the other hand, in FTWs, nutrients can be directly taken up from the water column. In the second place, the suspended root matrix of the FTW can provide a large surface area for the growth of biofilms where microbial transformations of pollutants (e.g., denitrification) can take place, and increase the sediment capture function by physically trapping suspended sediments and enhancing the flocculation and precipitation of particles. Despite these suspended roots occupying space on the water column, FTWs do not detract from the flood storage capacity of the pond, a main issue when designing wetlands within retention ponds [69]. In the third place, aerial and root tissues that have accumulated nutrients and pollutants can be easily harvested without damaging the benthic system of the pond. In the fourth place, shore-rooted vegetation suffers flooding stress and is prone to be flooded. However, the buoyant nature of the FTW allows water levels to change with no risk of threatening the plant's survival. This is especially relevant in the case of floodable parks and thus, it is suggested that it should be adopted in their design. One more advantage of FTWs is that they can be implemented inexpensively and on retrofitted parks, and therefore they can increase the performance of the urban ponds without significant land acquisition or earth removal [47,69].

In fact, one of the major challenges of urban planning nowadays is the lack of space for accommodating water and giving space to nature. One disadvantage of floodable parks is indeed their size. Other NbSs (e.g., green roofs and rain gardens) are often preferred to wet retention basins due to their easier and less costly implementation [27]. Therefore, retrofitting former urban parks into floodable parks should be considered as a way to optimize the available urban space. Of all 12 case studies, two parks, namely the Enghaven Park (Copenhagen) and Maple Park (Taiwan), were retrofitted from former urban parks, pioneering the upgrade of cities into more multifunctional spaces. Lourenço et al. (2020) [1] advocate for the multifunctionality of spaces in order to compact cities and increase their ecological performance. However, it is important to study the trade-offs that may arise between the different functions and services that these spaces provide.

Lastly, emphasis should be placed on the importance of water reuse, adding a layer to the multifunctionality of floodable parks. Floodable parks differ from other aNbSs in their intimate relation with water treatment facilities, as has been observed in this review paper. Tsatsou et al. (2023) [27] states that the circularity of NbSs can be enhanced through their integration with existing grey infrastructure, which treats water for its future reuse. Depending on its purpose of reuse, water quality standards vary. An interesting thought of this paper is to keep a certain concentration of nutrients in the water, since normally this reclaimed water is used for the irrigation of urban green spaces or urban agriculture, and thus the fertigation practices recycle these nutrients and reduce the need for additional fertilizer. However, this practice has the disadvantage that, since the fertilizers are not released slowly (as solid fertilizers are), there is an increased risk of the nutrients reaching the groundwater. The supply of water from floodable parks as a sustainable alternative of non-potable water is of high interest, but the practical details of such a connection (runoff areas, water quality standards, and technical design) should be better studied.

5. Conclusions

Urban areas are increasingly facing flooding which they are incapable of mitigating. The existing water management infrastructure is now frequently overflown (even under small precipitation events) due to the combined effects of climate change and urbanization. City design rarely incorporates permeable zones where runoff water can be infiltrated, nor gives enough space for water to be retained until it can be drained naturally. Floodable parks are unique solutions that combine stormwater management with natural processes to create a multifunctional space where water can be welcomed. This review of twelve floodable parks which were constructed for pluvial control gives insight on the rising needs for multifunctional recreational areas. The parks were constructed in low-land zones where both land modification and inadequate water infrastructure were exacerbating the impacts of floods. The parks were designed not only to serve for leisure, but also provide space for water retention and habitats for biodiversity. Remarkably, most of them were connected to treatment plants where the runoff water was treated for further reuse. This highlights the increasing importance of the utilization of all available water resources. However, it is of foremost importance to understand the trade-offs between the different functions provided by floodable parks in order to develop a design that can maximize its services and benefits. Our conceptual model revealed the need of incorporating vegetated aNbSs surrounding the main aNbS, the retention pond, in the design of floodable parks. Future research on the interdependency between these two types of aNbSs is needed in order to comprehend their mutual role on water runoff mitigation and purification. Cities have to adapt to the multiple challenges which global change is posing and will pose in the future, and floodable parks can help the transition towards more sustainable and resilient cities.

Author Contributions: S.M.M.: conceptualization, data curation, formal analysis, investigation, methodology, software, visualization, writing—original draft, writing-review and editing. S.E.: conceptualization, methodology, software, supervision, visualization. J.S. (Jan Staes): conceptualization, supervision, funding acquisition. J.S. (Jonas Schoelynck): supervision. All authors have read and agreed to the published version of the manuscript.

Funding: This research was developed within the project BiNatUr: Bringing nature back—biodiversity-friendly nature-based solutions in cities and funded through the 2020–2021 Biodiversa and Water JPI joint call for research projects, under the BiodivRestore ERA-NET Cofund (GA N°101003777), with the EU and the funding organizations the Academy of Finland (Finland), Bundesministerium für Bildung und Forschung (BMBF, Germany), Federal Ministry of Education and Research (Germany), National Science Centre (Poland), Research Foundation Flanders (fwo, Belgium), and Fundação para a Ciência e Tecnologia (Portugal).

Data Availability Statement: For the data supporting the reported results, contact the contributing author.

Conflicts of Interest: The authors declare no conflicts of interest.

Appendix A

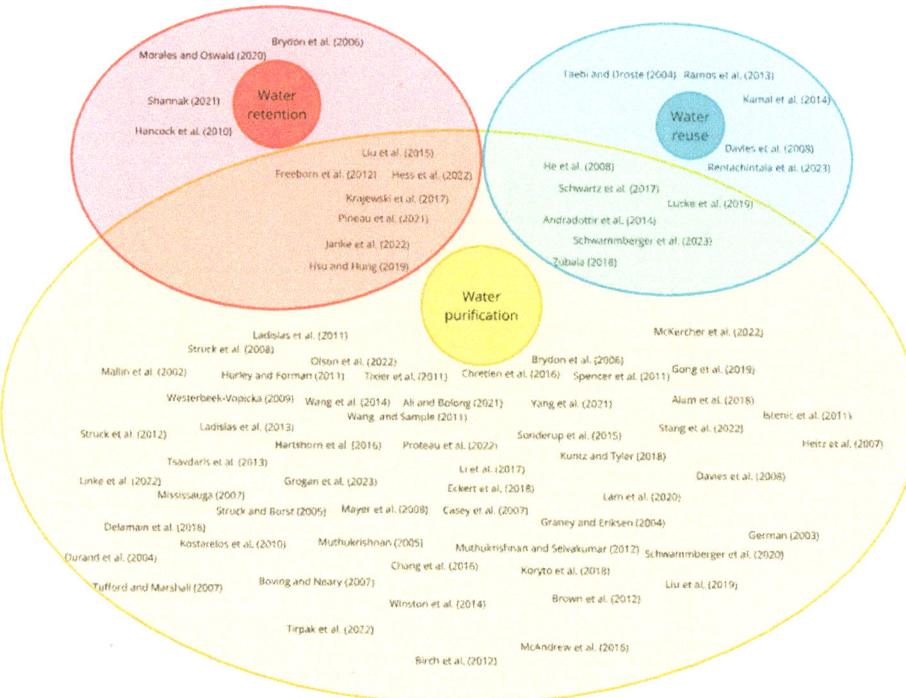

Figure A1. Scheme of reviewed papers according to the described water process. Papers between two circles belong to the two categories which these represent.

References

1. Lourenço, I.B.; Beleño de Oliveira, A.K.; Marques, L.S.; Quintanilha Barbosa, A.A.; Veról, A.P.; Magalhães, P.C.; Miguez, M.G. A framework to support flood prevention and mitigation in the landscape and urban planning process regarding water dynamics. *J. Clean. Prod.* **2020**, *277*, 122983. [CrossRef]
2. Matos Silva, M.; Costa, J. Flood Adaptation Measures Applicable in the Design of Urban Public Spaces: Proposal for a Conceptual Framework. *Water* **2016**, *8*, 284. [CrossRef]
3. Hirabayashi, Y.; Mahendran, R.; Koirala, S.; Konoshima, L.; Yamazaki, D.; Watanabe, S.; Kim, H.; Kanae, S. Global flood risk under climate change. *Nat. Clim. Change* **2013**, *3*, 816–821. [CrossRef]
4. Guerrero, P.; Haase, D.; Albert, C. Locating Spatial Opportunities for Nature-Based Solutions: A River Landscape Application. *Water* **2018**, *10*, 1869. [CrossRef]
5. Vörösmarty, C.J.; McIntyre, P.B.; Gessner, M.O.; Dudgeon, D.; Prusevich, A.; Green, P.; Glidden, S.; Bunn, S.E.; Sullivan, C.A.; Liermann, C.R.; et al. Global threats to human water security and river biodiversity. *Nature* **2010**, *467*, 555–561. [CrossRef]
6. Cremades, R.; Sanchez-Plaza, A.; Hewitt, R.J.; Mitter, H.; Baggio, J.A.; Olazabal, M.; Broekman, A.; Kropf, B.; Tudose, N.C. Guiding cities under increased droughts: The limits to sustainable urban futures. *Ecol. Econ.* **2021**, *189*, 107140. [CrossRef]
7. Pinho, P.; Haase, D.; Gebler, D.; Staes, J.; Martelo, J.; Schoelynck, J.; Szoszkiewicz, K.; Monaghan, M.T.; Vierikko, K. Urban Aquatic Nature-Based Solutions in the Context of Global Change: Uncovering the Social-ecological-technological Framework. In *Introduction to Designing Environments: Paradigms & Approaches*; Springer International Publishing: Cham, Switzerland, 2023; pp. 139–157. [CrossRef]
8. Fluhrer, T.; Chapa, F.; Hack, J. A Methodology for Assessing the Implementation Potential for Retrofitted and Multifunctional Urban Green Infrastructure in Public Areas of the Global South. *Sustainability* **2021**, *13*, 384. [CrossRef]
9. Markou, D. Sustainable Stormwater Management at Urban Expansion Areas of Potential Significant Flood Risk. *IOP Conf. Ser.* **2022**, *1123*, 012019. [CrossRef]
10. Martín Muñoz, S.; Schoelynck, J.; Tetzlaff, D.; Debbaut, R.; Warter, M.; Staes, J. Assessing biodiversity and regulatory ecosystem services in urban water bodies which serve as aqua-Nature-based Solutions. *Front. Environ. Sci.* **2024**, *11*, 1304347. [CrossRef]

11. Gil Alonso, F.; Bayona Carrasco, J.; López Villanueva, C.; Pujadas Rúbies, I. Diferencias geográficas de la fecundidad en España: Una perspectiva provincial. *Papeles de Geografía* **2017**, *63*, 21. [CrossRef]
12. Le, T.Q.; Devisch, O.; Trinh, T.A. Flood-resilient urban parks: Toward a framework. *Area* **2019**, *51*, 804–815. [CrossRef]
13. Hobeica, L.; Hobeica, A. Parques urbanos inundáveis: Articulando adaptação às inundações e sensibilização para o risco. *Territorium* **2018**, *25*, 143–160. [CrossRef] [PubMed]
14. Veról, A.P.; Battemarco, B.P.; Merlo, M.L.; Machado, A.C.M.; Haddad, A.N.; Miguez, M.G. The urban river restoration index (URRIX)—A supportive tool to assess fluvial environment improvement in urban flood control projects. *J. Clean. Prod.* **2019**, *239*, 118058. [CrossRef]
15. Bertilsson, L.; Wiklund, K.; de Moura Tebaldi, I.; Rezende, O.M.; Veról, A.P.; Miguez, M.G. Urban flood resilience—A multi-criteria index to integrate flood resilience into urban planning. *J. Hydrol.* **2019**, *573*, 970–982. [CrossRef]
16. Sánchez-Almodóvar, E.; Olcina-Cantos, J.; Martí-Talavera, J. Adaptation Strategies for Flooding Risk from Rainfall Events in Southeast Spain: Case Studies from the Bajo Segura, Alicante. *Water* **2022**, *14*, 146. [CrossRef]
17. Sánchez-Almodóvar, E.; Olcina-Cantos, J.; Martí-Talavera, J.; Prieto-Cerdán, A.; Padilla-Blanco, A. Floods and Adaptation to Climate Change in Tourist Areas: Management Experiences on the Coast of the Province of Alicante (Spain). *Water* **2023**, *15*, 807. [CrossRef]
18. Perrotti, D. Toward an agentic understanding of the urban metabolism: A landscape theory perspective. *Urban Geogr.* **2022**, *43*, 1–11. [CrossRef]
19. Nordic Built Cities Challenge. The Soul of Norrebro. 2019. Available online: https://www.nordicinnovation.org/sites/default/files/inline-images/Soul%20of%20Norrebro_booklet.pdf (accessed on 25 May 2024).
20. Jacob, A.C.P.; Rezende, O.M.; de Sousa, M.M.; de França Ribeiro, L.B.; de Oliveira, A.K.B.; Arrais, C.M.; Miguez, M.G. Use of detention basin for flood mitigation and urban requalification in Mesquita, Brazil. *Water Sci. Technol.* **2019**, *79*, 2135–2144. [CrossRef]
21. Hsu, C.-H.; Hung, C.-Y. Sustainable development of climate change resources: About recycling of water resources in Maple Park, Taichung, Taiwan. *E3S Web Conf.* **2019**, *117*, 00016. [CrossRef]
22. Braae, E.; Riesto, S. Designing Urban Natures. Ambiguities in open space design on the threshold of climate disaster. *Krit. Berichte* **2018**, *13*, 45–60.
23. A Commitment to Innovation and Sustainability. Available online: https://lanzine.com/ (accessed on 10 April 2024).
24. Peinhardt, K. *Resilience Through Placemaking: Public Spaces in Rotterdam's Climate Adaptation Approach (No. 1/2021)*; Discussion Paper; Deutsches Institut für Entwicklungspolitik (DIE): Bonn, Germany, 2021.
25. Quach, J. A Resilient Ecological Development of Ala Wai Golf Course: Symbiosis Between Buildings, Communities and Urban Waterways. Doctoral Dissertation, University of Hawai'i at Manoa, Honolulu, HI, USA, 2021.
26. Weber, J.-L. Classification of ecosystem services (EEA) (UNCEEA/5/7). In Proceedings of the Expert Meeting on Ecosystem Accounting, Copenhagen, Denmark, 11–13 May 2011.
27. Tsatsou, A.; Frantzeskaki, N.; Malamis, S. Nature-based solutions for circular urban water systems: A scoping literature review and a proposal for urban design and planning. *J. Clean. Prod.* **2023**, *394*, 136325. [CrossRef]
28. Maćkiewicz, B.; Asuero, R.P. Public versus private: Juxtaposing urban allotment gardens as multifunctional Nature-based Solutions. Insights from Seville. *Urban For. Urban Green.* **2021**, *65*, 127309. [CrossRef]
29. Urban Green Up. D1.1-NBS Catalogue. Available online: https://www.urbangreenup.eu/insights/deliverables/d1-1---nbs-catalogue.kl (accessed on 18 April 2024).
30. Janke, B.D.; Finlay, J.C.; Taguchi, V.J.; Gulliver, J.S. Hydrologic processes regulate nutrient retention in stormwater detention ponds. *Sci. Total Environ.* **2022**, *823*, 153722. [CrossRef] [PubMed]
31. EPA. International Stormwater BMP Database. Available online: https://bmpdatabase.org/ (accessed on 11 April 2024).
32. Minnesota Pollution Control Agency. *Minnesota Stormwater Manual-Retention Ponds*; Minnesota Pollution Control Agency: St Paul, MN, USA, 2008.
33. Hancock, G.S.; Holley, J.W.; Chambers, R.M. A Field-Based Evaluation of Wet Retention Ponds: How Effective Are Ponds at Water Quantity Control? *JAWRA J. Am. Water Resour. Assoc.* **2010**, *46*, 1145–1158. [CrossRef]
34. McAndrew, B.; Ahn, C.; Spooner, J. Nitrogen and Sediment Capture of a Floating Treatment Wetland on an Urban Stormwater Retention Pond—The Case of the Rain Project. *Sustainability* **2016**, *8*, 972. [CrossRef]
35. Hurley, S.E.; Forman, R.T.T. Stormwater ponds and biofilters for large urban sites: Modeled arrangements that achieve the phosphorus reduction target for Boston's Charles River, USA. *Ecol. Eng.* **2011**, *37*, 850–863. [CrossRef]
36. Morales, K.; Oswald, C. Water age in stormwater management ponds and stormwater management pond-treated catchments. *Hydrol. Process.* **2020**, *34*, 1854–1867. [CrossRef]
37. Proteau, K.; Binesh, N.; Duchesne, S.; Pelletier, G.; Lavoie, I. Urban runoff quality and quantity control: A functional comparison of various types of detention basins. *Urban Water J.* **2022**, *19*, 1080–1092. [CrossRef]
38. Sønderup, M.J.; Egemose, S.; Bochdam, T.; Flindt, M.R. Treatment efficiency of a wet detention pond combined with filters of crushed concrete and sand: A Danish full-scale study of stormwater. *Environ. Monit. Assess.* **2015**, *187*, 758. [CrossRef]
39. Maryland Department of the Environment Maryland Stormwater Design Manual, Volumes I and II. Available online: https://mde.maryland.gov/programs/water/stormwatermanagementprogram/pages/stormwater_design.aspx (accessed on 15 October 2024).
40. Kadlec, R.H.; Wallace, S. *Treatment Wetlands*; CRC Press: Boca Raton, FL, USA, 2008. [CrossRef]

41. Hess, K.M.; Sinclair, J.S.; Reisinger, A.J.; Bean, E.Z.; Iannone III, B.V. Are stormwater detention ponds protecting urban aquatic ecosystems? A case study using depressional wetlands. *Urban Ecosyst.* **2022**, *25*, 1155–1168. [CrossRef]
42. Chandrasena, G.I.; Shirdashtzadeh, M.; Li, Y.L.; Deletic, A.; Hathaway, J.M.; McCarthy, D.T. Retention and survival of E. coli in stormwater biofilters: Role of vegetation, rhizosphere microorganisms and antimicrobial filter media. *Ecol. Eng.* **2017**, *102*, 166–177. [CrossRef]
43. Vopicka, K.W. Sediment Assessment of Stormwater Retention Ponds within the Urban Environment of Calgary, Canada. *Water Qual. Res. J.* **2009**, *44*, 81–91. [CrossRef]
44. Istenič, D.; Arias, C.A.; Matamoros, V.; Vollertsen, J.; Brix, H. Elimination and accumulation of polycyclic aromatic hydrocarbons in urban stormwater wet detention ponds. *Water Sci. Technol.* **2011**, *64*, 818–825. [CrossRef] [PubMed]
45. Olson, E.; Hargiss, C.L.M.; Norland, J. *Escherichia coli* levels and microbial source tracking in stormwater retention ponds and detention basins. *Water Environ. Res.* **2022**, *94*, e1675. [CrossRef] [PubMed]
46. Pineau, B.; Brodeur-Doucet, C.; Corrivault-Gascon, J.; Arjoon, D.; Lessard, P.; Pelletier, G.; Duchesne, S. Performance of green infrastructure for storm water treatment in cold climate (Canada). *J. Environ. Eng. Sci.* **2021**, *16*, 185–194. [CrossRef]
47. Wang, C.-Y.; Sample, D.J.; Bell, C. Vegetation effects on floating treatment wetland nutrient removal and harvesting strategies in urban stormwater ponds. *Sci. Total Environ.* **2014**, *499*, 384–393. [CrossRef]
48. Heal, K.V.; Hepburn, D.A.; Lunn, R.J. Sediment management in sustainable urban drainage system ponds. *Water Sci. Technol.* **2006**, *53*, 219–227. [CrossRef]
49. Zubala, T. Technical and natural conditions and operating efficiency of a municipal stormwater treatment plant. *Environ. Sci. Pollut. Res.* **2018**, *25*, 952–962. [CrossRef]
50. Liu, F.; Vianello, A.; Vollertsen, J. Retention of microplastics in sediments of urban and highway stormwater retention ponds. *Environ. Pollut.* **2019**, *255*, 113335. [CrossRef]
51. EPA. *National Pollutant Removal Performance Database*; United States Environmental Protection Agency: Washington, DC, USA, 2007.
52. Hartshorn, N.; Marimon, Z.; Xuan, Z.; Cormier, J.; Chang, N.-B.; Wanielista, M. Complex interactions among nutrients, chlorophyll-a, and microcystins in three stormwater wet detention basins with floating treatment wetlands. *Chemosphere* **2016**, *144*, 408–419. [CrossRef]
53. Li, J.; Liu, H.; Chen, J.P. Microplastics in freshwater systems: A review on occurrence, environmental effects, and methods for microplastics detection. *Water Res.* **2018**, *137*, 362–374. [CrossRef] [PubMed]
54. Tixier, G.; Lafont, M.; Grapentine, L.; Rochfort, Q.; Marsalek, J. Ecological risk assessment of urban stormwater ponds: Literature review and proposal of a new conceptual approach providing ecological quality goals and the associated bioassessment tools. *Ecol. Indic.* **2011**, *11*, 1497–1506. [CrossRef]
55. Yang, Y.-Y.; Asal, S.; Toor, G.S. Residential catchments to coastal waters: Forms, fluxes, and mechanisms of phosphorus transport. *Sci. Total Environ.* **2021**, *765*, 142767. [CrossRef] [PubMed]
56. Kuntz, K.L.; Tyler, A.C. Bioturbating invertebrates enhance decomposition and nitrogen cycling in urban stormwater ponds. *J. Urban Ecol.* **2018**, *4*, juy015. [CrossRef]
57. Schwammberger, P.F.; Tondera, K.; Headley, T.R.; Borne, K.E.; Yule, C.M.; Tindale, N.W. Performance monitoring of constructed floating wetlands: Treating stormwater runoff during the construction phase of an urban residential development. *Sci. Total Environ.* **2023**, *865*, 161107. [CrossRef]
58. Greenway, M. The Role of Macrophytes in Nutrient Removal using Constructed Wetlands. *Environ. Bioremediation Technol.* **2007**, 331–351. [CrossRef]
59. Ladislas, S.; Gérente, C.; Chazarenc, F.; Brisson, J.; Andrès, Y. Performances of Two Macrophytes Species in Floating Treatment Wetlands for Cadmium, Nickel, and Zinc Removal from Urban Stormwater Runoff. *Water Air Soil Pollut.* **2013**, *224*, 1408. [CrossRef]
60. Awang Ali, A.N.; Bolong, N. Determination of water quality in Universiti Malaysia Sabah (UMS), Kota Kinabalu: The effectiveness of stormwater management systems. *Mater. Today Proc.* **2021**, *46*, 1848–1854. [CrossRef]
61. Rentachintala, L.R.N.P.; Mutukuru Gangireddy, M.R.; Mohapatra, P.K. Stormwater reuse for water-sensitive city—Integrated analysis of urban hydrology for efficient alternatives of Amaravati city, India. *Water Sci. Technol.* **2023**, *88*, 3151–3167. [CrossRef]
62. Li, Y.C.; Zhang, D.Q.; Wang, M. Performance Evaluation of a Full-Scale Constructed Wetland for Treating Stormwater Runoff. *CLEAN-Soil Air Water* **2017**, *45*, 1600740. [CrossRef]
63. Andersson, E.; Langemeyer, J.; Borgström, S.; McPhearson, T.; Haase, D.; Kronenberg, J.; Barton, D.N.; Davis, M.; Naumann, S.; Röschel, L.; et al. Enabling Green and Blue Infrastructure to Improve Contributions to Human Well-Being and Equity in Urban Systems. *BioSci.* **2019**, *69*, 566–574. [CrossRef] [PubMed]
64. Carsten, D.; Antje, W.; Martina, D. Approval Of Technical Suds In Germany. In Proceedings of the 36th IAHR World Congress, The Hague, The Netherlands, 28 June–3 July 2015.
65. Kamal, N.A.; Park, H.; Shin, S. Assessing the viability of microhydropower generation from the stormwater flow of the detention outlet in an urban area. *Water Supply* **2014**, *14*, 664–671. [CrossRef]
66. Ramos, H.M.; Teyssier, C.; Samora, I.; Schleiss, A.J. Energy recovery in SUDS towards smart water grids: A case study. *Energy Policy* **2013**, *62*, 463–472. [CrossRef]

67. Andradóttir, H.Ó.; Vollertsen, G.E.G. Temporal Variability of Heavy Metals in Suburban Road Runoff in a Rainy Cold Climate. *J. Environ. Eng.* **2015**, *141*, 04014068. [CrossRef]
68. Jabbar, M.; Yusoff, M.M.; Shafie, A. Assessing the role of urban green spaces for human well-being: A systematic review. *GeoJournal* **2022**, *87*, 4405–4423. [CrossRef]
69. Tirpak, R.A.; Tondera, K.; Tharp, R.; Borne, K.E.; Schwammberger, P.; Ruppelt, J.; Winston, R.J. Optimizing floating treatment wetland and retention pond design through random forest: A meta-analysis of influential variables. *J. Environ. Manag.* **2022**, *312*, 114909. [CrossRef]

Disclaimer/Publisher's Note: The statements, opinions and data contained in all publications are solely those of the individual author(s) and contributor(s) and not of MDPI and/or the editor(s). MDPI and/or the editor(s) disclaim responsibility for any injury to people or property resulting from any ideas, methods, instructions or products referred to in the content.

land

MDPI

Article

Making Hidden Sustainable Urban Planning and Landscape Knowledge Visual and Multisensorial

Carlos J. L. Balsas

School of Architecture and the Built Environment, Ulster University, Belfast BT15 1ED, UK; c.balsas@ulster.ac.uk

Abstract: Many cityscapes and landscapes create reactions of awe due to their unexpected characteristics or uniqueness. First glance observations may not reveal or enable the understanding of all natural, social, economic, environmental, and cultural processes taking place at a given moment or, particularly, how those places have evolved and been transformed into being what they are today. Research helps to uncover hidden layers of reality and make sense of relationships among elements, dimensions, and processes. Distinct research styles of variable intensity and comprehensiveness are required to respond to the needs of multiple stakeholders. Quantitative research methods have become increasingly sophisticated, and while they serve to enhance one's knowledge of complex phenomena, they can also be difficult to understand. To what extent can urban planners and landscape architects utilize visual and multisensorial literacies to bring about more transparency and significance to territorial transformation processes? This paper analyzes how four educational and research programs implemented at the University at Albany built upon visual and multisensorial research methods to augment educational outcomes, bolster scholarly significance, and provide up-to-date knowledge to empower concerned stakeholders to help change unsustainable urbanization practices, social inequities and injustices in cities, car dependence and sedentary lifestyles, protect natural resources and livelihoods, while reducing environmental and social harm. The methods comprised literature reviews, self-reflections on the implemented programs, and analyses of participant evaluations of various outreach activities. The paper concludes with remarks for other researchers interested in making their pedagogical and scholarly activities more effective and rewarding for themselves and their intended beneficiaries.

Keywords: visual and multisensorial literacy; citizen science; city center sustainability; working landscapes; place phenomenology

1. Introduction

The process of transforming data into information and information into useful knowledge is often convoluted and complex. The wicked nature of constantly changing socioeconomic phenomena complicates the research processes even further. How can research help to streamline and disentangle hidden relations of meaning, power, and forces to bring additional professional and scholarly value to society and the governance of common affairs?

Modern research tends to be dominated by two main research paradigms which have co-existed for generations, with the qualitative genre benefiting from new sociological, anthropological, and ethnographic methodologies within a post-modern context. However, visual literacy is needed in the design and social sciences professions [1–3]. Spatial, psychology, and cognitive approaches to transformational processes, interactive installations, sensory tours, visualization, and virtual reality techniques have helped to understand such concepts as place attachment, biophilia, topophilia, biomimicry, and wellbeing [4–7]. Nonetheless, there is still a lacuna in how sustainable urban planning and landscape knowledge and professional practice ought to help resolve societal complexities.

The impetus for this article came from three main sources: (i) Sandercock's edited volume entitled *Making the Invisible Visible* [8]; (ii) Gaber and Gaber's important contribution to qualitative research scholarship *Qualitative Analysis for Planning and Policy: Beyond the numbers* [9]; and (iii) Forsyth's article on *Alternative Cultures in Planning Research* [10]. The former two are discussed in the literature review and the analytical framework sections, while the latter source is also utilized to structure the article's empirical reflections.

Various planning research cultures tend to co-exist and answer different questions via different performative styles and purposes. It is argued that the specialization versus generalist dilemma in planning education and practice has precluded greater reliance on visual and multisensorial literacies, methods, and tools. The "universal man" paradigm is now being characterized as having different needs depending on gender, age, cultural upbringings, traditions, societal norms, and worldviews. The machine-like and automated approach to societal co-existence and affairs might work well with inanimate objects, but not necessarily with human beings and their likes and dislikes, preferences, needs, and wants. Automation and, increasingly, artificial intelligence are known to invisibly promote efficiency, but often at the cost of jobs, human happiness, and collective flourishing [11].

The transition from a rational and mostly positivist paradigm to a post-Fordist and post-Modernist plural reality is occurring at different paces in society, education, workplaces, and the built and natural environments. However, some societal trends, which impact specific societal groups disproportionately, are so ingrained in western world lifestyles, work, and recreational habits that they tend to invisibly perpetuate social injustices and negative environmental externalities. Apparently and independently of the degree of scientific evidence presented through different means and channels, in certain societies, established behaviors, rules, and transferred procedures tend to encourage increased consumption, which produce higher environmental footprints, more greenhouse gas emissions, and higher overall climate change pressures over finite natural resources and the carrying capacity of the planet [12–14].

Therefore, this paper's purpose is to analyze the extent to which four educational and research programs conducted at SUNY Albany in the U.S. Northeast [15] have contributed to ameliorate the abyssal divide between quantitative and qualitative research paradigms in the design, conduct, writing up, and dissemination of urban and landscape planning findings, as well as helping students and community members become more attuned to their own professional and societal interests and responsibilities, while enabling a wider breadth and more in-depth acquisition, comprehension, and deployment of job-ready skillsets and community-oriented attainments. In short, the paper aims to analyze how the four educational and research programs made invisible sustainable urban planning and landscape knowledge much more visual and multisensorial, to the point that it has influence in the lives of students, scholars, and community stakeholders.

It is argued that the four programs comprised the utilization, analyses, and the application of not only quantitative methods, but they also made greater use of visual and multisensorial literacies, typical of a qualitative methods genre, which helped to bolster greater understandings of the visual and multisensorial functions (i.e., sight, hearing, smell, taste, and touch) through direct and explicit engagements with the studied realities. The methods comprised mostly literature reviews, self-reflections on the four implemented programs, and participant evaluations of some of the activities undertaken. The paper concludes with remarks for other researchers interested in making their pedagogical and scholarly activities more effective and rewarding for themselves and their intended beneficiaries.

2. Literature Review

Planning scholarship and professional practice are the result of a myriad of academic and professional experiences, acquired knowledge, learned methodologies, personal and collective reflections, perceived contexts, and interpretations of reality. When one considers the length of a whole professional career, the planning knowledge acquired in college is just a small part of the accumulated information encountered during one's existence. However, the capacity to think and act to accomplish a task, or a series of tasks, has been put forward as one of the most important cognitive functions human beings will have to perform during their lifetime. Imagining a better world is central to assuring one's own survival in it. Within this utopian perspective, Vieira has identified these four modes of thinking: prospective thinking, critical thinking, holistic thinking, and creative thinking [16].

We contend that common to these four modes of thinking is the existence of means and ends needed to process a thought and, for instance, to accomplish a task or to decide not to act on a certain thought. A means-and-ends matrix was developed to help professional planners articulate their actions under conditions of uncertainty [17]. The existence of knowledge, or its lack thereof, tests how planners make decisions to deal with challenges and to resolve a specific problem. Depending on the means available and the methods employed in the pursuit of a specific goal, the anticipated ends are likely to be accomplished within a specific degree of certainty. Resourcefulness and the capacity to adjust goals will also tend to influence planning implementation. Given urban and regional planning's complex professional genealogy, evolution, and purposes in society, planning scholarships attempt to use a myriad of means to help accomplish consensual goals, which in the US are variously equated with the common interest or the public good.

To many community members, this concept is invisible and riddled with contradictions that are difficult to operationalize in ways that can be useful not only to public officials but also to the general population in their everyday tasks and decisions. Gaber and Gaber pioneered the use of visual resources in multicultural urban planning and qualitative methods [18], while Apostol et al. bolstered the study of scenic values in landscape architecture and planning [19]. Dühr has integrated the visual language of spatial planning via cartographic representations and visualizations of spatial planning processes in Europe [20].

Despite these notable contributions, and after about a hundred years of professional evolution, there appears to be a split between how planning and landscape knowledge is used in distinct career paths: (i) professional practice, and (ii) academic scholarship. Although research is equally valuable to both paths, it tends to be conducted in academia and utilized in academia and, to a limited extent, also in professional practice. Forsyth synthesized the following four main planning research cultures: (i) work at the scientific frontier; (ii) investigation of issues of practical relevance; (iii) reflection on the implications of practice; and (iv) attempts at answering the enduring questions of planning [10]. She also mapped correspondences between these four planning research cultures and four main research styles: empirical, logical argumentation/theory, critical/interpretive, and synthesis (see Figure 1) [10]. The various correspondences reflect different intellectual positionalities and scholarly identities.

The four research styles show a widening of the scope of topics being researched and the increasing number of theories construed to explain the interrelations put forward by those with a background in planning and/or who work in professional planning activities. Although rational planning theory has been utilized to allocate means to ends in the public sphere, in recent decades, other planning theories ranging from incrementalism, methodism, utopianism, and advocacy to collaborative and communicative planning and the just city theories have been formulated to encompass an ever-widening range of post Fordism territorial, economic, social, environmental, and cultural situations [21,22]. Planning's widening scope and movement away from blueprint comprehensive planning towards the socio-ecological realm and the right to the city and human flourishing worldviews of the twenty-first century have even resulted in more demanding professional accreditation requirements. As of last year, AICP-accredited planners ought to know not only the most recent professional developments in Law and Ethics in the US, as it was customary until the late 2010s, but they also now need to master the fundamentals of sustainability, resilience, and equity [23–25].

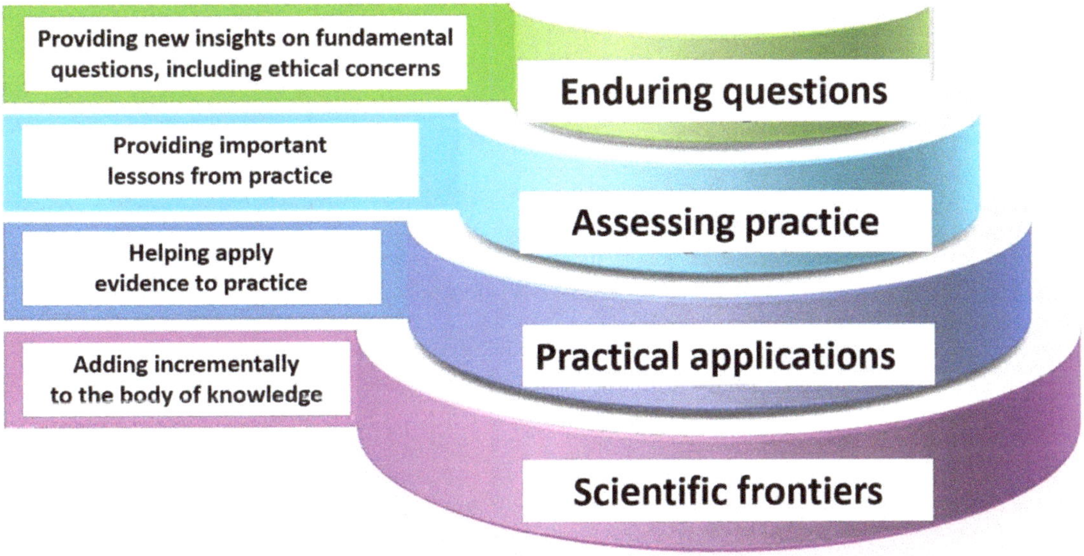

Figure 1. Forsyth's Planning Research Cultures in *Alternative Cultures in Planning Research* (illustration author's own) [10].

3. Analytical Framework

Social sciences have been shaped by two dominant research method paradigms: the quantitative and the qualitative. The former is associated with positivism and is traditionally perceived as being within the realm of the physical sciences. It aims to be objective and reduce subjective bias. Its explanatory power comes from the application of relatively immutable principles and postulates. The latter research paradigm has gained ground within the social sciences and makes use not only of mensuration and statistics but also of a myriad of other research methods, which excel at more in-depth and comprehensive characterizations of a social phenomenon. Said phenomena may be construed differently depending on regional and societal contexts; nonetheless, the research findings ought to be consensual according to (and potentially also replicable by) other researchers.

The qualitative research paradigm may require a higher degree of data, information, and comprehension for researchers to be able to make sense of the phenomenon being studied. Individual behaviors, such as the expression of feelings and emotions, may require careful interpretations and multiple observations, if not the triangulation of various data sources [26], before making definitive inferences about obtained results. As such, in-depth observations over extended periods of time and under distinct conditions may be required to assert a specific conclusion. While numbers and statistics tend to dominate the quantitative research paradigm, words, behaviors, rituals, drawings, esthetics, songs, and pictures all have value in the qualitative research paradigm [9,27].

Wicked (and often invisible) urban challenges are difficult to solve, if not impossible, as each attempt at resolving one issue is likely to cause a problem elsewhere in the system [28]. Nonetheless, there appears to be a relative consensus on the value of professional specializations and the frequent search for more appropriate solutions to pernicious problems. On the one hand, the struggle for adequate knowledge, together with the observation that knowledge ought to be utilized in ways that can bring about positive change, even outside of the traditional S.T.E.M. (science, technology, engineering, and mathematics) disciplines, is associated with the concept of "powerful knowledge" [29]. On the other hand, the notion of "public scholarship", meaning knowledge and scholarship deployed in ways that can ameliorate the virulence of wicked problems in the real world, and potentially also on some college campuses, and not necessarily only for knowledge's sake, such as research productivity, has gained considerable tracking in the social sciences and professional disciplines.

These two conceptions of knowledge have been applied mostly within educational settings to deal with educational curricula, institutional missions, career progression, and community embeddedness programs [30], which at their core aim to change not only societal behaviors and institutional arrangements, but above all aim to influence the territorial balance of forces in specific places, be they campuses, neighborhoods, cities, or countries. Recently, the literature on qualitative research has also recognized the importance of the visual medium and video in conveying more desirable courses of action, more inclusive and participatory public policy scenarios, and leading to transformative actions, instead of simply documenting a social phenomenon without eradicating its root causes [9,31,32].

We live in a visual era of ubiquitous YouTube and Vimeo video content streaming and image-centered social media platforms (Instagram, Facebook, and Flickr), which enables social media users to regularly check for updates, blog, and post selfies and other visual and audio material online. The visual culture of selfies in the age of social media is also becoming a powerful means for self-expression [33]. However, information and communication technology (ICT) can also be used to the advantage of institutions of higher education (IHE).

Specifically in the field of sustainable urban planning, it has been shown that YouTube videos can bring distinct city planning perspectives to "free-choice learners" [34]. Furthermore, videos can also be effectively utilized in community planning "to inspire students to make connections between complex urban theory and planning content and their everyday surroundings" [35] (p. 482). This finding extends two important earlier realizations that (i) "films communicate ideas about places that can sit heavily on people's emotions and sense of attachment to the represented locations and, additionally, to people's own ideas about the identity and meaning of the place" and (ii) "[l]ike maps, films are one more way of looking at the world, but they are more likely to evoke *[and make visible]* matters concerning power and contestation" [36] (p. 88) (our emphasis).

This is particularly relevant to this discussion because for generations, claims backed-up by mostly positivistic rational methods have worked to the advantage of powerful societal groups. Of a greater concern is the fact that some of those highly scientific methods have been employed to overrun advocacy claims that attempt to "speak truth to power" in hopes of legitimizing decisions [37,38]. It has also been claimed that social science methods can be utilized in uncovering hidden truths and turn invisible information into evidence to empower decision making [8,39].

In addition to the classical "*Episteme*" (scientific knowledge) and "*Techne*" (technical knowledge), Flyvbjerg has argued that social scientists ought to make use of "*Phronesis*" (practical wisdom) [39], which has a liberating effect in contexts of paralyzing power asymmetries. In this same vein, the notions of "Ecophronesis" and "Citizen Science" have been proposed to demonstrate reflective ecological practices, to further actionable science, and to legitimize and empower advocacy efforts [40,41].

Finally, recent advancements in qualitative research have demonstrated that visual literacy can be utilized in ways beyond illustration to also constitute its evidentiary corpus, and that multisensorial competencies and tools (e.g., participatory mapping, emotional mapping, and spatial perception analysis) can be utilized to create more comprehensive characterizations of complex situations [42–44]. In sum, mixed methods, visual literacy, and multisensorial competencies have potential to retool earlier emphasis on numbers and words to a visual and multisensorial reality capable of bridging knowledge gaps for experts and non-experts alike, while eliciting multiple representational opportunities [45,46].

4. Materials and Methods

The four educational and research programs are (i) UAlbany Documentary Series and Symposia; (ii) City Center Revitalization via walk-only precincts; (iii) Working Landscapes in transition; and (iv) Place Phenomenology. The programs are reviewed in the fashion of Wilson's *Reflective Planning Practice* [47]; the only difference being that the planning practice in this paper pertains to a combination of pedagogical programs and applied research in urban, community development, landscape, and wetland enhancement engagements in the US, Portugal, and China.

Materials from the various programs are reviewed and analyzed, including the documentaries and talks held in conjunction with the documentary and lecture discussion series as well as a documentary screened at the Land Use and Sustainability (LUS) Symposium; the symposia presentations and discussions; and participants' events evaluations. The research methods comprised mostly literature reviews and retrospective self-reflections on the advantages and limitations of the four programs under scrutiny.

The retrospective self-reflections made use of compare and contrast techniques of "before and after" or "then and now" analysis, even though the COVID-19 pandemic-related lockdowns precluded revisiting several of the original study sites. Nonetheless, various sites in Albany were revisited to unveil hints about their evolution since the programs' inception and community engagement with the Rapp Road Historical Association and the collaboration with UAlbany's S.T.E.P. program prospective study for Ezra Prentice in 2017 and 2018, respectively. Data for this paper also came from several papers on city center revitalization, working landscapes, social and cultural sustainability, as well as place phenomenology. The four programs were mapped onto Forsyth's four planning research cultures [10] (Table 1).

As such, the first educational and research program titled "*UAlbany Documentary Series and Symposia*" is categorized as work at the "scientific frontier" because it "add[ed] incrementally to the body of knowledge" [10]. The second program titled "*City Center*

Revitalization via *walk-only precincts*" was mapped as belonging to "practical application" because it "help[ed] apply evidence to practice" (p. 164). This research program applied the Commercial Revitalization Vibrancy (CRV) theory [48] to three pair-wise cases in Asia, Europe, and North America (China, Portugal, and USA).

Table 1. Mapping of Forsyth's planning research cultures [10] onto the four educational and research programs.

Planning Research Cultures	***Researcher Self-Description Internal Culture***	**Examples**	**Research Paradigm**	**Research Methods**
Scientific frontiers	Adding incrementally to the body of knowledge	UAlbany Documentary Series and Symposia (documentary and lecture discussion series; SUR in Japan; and LUS)	Communicating science (both quantitative and qualitative)	Documentary screening and lecture series; Japan Symposium with a field visit; LUS Symposium at a nature preserve and hike of a preserve's trail.
Practical applications	Helping apply evidence to practice; research that makes a difference	City Center Revitalization via walk-only precincts	Comparative research of walk-only precincts and their associated built environment transformations (qualitative)	Visits, data collection, and functional inventories in all 6 city centers on 3 continents (Asia, Europe, and North America).
Assessing practice	Providing important lessons from practice	Working Landscapes in transition (saltscapes in Portugal)	Ecosystem services and livelihoods (qualitative)	Visits to the three saltscapes locations (2 in central and 1 in southern Portugal).
Enduring questions	Providing new insights on fundamental questions, including ethical concerns	Place Phenomenology (town-gown citizen science in Albany)	Place phenomenology—placeless, place, and non-place (both qualitative and quantitative)	Visits to the three sites (affordable housing, suburban historic district, and landfill), interviews, guest lectures, focus group observations, and presentations.

The third program titled "*Working Landscapes in transition*" is an example of "assessing practice" because it "provid[ed] important lessons from practice" (p. 164). And finally, the fourth and last program on "*Place Phenomenology*" exemplifies some of the "enduring questions" of planning research culture because it "provid[ed] new insights on fundamental questions, including ethical concerns" (p. 164). This section is followed directly by an introduction of the four programs under review in terms of motivations, content, research methods, and assessments, where available, and by a discussion of results on how and why the programs constitute research innovations from visual and multisensorial literacy perspectives.

5. Ex-Poste Overview of the Educational and Research Programs

These educational and research programs are introduced in the order of lowest to highest complexity and potentially also societal relevance, not necessarily in chronological order of occurrence. On the one hand, complexity is here understood in terms of an increasing number of variables with a developing mesh of relationships among themselves as well as collective awareness, also an but unlikely ability by a small group of individuals to alter the intensity and pervasiveness of the societal problems. On the other, societal relevance is based on the motto "think globally, act locally". These two defining characteristics are well exemplified in terms of simply "communicating science", presenting research findings, and having discussions on professional practice in the first program versus attempting to answer "enduring questions" of environmental justice, zoning and siting of affordable housing, conservation of historic districts, and issues of sustainable consumption in the fourth program.

The programs' scope ranges from cityscape to landscape situations, from the most urban central business districts and walk-only precincts in both large metropolises and small cities to peri-urban neighborhoods. Even though of the three productive landscapes under scrutiny in Portugal—Mondego, Vouga, and Guadiana River valleys—the agricultural fields of the Mondego River valley are extremely well known for their rice and corn production, the third research program's emphasis is only on the three rivers' estuaries ancient salt production by solar evaporation landscapes, which have been in existence for many centuries.

5.1. UAlbany Documentary Series and Symposia

This educational and research program comprised one documentary screening and lecture discussion series and two symposia. *"Striving For Diverse Cities: A Documentary Film and Discussion Series"* was organized by two assistant professors and one graduate student at SUNY Albany with a grant from the Office of Diversity and Inclusion (ODI) in 2016–2017. The project was a cross-disciplinary collaboration among three UAlbany units: Geography and Planning; Latin American, Caribbean & U.S. Latino Studies; and Social Welfare. This series intended to highlight neighborhood transformation and adaptation processes in various North American cities. We sought to strengthen the study of the often invisible and rather painful issues of social injustice, racial and class segregation, and lack of inclusion, as well as the eradication of poverty, while endowing students and community members with the skills and awareness needed to identify, address, and prevent unjust and unfair societal practices (see Figure 2) [49].

The documentary series promoted engagement with a variety of themes central to the UAlbany's ODI mission, such as social justice, diversity, and inclusion. The documentary film screenings and discussions were open to everyone on campus as well as to community members. The sessions were scheduled in late afternoon and in an easily accessible location. The goal was to have a lecture and discussion with the documentary director(s), or someone who was directly involved in the making of each film, and who could speak inspirationally about the various issues featured in each documentary. We successfully screened the documentaries *MyBrooklyn*, *Los Sures*, and *When Strangers Become Friends*, and we canceled the screening of *The Throwaways* due to a major snowstorm that closed campus; Tom Angotti [50], Christopher Allen, and Leonie Sandercock [51], respectively, were brought to campus to discuss the making of the screened documentaries and their main themes. Attendance consisted mostly of the campus constituency; however many professional planners and community members also attended one or more events.

The symposium on *Sustainable Urban Regeneration (SUR) in Japan* was held at SUNY Albany on Saturday, February 11th, 2017. The symposium's goals were to understand and explore various sources of Japanese tradition, to acquire knowledge on Japanese urbanism, sustainable development, and urban regeneration, and to foster an appreciation for cultural development in Japan as well as to motivate interest in Japanese culture, urban development, and societal trends. The symposium consisted of presentations on the following specific topics: sustainable urban regeneration, consumption and urban form, mobility, walkability and public health, demographic challenges, and Japanese diaspora in the US. In addition to enabling one undergraduate and one graduate student to present their research papers, the symposium also created network opportunities between local and regional scholars of Japan Studies.

Figure 2. UAlbany documentary series and symposia posters (courtesy of C. Balsas, J. Londoño, and K. Hirokawa).

Albany's relative proximity to New York City enabled us to reflect on the parallels and differences between Japan and the U.S., with special emphasis on New York and Tokyo [52], in terms of countryside/rural preservation, small and medium-sized urbanization, de-industrialization trends, natural disaster preparedness and recovery, societal and demographic developments, urban regeneration, transportation, consumption, megalopolis transformations, urban and regional planning, and territorial infrastructure development [53,54]. This symposium also included an afternoon trip to the Clark Art Institute in Williamstown, Massachusetts, to visit the exhibit "Japanese Impressions: Color Woodblock Prints from the Rodbell Family Collection".

The 2018 *LUS Symposium* took place at the Albany Pine Bush Preserve on Friday, May 4th. The symposium's format was like the symposium on sustainable urban regeneration in Japan, with the main difference being that the whole event took place off-campus and, instead of a separate visit to a museum, we had a guided tour of one of the preserve's trails, a visit to the preserve's museum, and a roundtable discussion with scholars and practitioners. The symposium was co-organized with a professor of Environmental Planning at the Albany Law School. Although students were not invited to present their own research, we had a more encompassing range of assistant, associate, full, and distinguished professors from as far away as Boston, MA, USA, and Syracuse, NY, USA. The symposium's main speakers consisted of Neil Gifford, Albany Pine Bush Preserve; Jocelyn Gavitt and Richard Smardon [55], State University of New York ESF; Rik Scarce [56], Skidmore College; Robert Ryan [57], University of Massachusetts, Amherst; Justin Hollander [58], Tufts University; and Keith Hirokawa [59], Albany Law School.

Even though students did not present their own research at the LUS Symposium, students enrolled in the co-organizer's courses were given the option of either taking the semester's final exam or instead attending the symposium and explaining how the symposium related to their courses' materials. There was also an afternoon roundtable discussion on local, regional, and national planning and landscape "hot topics" after the guided tour of the Albany Pine Bush Preserve's museum and the hike of a preserve's trail.

The fields of research covered during this symposium ranged from land use planning, sustainability, landscape ecology, and landscape assessment and valuation to green infrastructure, environmental planning law, sense of place, social media, sustainability, cognitive architecture, campus planning, clean energy, and climate change [60,61], among others. We also screened portions of Rik Scarce's documentary on the Hudson River valley entitled *Sustaining this Place—Creating a new Hudson region landscape*. The Discovery Center at the Albany Pine Bush Preserve proved to be an excellent place to hold the symposium, given its state-of-the-art facilities and in loco opportunities for understanding nature conservation efforts in uptown Albany [62].

5.2. City Center Revitalization via Walk-Only Precincts

The second program under scrutiny comprised various research projects on city center revitalization, more specifically on walk-only precincts, which culminated with the publication of *Walkable Cities, Revitalization, Vibrancy and Sustainable Consumption* by State University of New York (SUNY) Press in 2019, recipient of The Society for the Study of Social Problems, Global Division 2020, and Independent Publisher 2021 Best Book awards [48], and the journal article titled "Exciting walk-only precincts in Asia, Europe and North-America" (Figure 3) [63]. The former comprised a comparative analysis of walk-only precincts in Portuguese and Spanish cities (Porto and Barcelona) as well as in the largest cities of their former Latin America colonies of Brazil and Mexico, São Paulo and Ciudad de Mexico, respectively. The latter extended the manuscript-length research by comparing walk-only precincts on three continents, Asia, Europe, and North America. In the latter case, city size was utilized as a broadly constructed proxy for urban agglomeration effect, in which the aim was to understand the main differences and similarities between walk-only precincts in large and small cities, respectively, in China (Shanghai and Macau SAR) [64], Portugal (Lisbon and Figueira da Foz) [65], and the US (New York City, NY, USA and Miami, FL, USA) [66].

5.3. Working Landscapes in Transition

The third program encompassed research on working landscapes (saltscapes) in central and southern Portugal. The main motivation for this research program was the important agricultural re-engineering public works associated with the correction of the path of the Mondego River between Coimbra and Figueira da Foz 1960–1990 to eliminate the risk of floods and to increase agricultural productivity. The much smaller area of saltscapes on the Mondego River delta near Figueira da Foz is of particular interest because of its functional decline when industrial-scale salt production and modern refrigeration techniques made traditional sea salt by solar evaporation obsolete. The research program involved visits to saltpans in Figueira da Foz and Aveiro in central Portugal and Castro Marim in the southern region of Algarve (Figure 4).

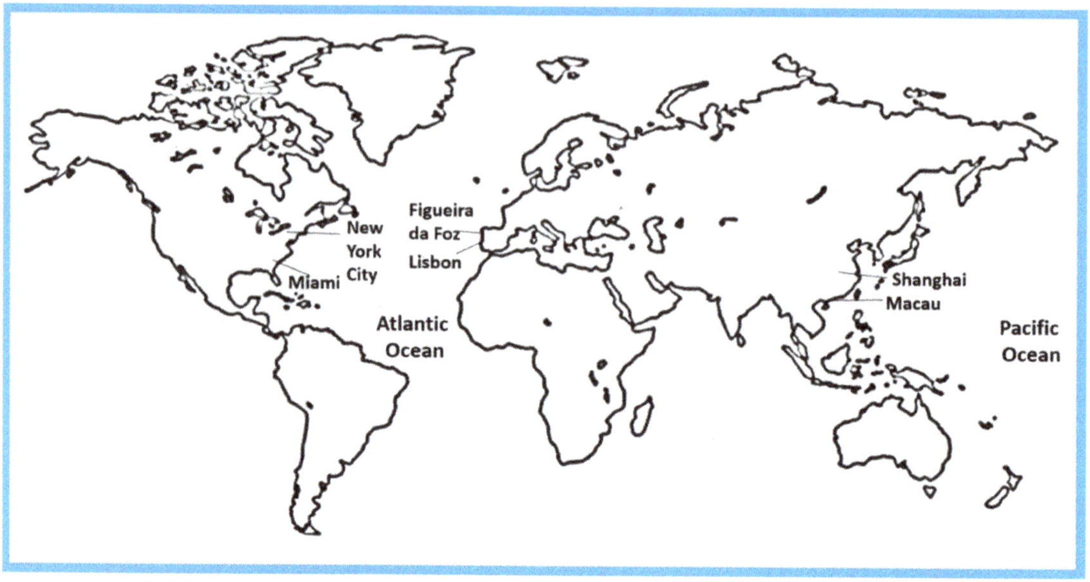

Figure 3. Location of the pairwise walk-only precincts in China, Portugal, and the United States [63].

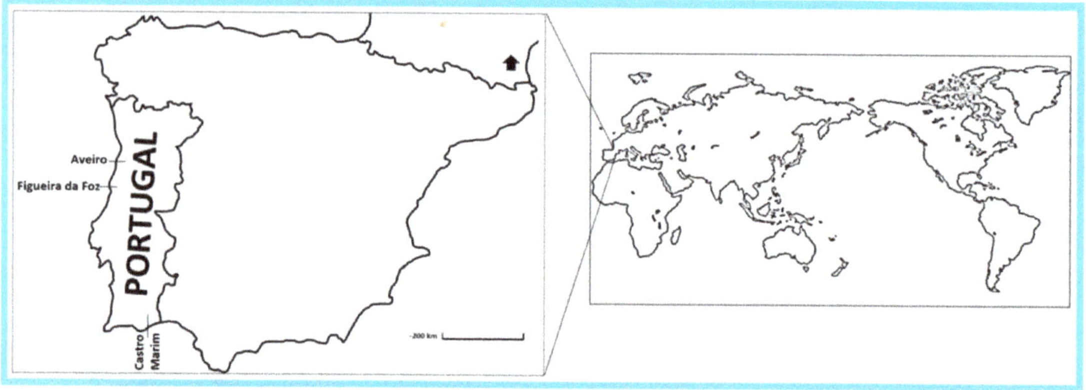

Figure 4. Location of the saltscape sites in Portugal.

Having studied on a university campus overlooking the extensive territory of the Ria de Aveiro and Vouga River salterns in the city of Aveiro for five years during the early 1990s, and with the knowledge that the salterns of Figueira da Foz were poorly studied, rendering the livelihoods of the saltpan workers socially and economically invisible compared to those of Aveiro, this compelled an analysis of the reasons for the fast decline of such an ecologically fragile territory. Recurrent visits to the salterns in the locality of Lavos, just south of Figueira da Foz, coupled with the strong interest of a former city mayor in the rehabilitation of those fragile ecosystems led to in-depth research about the solar salinas' decline as well as their transition to new uses such as eco-tourism, eco-museums, and nature preserve. Comparative studies involving extensive salterns in three different river estuaries have been conducted to date, with plans for further international comparisons of saltscapes and associated livelihoods [67].

5.4. Place Phenomenology

Thus far, the fourth program's purpose has been to find out how the university can become more relevant to underserved communities in the Capital Region of Upstate New York (Figure 5). In a way, this program aimed to analyze how to make invisible sustainable urban planning and landscape knowledge much more visual and multisensorial. This research program started with three requests for technical collaborations, initiated by (i) a student interested in participating in an awards competition involving the environmental impacts associated with the operation and eventual expansion of a suburban landfill; (ii) a supervisor of the UAlbany's S.T.E.M. program who was interested in demonstrating the applicability of a greenwall in a riverfront social justice community; and (iii) a motivation to bring forth (make more visible, visual, and multisensorial) notions of social injustice and ethnic exclusion in contexts of spatial and social development in uptown Albany in a Global Urbanism and Culture undergraduate course. Therefore, this research program comprised three study areas in the Capital Region: (i) a landfill site in Latham (Town of Colonie's landfill), (ii) a public housing complex in an industrial location adjacent to the Port of Albany, and (iii) an African American cultural heritage district in uptown Albany.

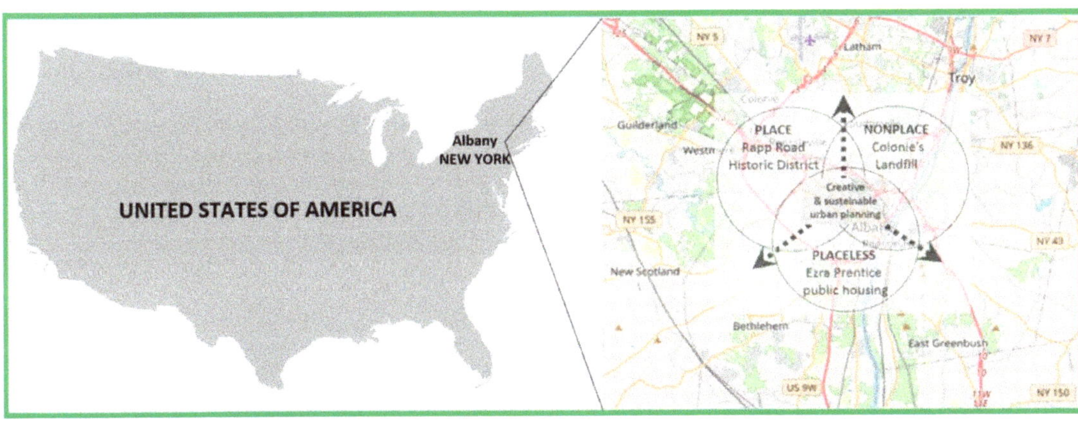

Figure 5. Place phenomenology sites in the Albany Capital Region of Upstate New York (openstreetmap.org and [68]).

At the beginning of summer 2018, we were alerted to the possibility of helping to provide some sustainable urban planning suggestions to the resolution of a very complex environmental health situation involving the residents at a public housing complex in an environmental justice community in Albany, New York, called Ezra Prentice. Such a housing complex is transversed by a busy state arterial road and is located adjacent to a railyard depot in the proximity of the Port of Albany. The residents, mostly low-income population of color, have been exposed to various types of air pollution generated by heavy roadway traffic and port-related operations such as crude oil transport by rail. Severe health problems such as high rates of asthma and other respiratory illnesses have been documented, most likely in connection with the poor air quality and noise pollution in the area. Prior environmental advocacy by the residents themselves and the leaders of various non-profit organizations in Albany, since at least 2014, served to alert various elected officials and decision makers to the severity of the problems in the area.

In collaboration with scientists from NYS DEC and a group of high-school students from SUNY Albany's S.T.E.P., we conducted visits to the study site to analyze the possibility of utilizing placemaking urban design strategies, including streetscape improvements and the installation of a greenwall to reduce the encroachment of air pollutants into the housing and playground areas located adjacent to the heavily traveled road through the housing complex [69]. The high school students had meetings with air quality scientists and planners, were taken on a guided tour of the air monitoring station, conducted their own air quality measurements, tested the feasibility of installing a greenwall pilot project, and then presented their findings at the SUNY Albany S.T.E.P. Summer Institute/Smart Scholars Summer Closing Ceremony.

Overall, the fourth research program has been guided by the following two main research and educational objectives: (i) to utilize engaged service learning, citizen science, and other community-based participatory methods in the resolution of very complex health, social, environmental, and consumption problems in the Capital Region, and (ii) to strengthen citizen science and design thinking in both high school and university education, respectively. The program's approaches consist of town–gown applied service learning and the phenomenology of place methodologies [70]. The applied town–gown service-learning activities aimed to resolve wicked territorial situations, such as those plaguing environmental justice communities and declining minority neighborhoods, and to properly retrofit landscape areas surrounding landfill sites. This program's actions require a strong intellectual motivation and creative ways to resolve the medium- to long-term commitments necessary to improve those particular places.

Some of these phenomenological sites have been studied according to the triad of Placeless (i.e., public housing complex), *Place* (i.e., African American cultural heritage district) and *Non*-place (i.e., landfill site) genius loci of socio-spatial town–gown interventions [68]. Upon completion, the program's expected results ought to include a better understanding of the proper siting and management of affordable housing developments and other sensitive infrastructure, the eradication of deeply imbedded racist behaviors, the celebration of diverse social and cultural heritage, and the accomplishment of more sustainable consumption practices at the metropolitan scale, which can lead to less material being sent to landfills and higher compost and recycling rates. Experiential curricular and pedagogical innovations aimed at strengthening citizen science and design thinking in both high school and university education have also been put forward [68].

6. Findings

This section examines the four programs according to the four research cultures and two educational and research innovations: (i) visual literacy and (ii) multisensory landscapes (Table 2). The results are presented with recourse to quotes from students' response papers and participants' remarks in their evaluations of the three events in the UAlbany Documentary Series and Symposia, together with visual materials produced for and gathered during the events.

The three research-based programs make use of quotes from referee's anonymous reports when available. With Table 2 and the in-depth analysis of the mostly research-oriented programs, we want to convey the complex nature of the "enduring questions" that have yet to be adequately and fully answered in terms of their substance and procedural ethics, as well as on the ground public policy improvements to the overall quality of life of those who reside in the neighborhoods adjacent to the study sites. It is important to note that the various programs represent distinct geographic scales and that climate change and sustainability responses have been influenced by distinct socio-economic, legal, and cultural contexts as well as planning policies and practices operationalized on at least one scale.

Table 2. Four educational and research programs and their visual literacy and multisensory landscapes.

Planning Research Cultures	Visual Literacy	Multisensory Landscapes
Scientific frontiers	Documentaries (e.g., filmed interviews, meeting observations, sense making, storytelling, narration, script writing, follow-up interpretations by directors and experts); Talks, presentations, and discussions (visit to the Clark Art Institute's museum exhibit); Talks, presentations, discussions, and film screening (e.g., guided visit to the preserve's museum, hike of a preserve's trail, roundtable panel)	Direct contact with documentary directors, formation of learning communities, acquisition of new knowledge, new learning community, visit to new spatial settings (Williamstown, peri-urban location in Albany), learning with past art tradition (Woodblock Prints), exposure to a different culture, in loco learning (land preservation and management, controlled fire), enhanced understanding of ecological functions (biodiversity, geomorphology, management practices, energy conservation) (lunch provided with registration)
Practical applications	Well-identifiable city center districts, commercial cityscapes, shopfront signs and window display, paved and landscaped public spaces, open spaces free of motorized vehicles, street furniture, stores' patronage	*Sight* (façade features assessment, window shopping, built environment characteristics, shoppers' demographic composition), *smell* (food, coffee roasting, proximity to the ocean); *touch* (landscape, salt, sand); *hearing* (people talking, music, stores); *taste* (food, drink, delicacies)
Assessing practice	Land-water interface, artisanal salterns technology, ecological value of working landscapes, material and cultural patrimony and traditions, salt barns	*Sight* (natural and man-made landscapes); smell (iodine), *touch* (salt; salt-based creams; spa treatment); *hearing* (*salineiros*, exhibits); *taste* (acquisition of salt-derived products, delicacies, salicornia)
Enduring questions	Road congestion, air pollution (tail-pipe fumes), noise from traffic and the port, urban design, and placemaking proposals (public space improvements); suburban decline, urban rehabilitation, and historic preservation measures; in loco understanding of landfill operations, visual and health impact assessments, landfill capacity dimensioning, and renewable energy production	*Sight* (esthetics of historic built heritage, traffic congestion, port operations, artificial mound, covering of solid waste with soil), *smell* (chemicals, solid waste), *touch* (playground equipment), *hearing* (street, industrial activity, heavy machinery trucks dumping waste, movement of topsoil)

6.1. UAlbany Documentary Series and Symposia

This public outreach set of initiatives took place in 2016–2018. It included three complementary initiatives centered on the Capital Region of Upstate New York and Western Massachusetts, and beyond. The program's main goal was to communicate, discuss, and shed light on complex societal topics such as social and environmental justice, immigration and diversity, gentrification, racism, crime, and urban poverty, as well as countryside/rural preservation, urbanization, de-industrialization, natural disaster preparedness and recovery, societal and demographic developments, urban regeneration, unsustainable transportation, consumption, megalopolis transformations, professional planning, territorial infrastructure development, urban sprawl, public health, land conservation, and climate change [71,72]. While these territorial problems are not necessarily new, their widespread existence throughout the US and abroad can be thought of as belonging to the "scientific frontier" planning research culture.

The use of documentaries with expressive visual stories in the US and Canada [73], in-person talks with movie directors and experts, and opportunities to engage in academic discussions with other community members were research innovations, which emphasized visual literacy and advocacy planning. The same can be said about the multisensorial learning experiences during the visit to the hands-on museum exhibits and the hiking of one of the local preserve's trails [74]. The impossibility of traveling to Japan was resolved with a tour to a carefully curated art exhibit at the Clark Art Institute in nearby Williamstown, MA. At the museum, we observed a timeline of Japanese art history, a variety of Woodblock Prints, artistic vases and jars, and specialty books.

The exhibit of Japanese Woodblock Prints represented three distinct generations of Japanese artists. The original artists worked in a tradition known as *ukiyo-e*, which utilized lots of color. The second and third generations comprised artists who attempted to revive the technique that had declined with the country's industrialization. The pieces evoked a sense of simplicity and grace and yet were captivatingly beautiful in their depictions of fashion and landscape scenes. The art on display portrayed a sense of fantasy while simultaneously revealing societal transformations through time from pre- to post-industrialization, potentially to convey increasing collective societal awareness.

The museum's exhibit related to some of the more cultural topics discussed in the morning, such as the multi-layered visual landscapes of most Japanese cities, issues of urban morphology and sustainable landscape planning, productive land uses, consumption, and agglomerations' shrinking tendencies. The LUS Symposium at the peri-urban nature preserve served to re-equate and extend the textbook knowledge acquired in, for instance, the Urban Design course [11], not necessarily in terms of land use planning, but more on urban sustainability and the proper design of public spaces in city centers and suburban cores, open spaces and parks, and city streets and linear greenways connecting multiple spaces in cities [75]. Having this symposium at the nature preserve also enabled community members to feel more at ease and to participate more actively in the roundtable discussions, which is integral to the "public scholarship" goals of most IHE, regardless of whether they are land grant universities or not.

It is believed that these three initiatives also excelled at fostering three well-recognized public outreach visions in Sustainable Urban and Regional Planning: (i) originality and transferability, (ii) effectiveness and results, and (iii) participation. The various activities attempted to establish parallels between local and regional developments in Upstate New York, the US, and internationally (mostly Latin America and Japan). Originality and transferability were well evident in the multitude of topics featured in the documentary screenings and subsequent discussions. The transferability element enabled participants to learn directly from local and out-of-town speakers, while facilitating the establishment and consolidation of academic and social bonds amongst themselves. Participants were academically stimulated by the film screenings and the various symposia presentations and discussions, to the point where they felt comfortable asking speakers to expand on some of the arguments made or to follow up with requests for answers to regional social and territorial problems. The events' effectiveness and results were well documented in the various initiatives' evaluations. Most participants recognized that they had gained something useful and learned new way(s) of thinking and acting on the subject matters presented and discussed in the documentaries and the symposia presentations.

The documentaries' themes, together with the participants evaluations are publicly available. Of those who returned the evaluations at the 2017 SUR in the Japan Symposium, 60% attended both the morning talks and discussions and the afternoon visit, while 40% attended only the morning talks and discussions. The tabulation of results demonstrated that 70% of participants strongly agreed and 30% moderately agreed that they gained something useful from attending the symposium. Furthermore, 80% strongly agreed and 20% moderately agreed that they had learned a new way to think about Urban Regeneration and Culture.

Of the 10% of participants who answered the short evaluation survey at the end of the 2018 LUS Symposium, all declared to have attended all symposium activities; fifteen participants answered "strongly agree" and one answered "moderately agree" to the question "I have gained something useful"; and thirteen participants answered "strongly agree" and three answered "moderately agree" to the question "I have learned a new way to think about or address issues relating to LUS".

The open-ended answers to elaborate on earlier responses or to add anything else they would like to share ranged from "fantastic group of presenters + engaged attendees", "I really enjoyed this [symposium], especially the end (open discussion) and the hike", "gave me access to many different perspectives and opportunities of how we can appreciate different problems", and "I enjoyed the legal aspect of planning, the presenters and the panel at the end was very interesting" to "definitely learned a lot of new points about these topics" and "would love to attend more seminars like this in the future". Table 3 shows correspondences between Bloom's Taxonomy and selected students' self-reflections about their participation in the 2018 LUS Symposium.

Finally, the participation criterion was evidenced by the attendance of not only members of the UAlbany community, but also by the participation of community members from throughout the Capital District and beyond. Many of the symposia participants were affiliated with the various Universities in Upstate New York, Western and Eastern Massachusetts, local and regional planning agencies, and professional consulting companies dedicated to providing services to local and regional authorities and to the State.

Members of local non-profit organizations were also among the participants in attendance at the three public outreach initiatives [76]. Various participants acknowledged that the 2018 Symposium provided valuable continuing education opportunities for volunteers serving on various city and town boards who are required to acquire a certain number of credits to stay current with the latest developments in the field of sustainable urban planning. The *Eventbrite's* webpage of the 2018 Symposium had 1115 views up to the event's date. The statistical significance of the symposium participants reporting continuing education credits to either maintain their professional accreditations or to continue serving on various city and town boards was calculated at 0.125, which is much higher than the probability of one in twenty (0.05), which was the convenient cutoff level to reject the null hypothesis in Fisher's seminal study on statistical methods for research workers [77].

Undergraduate and graduate students in the Department of Geography and Planning and the School of Social Welfare were integral to the implementation of these initiatives. The students' roles ranged from co-organizer to enrollee in an independent study and active symposia participants.

Undergraduate and graduate students were critical to reaching out to symposium speakers and participants and, above all, to learning in loco with symposia speakers and other participants, taking part in the two symposia's field visits, and writing reflection essays aimed at establishing relationships between the theories, concepts, developments, programs, and public policies learned mostly in the Global Urbanism and Culture and the Urban Design and Public Spaces courses, as well as the applied learning gained by taking part in the various film screenings, discussions, the symposia, and corresponding visits.

Table 3. Correspondences between Bloom's Taxonomy [78] and selected students' self-reflections about their participation in the 2018 LUS Symposium.

Bloom's Criteria	Students' Self-Reflections
I. Knowledge	"I enjoyed learning about the globally rare ecosystem located in the heart of the Capital District". "One of the lessons learned from the milfoil abatement project at Cazenovia Lake was that without properly informing the people adjacent to the lake and surrounding streams it fed, the algicide seemed a lot scarier than it actually was".
II. Comprehension	"He showed snippets of his documentary, including the interviews of different people from different towns along Hudson, that shared their feelings of care and appreciation for clean river". "From the documentary, I gathered that the motto is 'to save land when you consume the land.'"
III. Application	"This connects back to urban design because environmental sustainability is one of the factors in shaping the community". "In our Urban Design course, we talk about how pollution and cars affect rural places because of how much people use automobiles instead of conserving energy".
IV. Analysis	"During the hike, I felt a personal connection with nature. The lead hiker/tour guide fully educated us on the history of different species of wildlife residing in Albany Pine Bush Preserve". "As a student of urban studies and regional planning, it is my responsibility to integrate the information I have acquired in my course and module it in a way that can be productive to man and benevolent to the environment".
V. Synthesis	"After lunch, we regrouped and went for a hike of the preserve, which was one of my personal favorite parts of the day. I was so surprised to see a place like this existed in Albany". "It is inspiring to understand how community activists, local farmers, corporate executives, distillers, politicians, landscape architects all united their knowledge and resources to reclaim a subdued river ecosystem into thriving biodiversity".
VI. Evaluation	"In our class, we brainstormed ideas to improve our campus and one idea we came up with was the idea to possibly create murals to liven up the underground tunnel system". "The one that caught my attention the most that is relevant to me as well as talked about in class was the presentation of measuring sustainability using social media data".

6.2. City Center Revitalization via Walk-Only Precincts

Downtowns and city centers are among the most visually stimulating places in cities; this could have been one of the reasons for the journal editors to feature the "Exciting walk-only precincts in Asia, Europe and North-America" in the journal's Cities Profile section [63]. The existence of walk-only precincts in the selected cities helped to delimit the contours of their respective city centers. Their almost organic combination of functions, multiple uses, distinct architecture styles, building heights and typologies, busy sidewalks with people walking, seating, conversing, as well as the commercial landscape of retail establishments interspersed with restaurants, bars, coffee shops, public and private services such as post offices, city halls, and banks, street furniture, landscaping, vehicles, distinct levels of cleanliness, a high concentration of bus stops and transit hubs, parking garages, and subway stops all contribute to high degrees of accessibility and eased mobility, attractiveness, vibrancy, and livability [79].

Once streets are fully, or only partially, closed to vehicular traffic, pedestrians will surely feel a lot safer and comfortable strolling around [80–82]. One of the main findings of this research was that the smaller cities, although possessing other advantages, were hindered by not having a comparable level of accessibility to the walk-only shopping precincts as their large size counterparts [83].

This research program was conducted with field visits and in loco inventories in the studied cities on three continents. The visits, surveys, and visual documentation of commercial cityscapes through photography [84,85] enabled the identification of cues and unique features in the built environments, urban morphologies, planning functions, design deficiencies, and local economies that confirmed (and in other cases, discredited) tendencies, facts, and observations first encountered in interviews and desk research. The rich multisensory reality emerged via the following human senses: sight (façade features assessment, window shopping, built environment characteristics, shoppers' demographic composition), smell (food, coffee roasting, proximity to the ocean), touch (landscape, sand), hearing (people talking, music in public spaces, stores), and taste (food, drink, traditional delicacies).

6.3. Working Landscapes in Transition

This research program discovered that saltscapes are remarkable places in need of integrated public policy and scholarly attention. The three sets of solar salinas examined in central and southern Portugal have all been threatened by global economic transformations, changes in employment patterns, labor technologies, lifestyles, and the encroachment of urbanization, potentially leading to their irreplaceable destruction.

The recent reactivation of several solar salinas as part of pilot projects is a positive sign towards their transition to livelihood enhancing utilizations. The field work and subsequent desk research revealed visual cues ranging from their privileged locations at the land–water interface, artisanal technology and modus operandi, the esthetic and ecological value of their productive landscapes, the existence of material (salt barns, boats, and tools), cultural patrimony, and celebrated traditions (working journeys, theatrical performances, and festivals).

The multisensory landscapes are the result of powerful stimuli derived from the combination of visual cues (natural and humanmade landscapes); smell (iodine); touch (salt, salt-based creams, and minerally concentrated mud used in spa treatment); hearing (salineiros' business acumen and exhibits); and taste (acquisition of salt-derived products, delicacies, and salicornia). However, the continued existence of these fragile working landscapes is in danger if collaborative planning among stakeholders does not occur or is not fully embraced by those responsible for the planning and conservation of river watersheds and estuaries [86].

6.4. Place Phenomenology

The fourth research program initiated various studies on multiple sites in the Albany metropolitan area. The following are two examples of comments received from anonymous reviewers during the refereeing process of already published papers on this research program: (i) "This ambitious essay is committed to the pursuit of a larger, analytical, and somewhat retrospective view that embodies and pursues the social commitment and responsibility of discrete institutions/places, urban universities, which should be seen, it is argued, as embedded within and obligated towards local communities" and (ii) "the paper incorporates and reflects on a variety of public conversations about promoting sustainability in the Hudson River watershed, including college courses on planning and development, as well as public screenings of documentaries focusing on the theme of diversity".

This research program's emphasis was on the need to resolve the negative environmental externalities, public health, and quality of life issues plaguing an affordable housing complex sited near the port of Albany; the conservation of patrimonial heritage in an uptown Albany historic district; and the encouragement of more sustainable consumption practices aimed at reducing the region's ecological footprint.

The visual and multisensorial literacy component emerged during the research project in the form of road congestion, air pollution (tail-pipe fumes), protest yard signs against oil pipelines, noise from traffic and port operations, and urban design and placemaking proposals (potential greenwall deployment and public space improvements) [69,75]; suburban decline, urban rehabilitation, and historic preservation measures [87]; in loco understanding of landfill operations, visual and health impact assessments, landfill capacity dimensioning, and renewable energy production.

In aggregated forms, the main multisensorial landscapes comprised the following elements: sight (esthetics of historic built heritage, traffic congestion, port operations, artificial mound, and covering of solid waste with soil); smell (chemicals and solid waste); touch (playground equipment, gardening, and composting); and hearing (street, industrial activity, heavy machinery trucks dumping waste, and movement of topsoil). Exemplary research engagements in the Hudson River valley, and elsewhere [88–90], serve to demonstrate that modern town–gown citizen science can help establish the collaborative design, collection, and analysis of scientific data to answer "enduring questions" that will hopefully help to resolve, if not eradicate, pernicious societal challenges caused by some of the most relentless advances of uncontrollable market forces and inadequate professional planning.

7. Conclusions

This paper analyzed how four educational and research programs implemented at SUNY Albany at the turn of the decade built upon visual and multisensorial research methods to augment educational outcomes and provided up-to-date knowledge capable of empowering concerned stakeholders to help change a set of societal issues arising from a combination of perceived science-based decisions, uncontrolled market forces, and inadequate professional planning practices. These four programs potentially benefited from being implemented over an extended period and within an academic environment, which, to a certain extent, safeguarded real world politics, even though their deployment in partnership with other units on campus and other institutions in Albany still revealed power imbalances, entrenched administrative hierarchies, distinct funding priorities, disciplinary research agendas, and governance prejudices. Challenges and limitations were also encountered in terms of logistical constraints, participant biases, or a sense of disbelief with some of the sensory engagements.

The research presented in the article proved the extent to which urban planners and landscape architects can utilize visual and multisensorial literacies to bring about more transparency and significance to territorial transformation processes. The article tells the story of four educational and research town–gown programs in Albany, Upstate New York, where said venture was conducted in collaboration with others.

Similarly to the short-sighted siting of the affordable housing complex near the industrial area adjacent to the port of Albany five decades ago, professional planners and landscape architects ought to consider the negative consequences of exclusionary, nonetheless sustainably marketed, city center revitalization programs and initiatives, which unfortunately are wiping central city neighborhoods of their long-term residents and established businesses as a consequence of unrestrained green gentrification.

The recent retooling of research paradigms from numbers and statistics to words and a more fully immersed visual and multisensorial reality appears rather positive. More progressive and community-oriented research notions such as "powerful knowledge", "public scholarship", "ecophronesis", and "citizen science" have the potential to enable intervened realities to be more fully understood by not only the experts and decision makers, but also by the general public, who are likely to live with the transformed realities once experts find other potentially more rewarding market niches elsewhere.

In conclusion, the programs analyzed in this paper made a difference to the lives of students, scholars, local community members, and all those who somehow are at a higher risk of being disproportionately impacted by little understood and often invisible planning and landscape knowledge. Conducting science and communicating scientific findings ought to begin with the proper identification of the root causes of complex problems and their disentanglement in ways that bring more transparency, significance, relevance, and, above all, visuality and multisensorial understandings to future transformational processes [91].

As to the future practical implementation of teaching and learning strategies, one is reminded of Freire's four dialogical actions to overcome oppression in the educational system: cooperation (instead of conquest), unity for liberation (instead of divide and rule), organization (instead of manipulation), and cultural synthesis (as opposed to cultural invasion) [92]. Half a century after these suggestions were put forward in the context of major societal transformations in Latin America, and now conscious of a new, more connected, not necessarily more humane existence in the Global North, we are convinced that future learning opportunities ought to raise the community's awareness of its own natural, ecological, and human assets and to create place-based real-world opportunities for students and faculty to engage in socio-economic, environmental, and cultural sustainability studies. As argued earlier, these opportunities can also be further augmented with service-learning projects and programs, internships, and even fulltime jobs for recent graduates in local community development organizations.

This article has a few limitations. The author attempted to be as systematic with his data collection, analysis, and reporting as possible. However, future studies ought to make more extensive use of standardized analysis to increase the transparency and reproducibility of the findings. For instance, clarity can be enhanced with the inclusion of structured tools, such as predefined reflection prompts or coding frameworks for qualitative feedback. The reporting of the programs' evaluations likely ought to have accounted more extensively for the possibility of biases, including response biases or social desirability biases. Future program evaluation studies can minimize this limitation by providing comparative benchmarks or administer pre-event surveys to contextualize post hoc results, which can also guarantee higher representativeness. Striving towards a higher degree of quantification, in the form of empirical evidence, can potentially strengthen the relevance of sensorial and highly subjective interpretations of feelings about lived realities and experienced events in natural and built environments.

Zooming in and out on one's own "gratis" subscription of Google Earth software, we are conveyed the possibility of a total understanding of an infinitesimal myriad of planetary realities out there in the cosmos. However, research funding to study, analyze, and compare those distinct territories and associated relationships in a variety of geographical contexts remains a major limitation for most researchers. Future empirical studies ought to potentially augment the scope and comparability of the analysis undertaken to also account for the generalizability and usefulness of the discovered findings. The article cites only three of the author's, at least, ten research outputs directly related to the four programs discussed in the manuscript (see Appendix A). Except for one of them, nine were published in the last five years. While this does not account for the lack of long-term impact limitations of the scholarly work, it demonstrates the value attributed to the said embryonic body of work by the peers and editors of the outlets where those outputs were first recommended for publication and actually published.

Finally, the fact that the author is publishing this manuscript in an open access format while also paying the open access fees with vouchers earned by reviewing papers for M.D.P.I. journals, instead of choosing to publish it in a traditional outlet with a library subscription pay wall, which would make the work accessible mostly to students and scholars, is a genuine gesture at attempting to augment the transparency and reproducibility of the findings now being reported. While this publication option does not resolve external confounding factors that may influence the outcomes of the four educational and research programs analyzed in the article, at least it helps to clarify the distinctions amongst the four planning research cultures, which constitute the thrust of the article, while also increasing their pool of examples in North America, Southern Europe, and East Asia.

Funding: Various of the programs analyzed in the paper were funded by the University at Albany's Office of Diversity and Inclusion (ODI); the Japanese Chamber of Commerce and Industry of New York, Inc.; and the University at Albany and Albany Law School Collaboration Seminar and Conference Fund.

Data Availability Statement: Data are contained in the article.

Conflicts of Interest: The author declares no conflicts of interest.

Appendix A. Further Reading

Paper's Section | Planning Research Culture | | Educational or Research Program	Example of Additional Scholarly Outputs
Introduction	Enthralling Urban and Regional Planning forward. *Land*.
Scientific frontiers | | UAlbany Documentary Series and Symposia	Retaining social and cultural sustainability in the Hudson River watershed of New York, U.S.A. *J. Place Manag. Dev.* Historical and conceptual perspectives on urban regeneration: A prolog to a Special Issue. *J. Place Manag. Dev.* A collaborative geophilosophical perspective on creative territories: The Massachusetts-New York case. *Rev. Port. Filos.*
Assessing practice | | Working Landscapes in transition	Mediterranean Saltscapes: The need to enhance fragile ecological and cultural resources in Portugal. *ZARCH J. Interdiscip. Stud. Archit. Urban.* Revaluing saltscapes in Portugal: Ecotourism, ecomuseums and environmental education. In *Global Trends, Practices, and Challenges in Contemporary Tourism and Hospitality Management*; Batabyal D, Das D (Eds.) Engineering Science Reference: Hershey.
Enduring questions | | Place Phenomenology	Where does technology fit in the geospatial design and planning professions? In *5th International Conference on Connected Smart Cities.* Kommers P, Peng G, (Eds) IADIS & ISEP: Porto.

References

1. Yin, L. Integrating 3D visualization and GIS in planning education. *J. Geogr. High. Educ.* **2010**, *34*, 419–438. [CrossRef]
2. Inam, A. Meaningful urban design: Teleological/catalytic/relevant. *J. Urban Design* **2002**, *7*, 35–58. [CrossRef]
3. Shanken, A.M. The visual culture of planning. *J. Plann. Hist.* **2018**, *17*, 300–319. [CrossRef]
4. Guhathakurta, S. Visualizing planning for a sustainable world. In *Visualizing Sustainable Planning*; Steinebach, G., Guhathakurta, S., Hagen, H., Eds.; Springer: Berlin, Germany, 2009; pp. 37–50.
5. Portman, M.E.; Natapov, A.; Fisher-Gewirtzman, D. To go where no man has gone before: Virtual reality in architecture, landscape architecture and environmental planning. *Comput. Environ. Urban Syst.* **2015**, *54*, 376–384. [CrossRef]
6. Metze, T. Visualization in environmental policy and planning: A systematic review and research agenda. *J. Environ. Pol. Plan.* **2020**, *22*, 745–760. [CrossRef]
7. Spence, C. Senses of place: Architectural design for the multisensory mind. *Cogn. Res. Princ. Implic.* **2020**, *5*, 46. [CrossRef] [PubMed]
8. Sandercock, L. (Ed.) *Making the Invisible Visible: A Multicultural Planning History*; University California Press: Los Angeles, CA, USA, 1998.
9. Gaber, J.; Gaber, S. *Qualitative Analysis for Planning and Policy: Beyond the Numbers*; Routledge: New York, NY, USA, 2017.
10. Forsyth, A. Commentary: Alternative Cultures in Planning Research—From extending scientific frontiers to exploring enduring questions. *J. Plan. Educ. Res.* **2012**, *32*, 160–168. [CrossRef]
11. Montgomery, C. *Happy City: Transforming Our Lives Through Urban Design*; Penguin: London, UK, 2013.
12. Rapoport, E. Sustainable urbanism in the age of Photoshop: Images, experiences and the role of learning through inhabiting the international travels of a planning model. *Glob. Netw.* **2015**, *15*, 307–324. [CrossRef]
13. Kaika, M.; Tzaninis, Y.; Mandler, T.; Keil, R. (Eds.) *Turning Up the Heat: Urban Political Ecology for a Climate Emergency*; Manchester University Press: Manchester, UK, 2023.
14. Lyytimäki, J.; Teperi, A.-M.; Jylhä, K.M.; Vieira, R.S.; Mervaala, E. Dark side of resilience: Systemic unsustainability. *Front. Sustain.* **2023**, *4*, 1241553. [CrossRef]
15. Arribas-bel, D.; Gerritse, M. From manufacturing belt, to rust belt, to college country: A visual narrative of US urban growth. *Environ. Plan. A* **2015**, *47*, 1241–1253. [CrossRef]
16. Vieira, F. The four modes of thinking framed by utopian discursivity, or why we need Utopia. In *PHI Utopia(s)—Worlds and Frontiers of the Imaginary*; Monteiro, M., Kong, M., Neto, M., Eds.; CRC Press: London, UK, 2016; pp. 19–24.
17. Christensen, K. *Cities and Complexity: Making Intergovernmental Decisions*; Sage: Thousand Oaks, CA, USA, 1999.
18. Gaber JGaber, S.L. If you could see what I know: Moving planners' use of photographic images from illustrations to empirical data. *J. Archit. Plann. Res.* **2004**, *21*, 222–238.
19. Apostol, D.; Palmer, J.; Pasqualetti, M.; Smardon, R.; Sullivan, R. (Eds.) *The Renewable Energy Landscape: Preserving Scenic Values in our Sustainable Future*; Routledge: London, UK, 2017.
20. Dühr, S. *The Visual Language of Spatial Planning: Exploring Cartographic Representations for Spatial Planning in Europe*; Routledge: London, UK, 2007.
21. Beauregard, R.A. *Advanced Introduction to Planning Theory*; Edward Elgar Publishing: Northampton, UK, 2020.
22. Haughton, G.; White, I. *Why Plan? Theory for Practitioners*; Lund Humphries: London, UK, 2019.
23. Portney, K.E. *Taking Sustainable Cities Seriously: Economic Development, the Environment, and Quality of Life in American Cities*; MIT Press: Cambridge, MA, USA, 2013.
24. Ellin, N. Integral urbanism: A context for urban design. In *Resilience in Ecology and Urban Design: Linking Theory and Practice for Sustainable Cities*; Pickett, S.T., Cadenasso, M.L., McGrath, B., Eds.; Springer: New York, NY, USA, 2012; pp. 63–78.
25. Hexter, K.W.; Krumholz, N. (Eds.) *Advancing Equity Planning Now*; Cornell University Press: Ithaca, NY, USA, 2019.
26. Roth, M.; Hildebrandt, S.; Walz, U.; Wende, W. Large-area empirically based visual landscape quality assessment for spatial planning—A validation approach by method triangulation. *Sustainability* **2021**, *13*, 1891. [CrossRef]
27. Franklin, M.I. *Understanding Research: Coping with the Quantitative-Qualitative Divide*; Routledge: New York, NY, USA, 2013.
28. Dronova, I. Landscape beauty: A wicked problem in sustainable ecosystem management? *Sci. Total Environ.* **2019**, *688*, 584–591. [CrossRef]
29. Young, M. Bringing Knowledge Back. In *From Social Constructivism to Social Realism in the Sociology of Education*; Routledge: London, UK, 2007.
30. Bond, M.G.; Overholser, L.; Drinkwater, J.; Wilson, D.K.; Callahan, G. Toward improvement of community life: The history of arts and local development in cooperative extension. *Local Dev. Soc.* **2022**, *3*, 29–43. [CrossRef]
31. Rose, G. *Visual Methodologies: An Introduction to Researching with Visual Materials*, 4th ed.; Sage: London, UK, 2016.

32. Chen, Y.; Sherren, K.; Smit, M.; Lee, K.Y. Using social media images as data in social science research. *New Media Soc.* **2023**, *25*, 849–871. [CrossRef]
33. Million, A.; Machemer, P.; Kotval, Z. Engaging and educating young people in sustainable development: Communicating research findings and planning practice in an active learning environment. In *The Routledge Handbook of Planning Methods*; Silva, E., Healey, P., Harris, N., Broeck, P.V., Eds.; Routledge: Oxon, UK, 2015; pp. 502–518.
34. Amos, D. Planning education and "free-choice" learners: Teaching the YouTube Classroom. *J. Plan. Educ. Res.* **2021**, *44*, 565–575. [CrossRef]
35. Grodach, C. Video learning in community planning. *J. Plan. Educ. Res.* **2020**, *40*, 482–490. [CrossRef]
36. Tewdwr-Jones, M. Modern planning on film: Re-shaping space, image and representation. *Berkeley Plan. J.* **2013**, *26*, 86–106. [CrossRef]
37. Pielke, R.A., Jr. *The Honest Broker: Making Sense of Science in Policy and Politics*; Cambridge University Press: Cambridge, UK, 2007.
38. Grundmann, R. The problem of expertise in knowledge societies. *Minerva* **2017**, *55*, 25–48. [CrossRef]
39. Flyvbjerg, B. *Making Social Science Matter: Why Social Inquiry Fails and How It Can Succeed Again*; Cambridge University Press: Cambridge, UK, 2001.
40. Xiang, W.-N. Ecophronesis: The ecological practical wisdom for and from ecological practice. *Lands. Urban Plan.* **2016**, *155*, 53–60. [CrossRef]
41. von Gönner, J.; Herrmann, T.M.; Bruckermann, T.; Eichinger, M.; Hecker, S.; Klan, F.; Lorke, J.; Richter, A.; Sturm, U. Citizen science's transformative impact on science, citizen empowerment and socio-political processes. *Socio-Ecol. Pract. Res.* **2023**, *5*, 11–33. [CrossRef]
42. Powell, K. Making sense of place: Mapping as a multisensory research method. *Qual. Inq.* **2010**, *16*, 539–555. [CrossRef]
43. Goharipour, H.; Gibson, H. Reading the represented city and society: Signs, theory, and the dynamic interpretativeness of Peircean semiotics. *Qual. Res.* **2021**, *23*, 962–980. [CrossRef]
44. Koegst, L.; Kühne, O.; Edler, D. (Eds.) *Multisensory Landscapes Theories and Methods*; Springer: Wiesbaden, Germany, 2023.
45. Bergman, M.M. *Advances in Mixed Methods Research: Theories and Applications*; Sage: Los Angeles, CA, USA, 2008.
46. Roehr, D. *Multisensory Landscape Design: A designer's Guide for Seeing*; Routledge: Oxon, UK, 2022.
47. Willson, R. (Ed.) *Reflective Planning Practice: Theory, Cases, and Methods*; Routledge: New York, NY, USA, 2020.
48. Balsas, C. *Walkable Cities: Revitalization, Vibrancy and Sustainable Consumption*; SUNY Press: Albany, NY, USA, 2019.
49. Jackson, A.; Garcia-Zambrana, I.; Greenlee, A.J.; Lee, C.A.; Chrisinger, B. All talk no walk: Student perceptions on integration of diversity and practice in planning programs. *Plan. Pract. Res.* **2018**, *33*, 574–595. [CrossRef]
50. Angotti, T. What will a non-capitalist city look like? In *Cities Without Capitalism*; Sadri, H., Zeybekoglu, S., Eds.; Routledge: New York, NY, USA, 2022; pp. 187–199.
51. Sandercock, L. When strangers become neighbours: Managing cities of difference. *Plan. Theory Pract.* **2000**, *1*, 13–30. [CrossRef]
52. Njeru, A.M. Carfree streets as negotiated play spaces in Tokyo. *J. Urban Hist.* **2022**, *48*, 896–912. [CrossRef]
53. Ursic, M.; Imai, H. *Creativity in Tokyo: Revitalizing a Mature City*; Palgrave: Singapore, 2020.
54. Almazán, J. *Emergent Tokyo: Designing the Spontaneous City*; ORO Editions: Novato, CA, USA, 2023.
55. Gavitt, J.M.; Smardon, R.C. Calculating cultural ecosystem services as part of greenspace management? *Int. J. Res. Mark.* **2019**, *4*, 16–21. [CrossRef]
56. Scarce, R. *Sustaining this Place: Creating a New Hudson Region Landscape—Documentary*; G. Zero: Saratoga Springs, NY, USA, 2015.
57. Stacy, J.; Ryan, R.L.; Roy, A.H.; Milman, A. Homeowners' willingness to adopt environmentally beneficial landscape practices in an urbanizing watershed. *Cities Environ.* **2021**, *14*, 2. [CrossRef]
58. Hollander, J.B.; Renski, H.; Foster-Karim, C.; Wiley, A. Micro quality of life: Assessing health and well-being in and around public facilities in New York City. *Appl. Res. Qual. Life* **2020**, *15*, 791–812. [CrossRef]
59. Hirokawa, K.H.; Rosenbloom, J. Climate change adaptation and land use planning law. In *Research Handbook on Climate Change Adaptation Law*; Verschuuren, J., Ed.; Edward Elgar Publishing: Northampton, MA, USA, 2022; pp. 300–331.
60. Daniel, T.C.; Muhar, A.; Arnberger, A.; Aznar, O.; Boyd, J.W.; Chan, K.M.; Costanza, R.; Elmqvist, T.; Flint, C.G.; Gobster, P.H.; et al. Contributions of cultural services to the ecosystem services agenda. *Proc. Natl. Acad. Sci. USA* **2012**, *109*, 8812–8819. [CrossRef] [PubMed]
61. Ruzow Holland, A.H. If all planning is local, how are we going to save tomorrow? Ten pragmatic lessons from the field. *J. Environ. Stud. Sci.* **2022**, *12*, 177–192. [CrossRef] [PubMed]
62. Naiman, S.M.; Allred, S.B.; Stedman, R.C. Protecting place, protecting nature: Predicting place-protective behaviors among nature preserve visitors. *J. Environ. Stud. Sci.* **2021**, *11*, 610–622. [CrossRef]
63. Balsas, C. Exciting walk-only precincts in Asia, Europe and North-America. *Cities* **2021**, *112*, 103129. [CrossRef]

64. Wang, L.; Tosi, M.C.; Zordan, M.; Villani, C.; Maroso, S.; Pellizer, A.; Aymonino, A.; Talamini, G.; Turvani, M. *Walkable Cities in High Density China: Livable, Healthy, and Sustainable*; Tongji University Press: Shanghai, China, 2017.
65. Moura, F.; Cambra, P.; Gonçalves, A.B. Measuring walkability for distinct pedestrian groups with a participatory assessment method: A case study in Lisbon. *Landsc. Urban Plan.* **2017**, *157*, 282–296. [CrossRef]
66. Topal, H. Pedestrian Plazas: A new type of neighborhood space in New York City. In *Routledge Handbook of Urban Public Space: Use, Design, and Management*; Franck, K., Huang, T.-S., Eds.; Routledge: New York, NY, USA, 2023.
67. Kortekaas, K.H. Artisanal Saltscapes in Europe: From mining heritage to living cultural landscapes. *Ident. Terr. Cult. Patr.* **2019**, *8*. [CrossRef]
68. Balsas, C. Creative and sustainable town-gown a place triad *genius loci*. *Environ. Space Place* **2021**, *13*, 76–92. [CrossRef]
69. Tang, K.H.D. Green walls as mitigation of urban air pollution: A review of their effectiveness. *Res. Ecol.* **2023**, *5*, 1–13. [CrossRef]
70. Matthews, T.; Smith, G. 'Town + Gown' and CCNY's sustainability in the urban environment program. *Curr. Opin. Environ. Sustain.* **2015**, *17*, 42–47. [CrossRef]
71. Hosler, A.S.; Dharssi, A. Identifying retail food stores to evaluate the food environment. *Am. J. Prev. Med.* **2010**, *39*, 41–44. [CrossRef]
72. Roth, J.H. Kamikaze truckers in postwar Japan. *Transfers* **2019**, *9*, 1–19. [CrossRef]
73. Sandercock, L.; Attili, G. Digital ethnography as planning praxis: An experiment with film as social research, community engagement and policy dialogue. *Plan. Theory Pract.* **2010**, *11*, 23–45. [CrossRef]
74. Steinwald, M.; Harding, M.A.; Piacentini, R.V.; Siyi, W. Multisensory engagement with real nature relevant to real life. In *The Multisensory Museum: Cross-Disciplinary Perspectives on Touch, Sound, Smell, Memory, and Space*; Levent, N., Pascual-Leone, A., Eds.; Rowman & Littlefield: Lanham, MD, USA, 2014; pp. 45–60.
75. Beatley, T. Biophilic urbanism: Inviting nature back to our communities and into our lives. *Wm. Mary Envtl. L.* **2009**, *34*, 209–238.
76. Larson, L.R.; Lauber, T.B.; Kay, D.L.; Cutts, B.B. Local government capacity to respond to environmental change: Insights from towns in New York State. *Environ. Manag.* **2017**, *60*, 118–135. [CrossRef]
77. Fisher, R.A. *Statistical Methods for Research Workers*, 5th ed.; Oliver and Boyd: Edinburgh, UK, 1934.
78. Bloom, B. *Taxonomy of Educational Objectives, the Classification of Education Goals*; Longman: New York, NY, USA, 1956.
79. Sadik-Khan, J.; Solomonow, S. *Streetfight: Handbook for an Urban Revolution*; Penguin: New York, NY, USA, 2017.
80. Zacharias, J. Pedestrian behavior pedestrian behavior and perception in urban walking environments. *J. Plan. Lit.* **2001**, *16*, 3–18. [CrossRef]
81. Carmelino, G.; Hanazato, T. The built environment of Japanese shopping streets as visual information on pedestrian vibrancy. *Front. Arch. Res.* **2019**, *8*, 261–273. [CrossRef]
82. Mehta, V. What commerce can do: The case of streets in the US. In *Routledge Handbook of Urban Public Space: Use, Design, and Management*; Huang, T.-S., Franck, K.A., Eds.; Routledge: Oxon, UK, 2023.
83. Villani, C.; Talamini, G. Pedestrianised streets in the global neoliberal city: A battleground between hegemonic strategies of commodification and informal tactics of commoning. *Cities* **2021**, *108*, 102983. [CrossRef]
84. Hunt, M.A. Urban photography/cultural geography: Spaces, objects, events. *Geogr. Compass* **2014**, *8*, 151–168. [CrossRef]
85. Altamirano-Allende, C.; Selin, C. Seeing the City: Photography as a place of work. *J. Environ. Stud. Sci.* **2016**, *6*, 460–469. [CrossRef]
86. Cannavò, P.F. *The Working Landscape: Founding, Preservation, and the Politics of Place*; MIT Press: Cambridge, MA, USA, 2007.
87. Cooke, S.; Buckley, K. Visual research methodologies and the heritage of 'everyday' places. In *People-Centred Methodologies for Heritage Conservation: Exploring Emotional Attachments to Historic Urban Places*; Madgin, R., Lesh, J., Eds.; Routledge: Oxon, UK, 2021; pp. 143–155.
88. Michails, M. Toward an ecological urbanism: Public engagement in contemporary art practice. In *Media Art and the Urban Environment—Engendering Public Engagement with Urban Ecology*; Marchese, F.T., Ed.; Springer: Cham, Switzerland, 2015; pp. 1–46.
89. Kimura, A.H.; Kinchy, A. Citizen science: Probing the virtues and contexts of participatory research. *Engag. Sci. Technol. Soc.* **2016**, *2*, 331–361. [CrossRef]
90. Schneller, A.; Hannan, S.; Bah, H.I.; Livecchi, S.; Pettigrew, S. Environmental justice is exhausting: Five decades of air pollution and community advocacy at Ezra Prentice Homes in Albany, New York. *Local Environ.* **2022**, *27*, 1514–1535. [CrossRef]
91. Kher Kaw, J.; Lee, H.; Wahba, S. (Eds.) *The Hidden Wealth of Cities: Creating, Financing, and Managing Public Spaces*; World Bank: Washington DC, USA, 2020.
92. Freire, P. *Pedagogy of the Oppressed*; Herder and Herder: New York, NY, USA, 1970.

Disclaimer/Publisher's Note: The statements, opinions and data contained in all publications are solely those of the individual author(s) and contributor(s) and not of MDPI and/or the editor(s). MDPI and/or the editor(s) disclaim responsibility for any injury to people or property resulting from any ideas, methods, instructions or products referred to in the content.

land

MDPI

Article

The Effects of Low-Impact Development Best Management Practices on Reducing Stormwater Caused by Land Use Changes in Urban Areas: A Case Study of Tehran City, Iran

Sajedeh Rostamzadeh [1], Bahram Malekmohammadi [1], Fatemeh Mashhadimohammadzadehvazifeh [2] and Jamal Jokar Arsanjani [3,*]

1 Department of Environmental Planning, Faculty of Environment, University of Tehran, Tehran 1417853111, Iran; rostamzade.sajede@ut.ac.ir (S.R.); malekb@ut.ac.ir (B.M.)
2 Washington-Alexandria Architecture Center, School of Architecture, Virginia Polytechnic Institute and State University, Alexandria, VA 22314, USA; fatemehmmv@vt.edu
3 Geoinformatics Research Group, Department of Planning and Development, Aalborg University Copenhagen, A.C. Meyers Vænge 15, DK-2450 Copenhagen, Denmark
* Correspondence: jja@plan.aau.dk

Abstract: Urbanization growth and climate change have increased the frequency and severity of floods in urban areas. One of the effective methods for reducing stormwater volume and managing urban floods is the low-impact development best management practice (LID-BMP). This study aims to mitigate flood volume and peak discharge caused by land use changes in the Darabad basin located in Tehran, Iran, using LID-BMPs. For this purpose, land use maps were extracted for a period of 23 years from 2000 to 2022 using Landsat satellite images. Then, by using a combination of geographic information system-based multi-criteria decision analysis (GIS-MCDA) method and spatial criteria, four types of LID-BMPs, including bioretention basin, green roof, grass swale, and porous pavement, were located in the study area. Next, rainfall–runoff modeling was applied to calculate the changes in the mentioned criteria due to land use changes and the application of LID-BMPs in the area using soil conservation service curve number (SCS-CN) method. The simulation results showed that the rise in built-up land use from 43.49 to 56.51 percent between the period has increased the flood volume and peak discharge of 25-year return period by approximately 60 percent. The simulation results also indicated that the combined use of the four selected types of LID-BMPs will lead to a greater decrease in stormwater volume and peak discharge. According to the results, LID-BMPs perform better in shorter return periods in a way that the average percentage of flood volume and peak discharge reduction in a 2-year return period were 36.75 and 34.96 percent, while they were 31.37 and 26.5 percent in a 100-year return period.

Keywords: LID-BMP; LULC change; urban stormwater; GIS-MCDA; urban growth

1. Introduction

The ever-increasing growth in the world population, especially in cities, and global warming, on the one hand, have caused the growth in urbanization, increasing the impermeability of urban surfaces, significantly reducing the ratio of infiltration capacity, raising the intensity of surface stormwater, decreasing the time of concentration, and declining rainfall [1], and on the other hand, it has caused an increase in atmospheric humidity, evaporation, and transpiration and related changes in rainfall patterns [2]. Climate change, even on a small scale, can increase the intensity of rainfall in urban areas [3]. The physical growth of cities (especially metropolitan cities), rapidly shifting patterns of land use, and demographic increase cause the natural vegetation cover and the capacity of rainfall infiltration to decrease on the surface of the earth; therefore, the risk of flooding in megacities has raised [4,5]. Although most metropolitan cities now use pumping stations and sewage systems to facilitate flood evacuation from urban areas, when the duration and intensity of a flood event exceed the maximum loading capacity of the urban sewage system, the residents of the surrounding areas may suffer from financial and life losses caused by the flood [6,7].

One of the effective ways to reduce the damage caused by floods in urban areas is to upgrade and expand the capacity of drainage systems; however, it has been proven in various studies that increasing the capacity of drainage systems can be costly, unsustainable, and even impractical, especially in dense urban areas [8–10]. In addition, drainage systems that have already been designed in urban areas are usually based on old climatic data and do not consider changes in the nature of rainfall over time [11,12]. Therefore, new techniques such as bioretention basin, green roof, grass swale, and porous pavement have been developed to deal with the problem of urban stormwater during floods. All of these techniques are called low-impact development best management practices (LID-BMPs) [13]. LID-BMPs, as innovative approaches in environmental management and urban improvement, play a vital role in strategic planning and urban sustainability. These strategies, including permeable pavements, green roofs, rain gardens, and other similar solutions, significantly enhance groundwater recharge, reduce runoff [14], and mitigate urban heat islands while integrating ecological [15] and social equity into urban design [16]. They strengthen cities' resilience to climate change by effectively managing stormwater, reducing flooding risks [17], and benefiting underserved areas through equitable urban development [18]. Additionally, these practices improve public health by creating green spaces, lower long-term infrastructure costs by reducing reliance on traditional systems, and foster sustainable economic growth aligned with long-term development goals [19]. Although LID-BMPs offer numerous advantages, their implementation is not without challenges. In some cases, the costs of extending stormwater systems with devices aimed at restoring natural water conditions can exceed those of traditional drainage systems [20]. Additionally, space limitations, especially in densely populated urban areas, may hinder their development [21]. Factors such as soil permeability and potential contamination risks also significantly influence the feasibility of these systems. Addressing these considerations ensures the effective and sustainable integration of LID-BMPs into urban environments.

Overall, LID-BMP serves as a multifaceted tool within strategic urban planning frameworks, providing environmental, social, and economic benefits that are essential for sustainable development. LID-BMP is one of the stormwater management strategies to increase permeability, reduce surface stormwater, and reuse water [22], which has been implemented in recent years to boost urban resilience and diminish flood risk [23,24]. In various studies, the efficiency of LID-BMPs has been reported as an effective method to lessen the effects of urban floods [2,18,25–27]. In other words, these studies show that different LID-BMP strategies can have different benefits for urban ecosystems in different geographical locations. In addition to decreasing the effects of urbanization on the basin's hydrology, LID-BMPs increase the level of groundwater, reduce stormwater, and preserve the natural environment. For instance, a study concluded that a green roof can reduce 60 to 70% of the volume of stormwater compared to conventional roofs [23]. Similarly, an experiment conducted found that green roofs can reduce stormwater volume by 30–78% in comparison with conventional roofs, in addition to reducing discharge time [28]. However, the optimal combination of LID-BMP measures varies in different severity zones, highlighting the need for a balanced approach to achieve sustainable development in sponge cities [29]. LID-BMPs can also reduce or eliminate the need to make ponds that are created to store rainwater in urban areas and usually occupy a large area from the occupancy level [2]. They provide the possibility of high-density urbanization with the least negative hydrological effects, and as a result, they can be considered as a positive measure in urban development in the direction of infrastructure development [30]. Another important feature of LID-BMPs is their affordability. For example, the economic importance of LID-BMPs was investigated in Hong Kong by presenting a cost–benefit model and concluded that USD 1.2 billion and 5.3 billion of environmental and economic benefits have been saved in a 30-year period [31]. According to the mentioned studies, LID-BMPs can be considered as an alternative to the old and traditional ways of managing stormwater at the city level and provide suitable solutions for the coping capacity against climate change and the sustainable development of cities.

Tehran, receiving considerable rainfall, is located at the foot of the Alborz Mountain range. Many rivers drain the water from the rain and transfer it to the downstream plains. The general direction of the rivers and canals is mainly towards the urban area, and this has caused the residents to witness waterlogging of the streets due to torrential rains and floods in case of heavy and continuous rains. At the same time, the urban area of Tehran is the passageway of many channels and rivers that are responsible for the water drainage of the upstream basins and also the collection of rainwater in the urban area. In some parts of the city where the cost of land is high, beds of rivers and canals inside the city have been encroached upon and their cross-section has been reduced and limited. This issue can affect the natural regime and flow of the river and cause irreparable damages in the time of heavy rainfalls. Therefore, the purpose of this study is to use LID-BMPs to compensate for the decline in permeability of the surfaces due to the rise in urban land use and the escalation in stormwater due to rainfall in a basin located in the northeast of Tehran. Among the innovations of this study is an integrated investigation of the cause (change in land use) to the solution (use of LID-BMPs) for urban stormwater management all in one study. Another novelty of this study is the use of a combination of the GIS-MCDA method and spatial criteria for the site selection of LID-BMPs. In addition, eight criteria, including flooding potential, were used to select the optimal locations for installing the LID-BMPs. In this study, four proposed scenarios of the use of bioretention basin, green roof, grass swale, and porous pavement are investigated.

2. Materials and Methods

2.1. Study Area

The Darabad basin, situated in the northeast of Tehran (35°49′31″ N, 51°30′12″ E), covers an area of 3987.39 hectares and receives an average annual rainfall of 380 mm. Tehran, the capital of Iran, is a highly populated city, with a population of 15.7 million, accounting for one-fifth of the country's total population, based on the 2016 census [32]. The northern area of Tehran city is connected within the Alborz Mountain ranges and has a length of about 40 km. The Darabad river originates from the Alborz Mountain ranges at an altitude of 3160 m above the mean sea level (amsl), and then, it is directed toward the inner city channels, and after the city sewage enters it, it finally terminates in the Kavir plain at an elevation of about 1700 m amsl. The lack of proper land use planning in the region can be deduced from the following: i. soil erosion, ii. inadequate urban drainage system, iii. building urban settlements up to the boundary of the protected area, iv. The poor management of public green spaces, and v. the absence of LID-BMPs and BMPs for managing stormwater [33]. Human activities along with natural influences (intense extreme events) have aggravated the vulnerability of the basin [34]. In recent years, as urbanization progresses, the basin has witnessed the transformation of large areas of gardens, grass land, and bare land into construction land, which has increased the amount of impervious area. A display of the study area is shown in Figure 1.

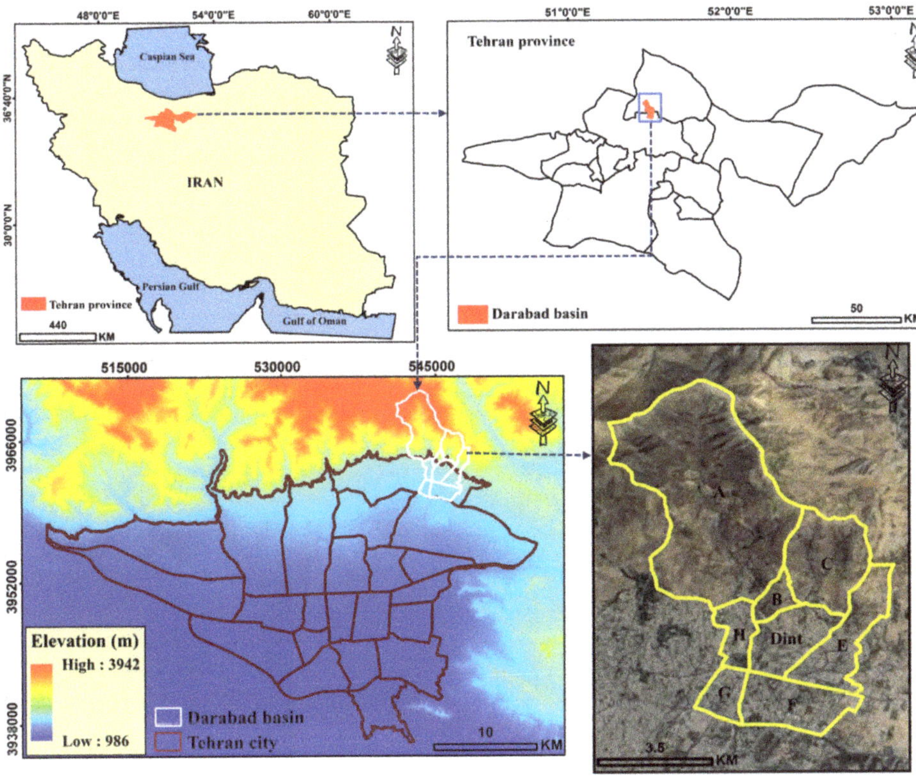

Figure 1. The map of the geographical location of the study area in Iran and Tehran.

2.2. Data

The materials used in this study mainly consist of land use data, meteorological data, and basic geographic information. In terms of land use data, two sets of temporal multi-spectral satellite images of 30 m resolution for the years 2000 and 2022 were acquired from the United States Geological Survey (USGS) website (https://earthexplorer.usgs.gov/, accessed on 24 December 2024). The average annual rainfall and the average maximum rainfall of 24 h of the Darabad basin were obtained from I.R. of Iran Meteorological Organization (https://www.tehranmet.ir, accessed on 24 December 2024). In addition, the maximum instantaneous discharge statistics of Qalak rain gauge, which is located inside the basin, were acquired from Regional Water Company of Tehran (https://thrw.ir/, accessed on 24 December 2024). The digitized topographic map of the basin with a spatial resolution of 5 m was taken from the Municipality of District 1 of Tehran city.

2.3. Methodology

The general procedure carried out in this research is depicted in a flowchart in Figure 2.

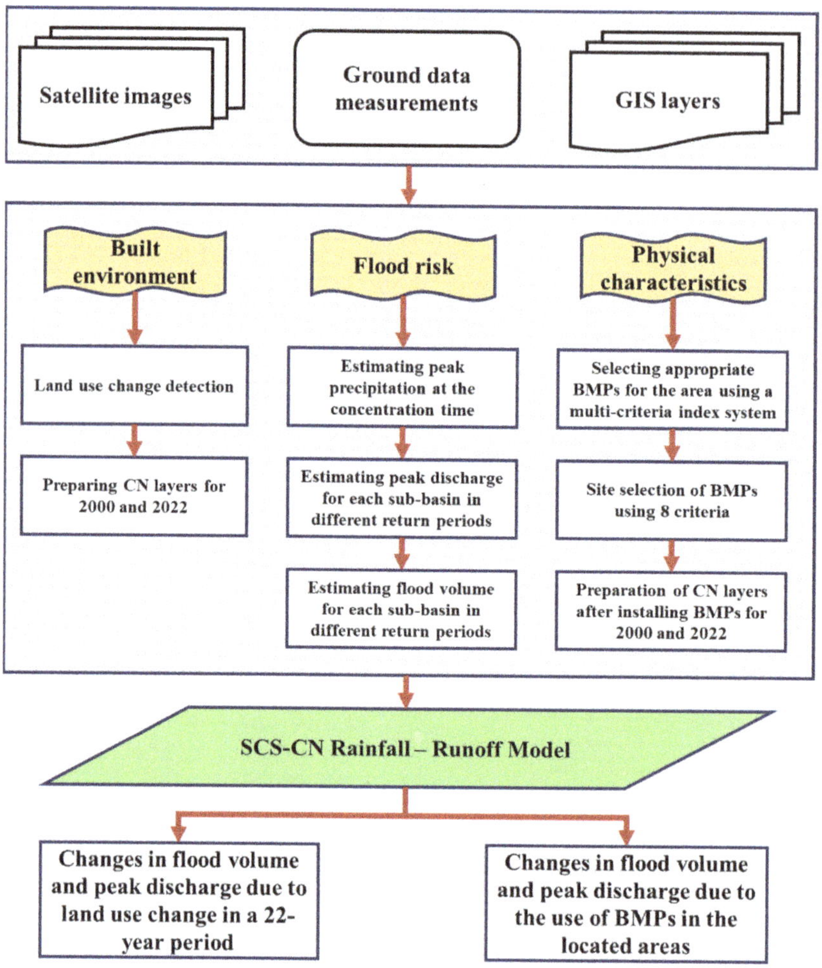

Figure 2. The flowchart of the main research steps.

2.3.1. Land Use Land Cover (LULC) Change

Recognizing LULC is a crucial component of environmental analysis, planning, and management [35], since these changes are regarded as one of the most significant environmental problems facing the world today [36]. Meanwhile, satellite imagery has been widely utilized to track environmental degradation and changes in LULC over time [37]. The time range of the Landsat series of satellites, from 1975 to the present time, has made it the best option for studying urban physical development [38]. The details corresponding to the satellite data are given in Table 1. The screening and processing of enormous amounts of remote sensing data are essential in remote sensing technology-based research and remote sensing image analysis [39]. After applying radiometric and atmospheric correction to the images, a total of five LULC classes, i.e., built-up (residential, commercial, industrial, and traffic land), barren land (bare soil without any cover), vegetation (parks and gardens), rangeland (shrubs and grasses), and ridge (mountainous lands), were identified. The LULC classification and mapping process includes the post-classification accuracy assessment, which is used to evaluate the accuracy of the classified maps. The accuracy of a classified map should not be less than 80% for accurate interpretability and identification [40]. Utilizing 400 randomly chosen points, the Kappa coefficient technique is employed in this study to assess the precision of LULC maps of the Darabad basin. In the selection of the points, it was performed in a way that covers all the study area and all the classes. Since field data were not available for 2000, the ground observations for this year were acquired from the Google Earth Pro platform. Additionally, for 2022, the ground observations were obtained from a combination of field visits and utilizing the Google Earth Pro platform. The accuracy assessment results indicate that the overall accuracy level was 93.56 percent and 96.80 percent for the years 2000 and 2022, respectively. The corresponding Kappa statistics were 0.8884 and 0.9441, respectively.

Table 1. Details of the satellite data.

Satellite	Sensor	Spectral Bands	Month/Year	Path/Row
Landsat 7	ETM+	8	July 2000	164/35
Landsat 8	OLI/TIRS	11	July 2022	164/35

2.3.2. Hydrology

Hydrological Divisions of the Area

In order to extract more accurate and uniform information, the study area was divided into 7 hydrological units and 1 non-hydrological unit. A non-hydrological unit is a basin that does not have an independent stream, and water enters it from adjacent basins. The division of sub-basins along with the digital elevation model (DEM) of the Darabad basin is shown in Figure 3.

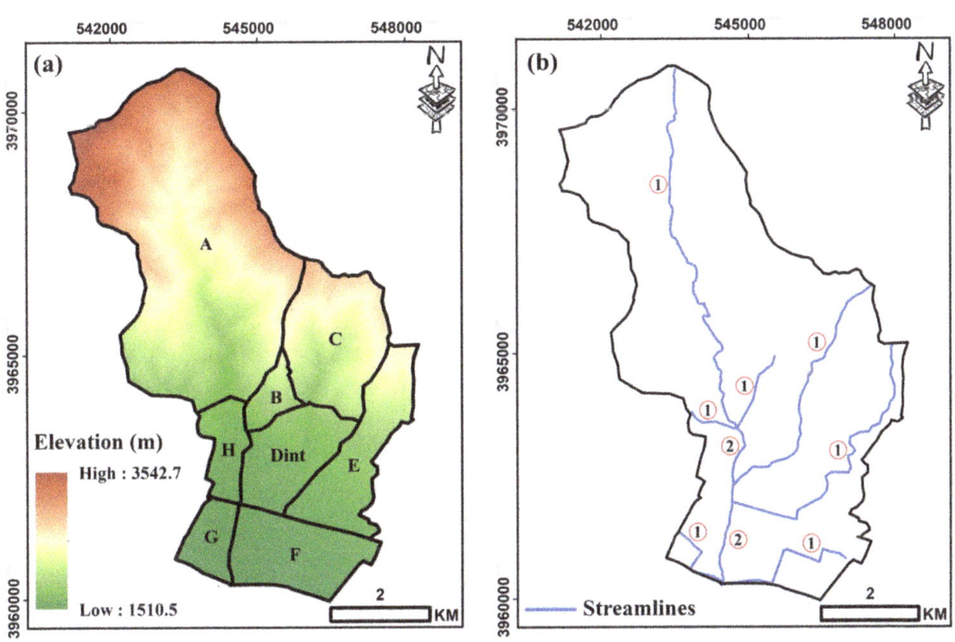

Figure 3. (**a**) Sub-basin divisions with the DEM of the Darabad basin and (**b**) streamlines and stream order of the Darabad basin.

Longitudinal Profile of the Main Stream and Slope Calculation

Knowing the slope of different points of a river's path can give a clear picture of the river's destructive power in different places. In addition, in order to calculate the time of concentration (TOC) of streams in the sub-basins, it is necessary to calculate their slope first. For this purpose, the longitudinal profile of streams was calculated and drawn separately for each sub-basin. The longitudinal profile of the Darabad river is obtained by transferring the length of the river on the X-axis and the height of its different points on the Y-axis and with the help of contour lines from topographic maps. In drawing the longitudinal profile of the main river of the study area and its hydrological units, the digitized topographic map (with a spatial resolution of 5 m) of the area was used. According to the profile, sub-basin A has the highest and sub-basin G has the lowest slope with 1.09 and 36 percent, respectively.

Calculation of Stormwater Using the Soil Conservation Service Curve Number (SCS-CN) Method

One approach for calculating the direct stormwater depth for a rainfall event is SCS-CN. Equations (1) and (2) are used in surface stormwater prediction equations based on the SCS-CN approach [41]. CN is determined based on the soil type, land use and land cover, hydrological conditions, and antecedent moisture conditions of the soil acquired from the SCS [42]. In this way, after preparing the map of the hydrological groups of the basin and integrating it with the land use map, the value of CN and then its weighted average were calculated using Equations (1) and (2).

$$Q_{SCC-CN} = \frac{(P - 0.2S)^2}{P + 0.8S} \tag{1}$$

$$S = \frac{25400}{C_N} - 254 \tag{2}$$

where Q—direct stormwater (mm); P—total is precipitation (mm); S—maximum retention capacity obtained as per the SCS-CN method (mm); and C_N—curve number that depends on the soil type [43].

TOC

One of the most pivotal and widely used physical parameters of a basin, especially in hydrological equations, is the TOC. TOC is the interval of time between the centroid of rainfall excess and the recession's inflection point for direct stormwater [44]. The time needed for a stormwater particle to move from the watershed boundary that is hydraulically farthest away along the longest watercourse to the basin outlet is the physical definition of TOC [45]. The TOC in each basin depends on several factors, including the slope of the basin, the slope of the waterway, the geological structure, the type and amount of vegetation, the intensity of rainfall, and the amount and distribution of rainfall in time and place. This study employed the Kirpich method (Equation (3)) for calculating the TOC.

$$T = 0.0663L^{0.77}S_o^{-0.305} \tag{3}$$

where T_c is time of concentration by the Kirpich method in hours, L is the main channel length in kilometers, and S_o is the main channel slope in meters per meters [46]. In this study, the watershed was divided into eight distinct sub-basins to enhance the precision of the hydrological analysis. For each sub-basin, both the basin slope and the main channel slope were computed. The results revealed minimal differences between the two parameters within most sub-basins, with the exception of sub-basin A (Table 2). As a result, the main channel slope was chosen as the primary parameter for calculating the specific TOC for each sub-basin. This approach was deemed both practical and representative of the flow dynamics within the sub-basins, reflecting the key characteristics driving water movement.

Table 2. Comparison of the average channel slope and basin slope for sub-basins.

Sub-Basin	A	B	C	Dint	E	F	G	H
Basin slope (%)	47.79	41.11	45.03	20.21	24.14	6.30	4.31	13.16
Channel slope (%)	44.93	39.87	43.67	19.30	23.34	6	3.93	12.38
Difference (%)	2.86	1.24	1.36	0.91	0.80	0.30	0.38	0.78

2.3.3. Meteorology

In order to implement the SCS-CN rainfall–runoff model, the maximum rainfall at the TOC is required, and its calculation requires the average annual rainfall of the basin and the average maximum rainfall of 24 h. The minimum, average, and maximum annual rainfall statistics of the basin were calculated from three meteorological stations around the basin from 1990 to 2022 and can be seen in Figure 4.

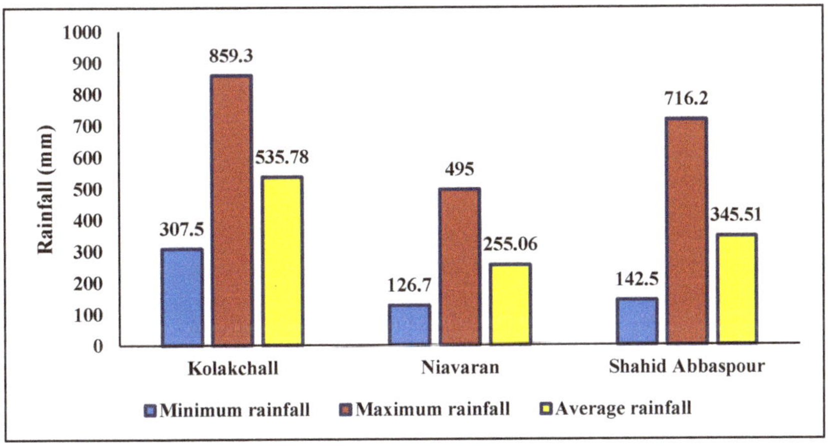

Figure 4. The rainfall statistics of meteorological stations in the basin in mm (Source: Tehran Regional Water Company).

Rainfall Gradient in the Area

In a general sense, the gradient represents the change in a phenomenon. In the study of rainfall gradient, a logical relationship is established between rainfall as the dependent variable and altitude or longitude and latitude as independent variables, to estimate the amount of rainfall in different parts of the basin. To determine the rainfall gradient in the area using meteorological station data, the relationship between rainfall and altitude in the basin and sub-basins was investigated. Due to the absence of stations within the study area, data from the nearest available stations were used. These stations, as shown in Figure 5, are located at varying elevations. The presence of meteorological stations at different elevations helps improve the accuracy of analyses, covers diverse areas, and enhances hydrological modeling. Although the trend line is based on three points, these points were selected as the most representative data available for the study area to ensure the reliability of the analysis within the given constraints. The results are presented separately below. Using the 33-year average rainfall statistics, the relationship between rainfall and altitude in the study area was analyzed, and the rainfall gradient was finally obtained. In Figure 5, the relationship and correlation diagram between rainfall amount and the elevation of meteorological stations are shown. The coefficient of determination obtained for the rainfall gradient equation in the study area is 0.95, which is acceptable.

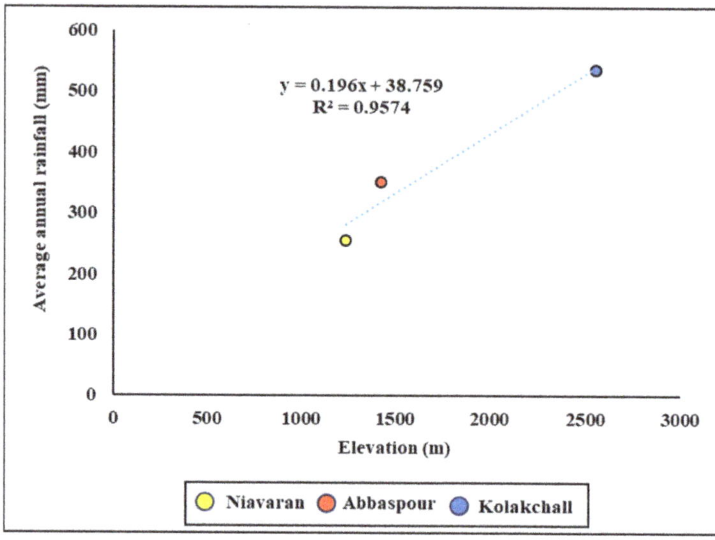

Figure 5. Annual rainfall gradient diagram of study area.

Average Maximum 24-Hour Rainfall in Different Return Periods

This parameter, which is very important in the design of structures and calculations related to the maximum instantaneous discharge, can be analyzed using the data recorded in nearby meteorological stations. Kolakchall (28 years), Niavaran (30 years), and Shahid Abbaspur (24 years) stations were the stations with the data of average maximum 24-hour rainfall around the Darabad basin, and all their data were used in this research. Using the EasyFit 5.6 software, the average maximum 24-hour rainfall was calculated in different return periods. EasyFit is a software tool used for fitting probability distributions to data. It provides a wide range of statistical distributions and helps in selecting the best-fitting distribution based on the data. In Table 3, the selected statistical distributions of average maximum 24-hour rainfall for the meteorological stations are given. The Wakeby and LogLogistic distributions are statistical models often used in hydrology and other fields to analyze extreme events like floods or rainfall. The Wakeby distribution is a five-parameter model defined by ξ (location), α (shape), β (shape), γ (scale), and δ (shape). The LogLogistic distribution, also used for skewed data such as rainfall or survival times, has parameters such as α (scale), β (shape), and γ (location). In EasyFit software, these parameters are derived from dataset and displayed with their values to help model the data accurately. In Table 4, the average maximum 24-hour rainfall (mm) in different return periods for the selected stations along with the gradient of annual rainfall frequency has been calculated.

Table 3. Selected statistical distributions used for calculating average maximum 24-hour rainfall (along with distribution parameters) for index stations.

Meteorological Station	Average Annual Rainfall	Statistical Distribution
Kolakchall	511.7	Wakeby (ξ: 189.15; α: 11.046; β: 8.697; γ: 0.19457; δ: 16.588)
Niavaran	243.3	LogLogistic (α: 3.7741; β: 19.409; γ: 7.1402)
Abbaspour	334	LogLogistic (α: 2.558; β: 15.643; γ: 17.546)

Table 4. The average maximum 24-hour rainfall (mm) in different return periods for the selected stations along with the gradient of annual rainfall frequency.

Meteorological Station		Average Maximum 24-Hour Rainfall with Return Period (Year)					
		2	5	10	25	50	100
Kolakchall		40.2	50.2	59.0	72.6	84.7	98.5
Niavaran		26.6	35.2	41.9	52.2	61.6	72.7
Abbaspour		33.2	44.4	54.5	71.7	89.2	111.8
Rainfall gradient	R2	0.99	0.95	0.90	0.78	0.65	0.50
	Gradient of a line	0.0491	0.0526	0.0584	0.0663	0.0708	0.0724
	Intercept	15.47	24.17	30.58	41.44	52.78	68.06

Analysis of Short-Term Rainfall and the Intensity, Duration, and Frequency (IDF) Curve

The intensity of rain in a rain storm is the amount of rain per unit of time, which is usually expressed in millimeters per hour or inches per hour. In general, the shorter the duration of the rainfall, the greater its intensity and vice versa. Rainfall intensity–duration equations change with frequency or return period. The longer the return period, the more intense rains are expected. Due to the lack of suitable stations containing information near the study area, IDF diagrams in this report were drawn using experimental and calibrated formulas. Risk management requires a precise and trustworthy estimation of the amount of rainfall indicated by the points on the IDF curves [47]. By analyzing the relationship between the intensity, duration, and frequency of observed rainfall data, it is possible to estimate the intensity of extreme rainfall events with a particular duration and return period [48]. There are diverse equations to calculate the intensity of short-term rainfall with different return periods, and in this report, Ghahraman and Abkhezr method [49], which is a generalized Bell method for Iran, has been used. In this method, Iran is divided into six regions, and for each of these regions, specific equations are provided to calculate one-hour rainfall with a return period of 10 years. Based on this zoning, the Darabad basin is located in District 2. The relation between one-hour rainfall calculation and 10-year return period in this area is given in Equation (4):

$$P_{60}^{10} = 9.99 + 0.212P_{1440}^{2} \tag{4}$$

where P_{60}^{10} is 60-minute rainfall with a return period of 10 years, and another parameter of the equation is the average maximum daily rainfall in the study area. To determine the amount of rainfall in the return period and the duration of different rainfall, Equations (5) and (6) are used for the duration of rainfall of less and more than 2 h:

$$P_t^T = 0.1372t^{0.4778}[0.4608 + 0.2349Ln(T - 0.62)]P_{60}^{10} \tag{5}$$

$$P_t^T = 0.2009t^{0.3937}[0.5565 + 0.1948Ln(T - 0.8)]P_{60}^{10} \quad (6)$$

Equation (5) is used for rainfalls of up to 2 h, and Equation (6) is used for rainfalls of more than 2 h. In the above equations, t is the duration of rainfall in minutes, T is the return period in years, and t is minutes of rainfall with a return period of T years. After calculating the TOC using the Kirpich method and placing it in the above equations, the amount of rainfall and its intensity in different return periods were obtained separately for each sub-basin. The diagram related to the IDF of the basin is given in Figure 6.

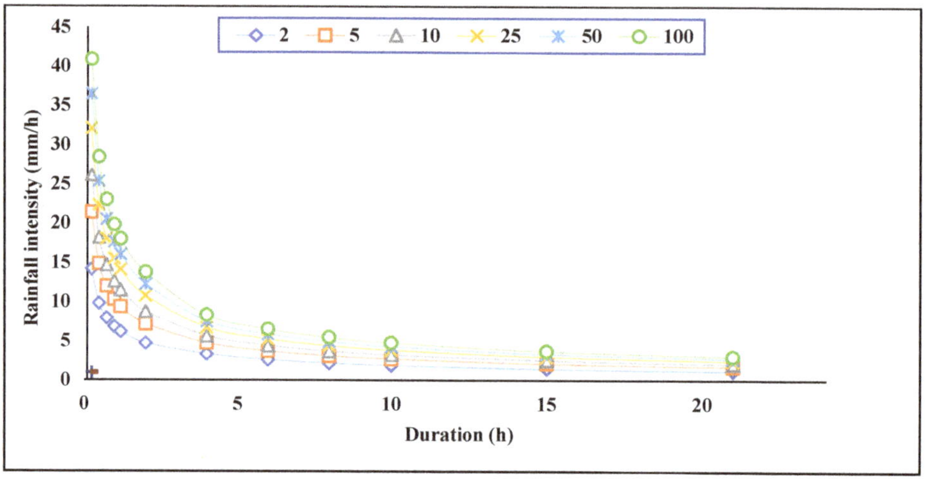

Figure 6. The intensity–duration–frequency curve of short-term rainfall by Ghahraman's method in the basin.

Unit Hydrograph of Sub-Basins by SCS-CN Method

One of the most important steps in hydrological analysis is the calculation of the unit hydrograph. In this research, the SCS-CN empirical method and the following relations are used to calculate the unit hydrograph.

$$\begin{cases} Q_p = \frac{0.2083AQ}{T_p} \\ T_p = \frac{D}{2} + T_i \\ D = 0.1333T_c \\ T_i = 0.6T_c \\ T_b = 2.67T_p \end{cases} \quad (7)$$

In above equations, Q_p is the peak discharge of the unit hydrograph (m^3/s), A is the area of the basin (km^2), Q is the surface stormwater height (mm), T_p is the time to reach the highest point of the unit hydrograph (h), D is the duration of effective rainfall of unit hydrograph (h), T_l is the lag time of the basin from the center of gravity of rainfall to the peak of the hydrograph (h), T_b is the base time of the unit hydrograph (h), and T_c is the time of concentration of the basin (h) [50].

To obtain unit hydrograph dimensions, the value of TP and QP must be multiplied in the SCS-CN dimensionless table; in this way, the dimensions of the hydrograph corresponding to each hydrological unit can be obtained. The dimensions of the unit hydrograph of each sub-basin, as well as the entire basin, are given in Table 5. Using the dimensions of the unit hydrograph obtained by the SCS-CN method and also the amounts of rainfall excess for different return periods, the dimensions of the flood hydrograph and the maximum values of the flood discharge were estimated.

Table 5. Unit hydrograph dimensions of sub-basins for one centimeter of stormwater by SCS-CN method.

Sub-Basin	Area (km^2)	T_c (h)	D (h)	T_i (h)	T_p (h)	Q_p (m^3/s)
A	19.81	0.66	0.09	0.40	0.44	9.31
B	0.96	0.15	0.02	0.09	0.10	1.98
C	5.13	0.25	0.03	0.15	0.17	6.37
Dint	3.24	0.38	0.05	0.23	0.26	2.63
E	3.99	0.45	0.06	0.27	0.30	2.75
F	3.78	0.62	0.08	0.37	0.41	1.91
G	1.40	0.54	0.07	0.32	0.36	0.81
H	1.57	0.19	0.03	0.11	0.13	2.60
Darabad basin	39.87	0.93	0.12	0.56	0.62	13.39

Table 5. *Cont.*

2.3.4. Selection of Suitable LID-BMPs

There are many types of LID-BMPs and possible sites for their installation, and the process of selecting LID-BMPs is a complex one, given the diversity in site conditions, performance, and cost of LID-BMP [51]. In this research, the multi-criteria index system (MCIS) method [52] was used to select LID-BMPs. The evaluation process was conducted based on three main criteria, i.e., site suitability, stormwater control benefits, and cost and maintenance, along with a set of sub-criteria. The sub-criteria under site suitability included land availability, groundwater characteristics, site conditions, slope, and soil type, while stormwater control benefits were evaluated based on runoff reduction efficiency, water quality improvement, and flood risk mitigation. The cost and maintenance criterion included construction costs, maintenance requirements, and lifespan.

To ensure a robust evaluation, a panel of experts, including water resource engineers, urban planners, environmental civil engineers, GIS experts, and environmental engineers, was formed to score each LID-BMPs against the defined criteria and sub-criteria. A range of potential LID-BMPs was initially considered, including constructed wetlands, infiltration basins, wet detention ponds, dry detention ponds, bioretention basins, green roofs, grass swales, and porous pavements. However, certain options, such as constructed wetlands, infiltration basins, wet detention ponds, and dry detention ponds, were excluded from the final list due to the need for a large area, which is practically not feasible in the Darabad basin due to the high price of land.

The scoring process was performed using a weighted average approach, where weights were assigned to each criterion based on their importance to the study objectives. As a result, four LID-BMPs with the highest scores were selected as suitable options for the study area: 1—bioretention basin; 2—green roof; 3—grass swale; and 4—porous pavement.

2.3.5. Optimal Location Identification for LID-BMPs

In order to determine the optimal locations for LID-BMPs, 8 criteria of slope, land use, rainfall, flooding potential, groundwater level, distance from the waterway, distance from the street, and distance from the fault line were used. Since the unit of measurement of each of these criteria and their valuation range is different, it is necessary to compare all the criteria with the same unit of measurement; fuzzy logic was used for this purpose. A fuzzy logic system is distinctive in that it can manage linguistic information and numerical data at the same time. To use fuzzy logic in ArcGIS, it is necessary to assign a weight to each criterion (layer), and the greater weight indicates the greater importance of that criterion. The assigned weights were extracted by the analytical hierarchy process (AHP) using Expert Choice 11 software. To derive these weights, a structured questionnaire was designed, which was completed by 25 experts specializing in water resource engineering, urban planning, environmental civil engineering, GIS, and environmental engineering. The experts were selected based on their professional experience and familiarity with LID-BMP implementation in urban basins.

Each expert provided pairwise comparisons of criteria and sub-criteria based on their judgment, and the final results were obtained by aggregating these individual judgments using the geometric mean method. This approach ensures a balanced and representative evaluation, reflecting the collective expertise of the panel [53]. The consistency ratio (CR) was also calculated to confirm the reliability of the judgments, with all values falling below the acceptable threshold of 0.1, indicating consistent responses [54,55]. To assign optimal locations to the LID-BMPs using the fuzzy-AHP method, it is necessary to fuzzify each criterion according to its characteristics, with fuzzy membership functions. It should be noted that the type of membership function for each criterion has been determined using a questionnaire and experts' opinions. Table 6 shows the fuzzy functions and weight of each criterion.

Table 6. Fuzzy membership functions and weights of criteria.

Criterion	Fuzzy Function		Weight			
			Bioretention Basin	Green Roof	Grass Swale	Porous Pavement
Slope	Linear (decrease)	Min: 20 Max: 0	0.217	0.097	0.243	0.297
Groundwater level	Linear (increase)	Min: 1.22 Max: 97.3	0.062	0.04	0.138	0.115
Distance from waterway	Linear (increase)	Min: 0 Max: 2220	0.04	0.041	0.057	0.035
Distance from street	Linear (decrease)	Min: 6556.7 Max: 0	0.122	0.045	0.303	0.224
Distance from fault line	Linear (increase)	Min: 0 Max: 5.790	0.083	0.068	0.064	0.057
Flooding potential	Linear (increase)	Min: 0.66 Max: 1.28	0.073	0.14	0.053	0.078
Rainfall	Linear (increase)	Min: 432 Max: 555.12	0.032	0.231	0.034	0.056
Land use	User defined		0.371	0.338	0.108	0.138

For the land use criterion, since 4 different types of LID-BMPs have been examined in this research and the priority of use for each of them is different, the land use scoring map for each LID-BMP was given separately. It should be noted that a land use type with a lower priority number in Table 7 has a higher number of points in the land use scoring map. To prepare Table 7, a comprehensive questionnaire was designed and distributed among experts in water resource engineering, urban planning, environmental civil engineering, GIS, and environmental engineering. The responses were analyzed using statistical methods to extract the priority of different land use types for each LID-BMP. The resulting priorities were then standardized on a scale from 1 to 10, where a higher score indicates better compatibility of a specific land use type with the corresponding LID-BMP. This table serves as the basis for fuzzifying the land use layer, with each priority level reflecting its contribution to the overall suitability analysis.

After preparing the fuzzy layers of criteria, it is necessary to multiply them by their obtained weights using the AHP method to obtain weighted fuzzy layers. According to the different functions of LID-BMPs, the relative weight of each criterion is also different in these systems. For example, for grass swale, the distance from the street is the first priority, and then, the slope criterion has the highest relative weight, while for the bioretention basin, due to the possibility of using the stored stormwater in the irrigation of green spaces or the car wash, the land use criterion holds the highest importance [56]. Table 6 shows the fuzzy membership functions and the resulting weight for each criterion separately for 4 types of LID-BMPs.

Table 7. Locative priority for each type of LID-BMP.

Land Use	Location Priority			
	Bioretention Basin	Green Roof	Grass Swale	Porous Pavement
Public services	5	2	4	4
Pertaining to the police	10	10	10	10
Residential garden	3	5	5	9
Residential	4	1	6	8
Green space	1	-	2	1
Recreational	2	-	3	2
Religulous	5	4	7	4
Parking	6	6	7	3
Bare land	6	-	1	-
Rocky land	5	-	5	-
Commercial–industrial–therapeutic–infrastructure	8	3	9	5

To fuzzify the land use layer, since the priority of each kind of land use is different for each type of LID-BMP, the User Defined function is used in the TerrSet 19.0 software. TerrSet is an integrated geospatial software designed for land planning, monitoring, and analysis, widely used for environmental management and urban planning applications [57]. The User Defined function in TerrSet allows for customized fuzzy membership functions to be applied, ensuring the accurate representation of expert-derived priorities.

In Table 7, the priority of each kind of land use for each LID-BMP is shown, which was extracted using the aforementioned questionnaire and expert opinions. The values in this table range from 1 to 10, where a higher number indicates a better compatibility between the land use and the corresponding LID-BMP.

In the next stage of locating the LID-BMPs, it is necessary to overlay the weighted fuzzy layers of the criteria using one of the fuzzy overlay operators. The common areas of the 8 criteria will indicate the appropriate place for placing the LID-BMP. In this research, gamma operator is used to overlay the layers. The gamma operator, applied in this study, is a powerful fuzzy aggregation operator used to combine multiple criteria layers. By adjusting the gamma value, it balances the trade-off between the minimum (conservative) and maximum (optimistic) aggregation methods, providing a more flexible approach to decision making [58,59]. This operator was instrumental in generating the final suitability maps, as it ensured that the specific characteristics and priorities of each LID-BMP type were appropriately integrated.

2.3.6. Modeling of the Selected LID-BMPs

Since the SCS-CN rainfall–runoff model calculates the infiltration loss using CN, LID-BMP can be defined as a permeable sub-basin (with a change in the infiltration rate) such that the rainfall excess in this unit is less compared to other units and its impact is observed in the outflow. In this method, each modeled sub-basin consists of several different parts of the CN (places with different permeability), and by calculating the new CN obtained cumulatively, the amount of new stormwater is calculated. The estimation of the new CN for LID-BMP is also based on its type as well as the review of previous research. A study by Jia, Yao, Tang, Yu, Zhen, and Lu [51] was conducted on the development of an MCIS to select the best urban stormwater management practices and concluded that the LID-BMP was effective in controlling the quantity of stormwater as shown in Table 8. The LID-BMPs used in this research, which include bioretention basin, green roof, grass swale, and porous pavement, are all effective in reducing the quantity of rainfall.

Table 8. LID-BMP's stormwater quantity control effectiveness.

LID-BMP	Stormwater Quantity Control			
	Stormwater Volume	Peak Flow	Flow Rate	Total
Bioretention basin	16.25	18.3	14.3	48.75
Green roof	14.75	16.8	14	45.5
Grass swale	15.75	17.8	13.8	47.25
Porous pavement	16	18.5	13.3	47.75

3. Results

3.1. Spatio-Temporal Land Use Change

The analysis of land use changes using remotely sensed data helps us gain a deeper understanding of the historical interactions between human activities and the environment. It also enables us to predict potential landscape conversions, which in turn can aid in the creation of realistic and multi-dimensional scenarios for developing sustainable environmental policies and management practices. As it is shown in Table 9, the most considerable change pertains to the transformation of barren land to built-up areas. Thus, during the time period of the study, 134 ha of barren land were converted into urban land use. The second significant change relates to 128 ha increase in built-up area, which was made possible by the conversion of barren land and vegetation areas. The considerable decrease in vegetation area in the urban part of the basin was moderated to some extent by the creation of gardens in its mountainous part. Gardens were also responsible for the decline in rangeland area; however, it mostly decreased naturally, which was also the reason for the increase in mountainous areas. The overall spatio-temporal LULC changes in the Darabad basin are shown in Figure 7.

Table 9. Changes in all land use classes during a period of twenty-two years in hectare (ha).

Year	Built-Up		Barren Land		Vegetation		Rangeland		Ridge	
	ha	%	ha	%	ha	%	ha	%	ha	%
2000	662	43.49%	262	67.17%	357	52.19%	1332	52.44%	1380	48.43%
2022	860	56.51%	128	32.83%	327	47.81%	1208	47.56%	1469	51.57%
Total change	+198	+13.02%	−134	−34.34%	−30	−4.38%	−124	−4.88	−89	+3.14

3.2. Annual Rainfall Values and Average Maximum 24-Hour Rainfall

After calculating the rainfall gradient in the study area, the average annual rainfall in the Darabad basin and its sub-basins was obtained, as shown in Figure 8. As observed in Figure 8, the Darabad basin has the highest average height (2617.9 m) and annual rainfall (527.2 mm) among all sub-basins. In contrast, sub-basin E exhibits the lowest values, with an average height of 1704.5 m and annual rainfall of 354.9 mm. Significant differences in annual rainfall are evident between the sub-basins. For example, sub-basin C, with an average height of 2725.6 m, receives an annual rainfall of 463.4 mm, highlighting the direct relationship between elevation and rainfall in the region. Additionally, sub-basins with lower elevations, such as E and F, experience annual rainfall below 400 mm.

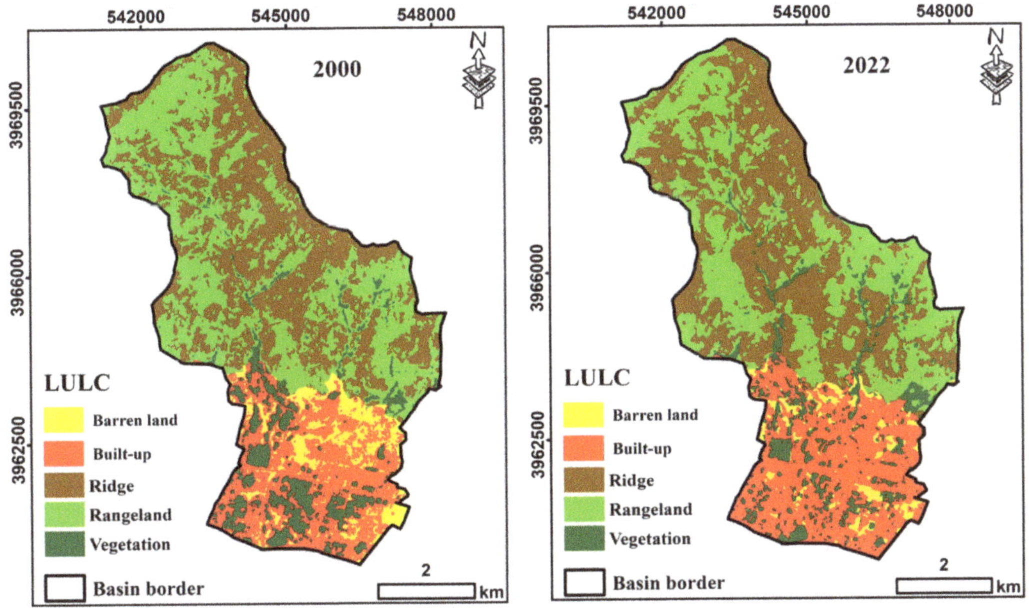

Figure 7. Spatio-temporal LULC change in Darabad basin.

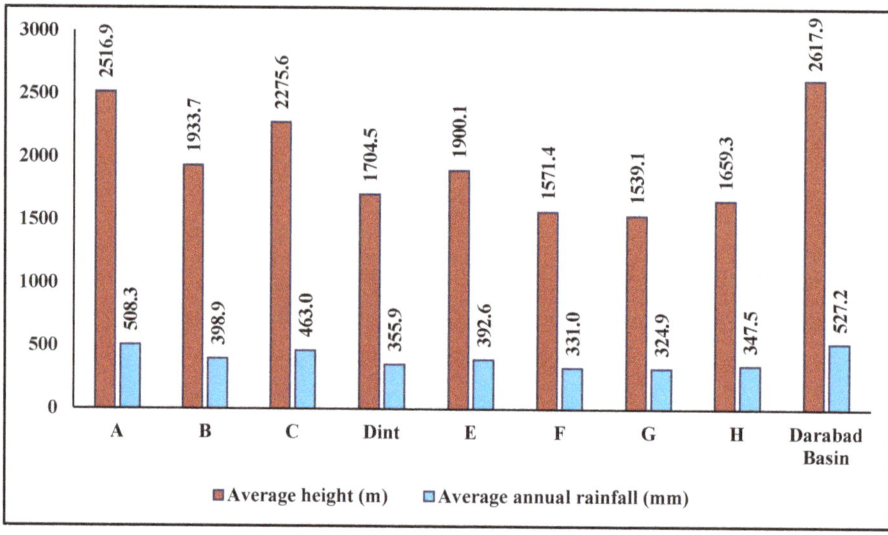

Figure 8. Average annual rainfall of Darabad basin and its sub-basins.

To calculate the maximum 24-hour rainfall in different return periods for the sub-basins, linear relationships were established between the maximum 24-hour rainfall of the stations in each return period and their average annual rainfall. Figure 9 illustrates the maximum 24-hour rainfall values of the sub-basins for return periods of 2, 5, 10, 25, 50, and 100 years.

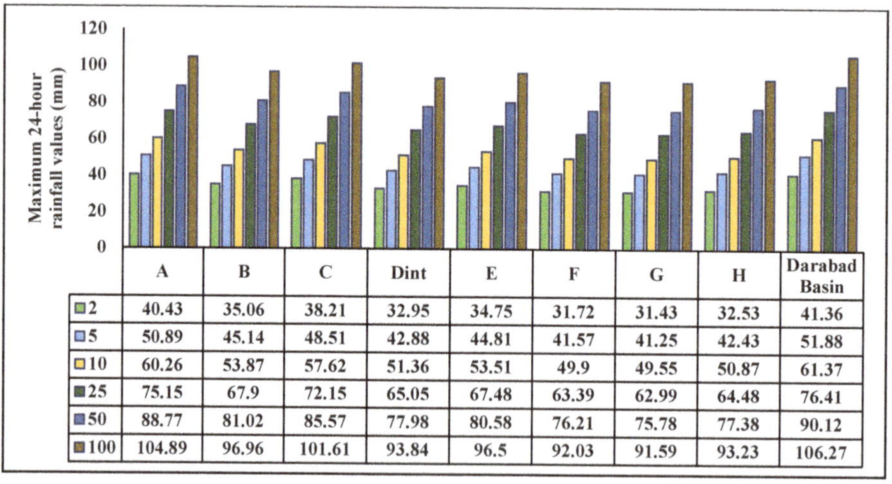

	A	B	C	Dint	E	F	G	H	Darabad Basin
2	40.43	35.06	38.21	32.95	34.75	31.72	31.43	32.53	41.36
5	50.89	45.14	48.51	42.88	44.81	41.57	41.25	42.43	51.88
10	60.26	53.87	57.62	51.36	53.51	49.9	49.55	50.87	61.37
25	75.15	67.9	72.15	65.05	67.48	63.39	62.99	64.48	76.41
50	88.77	81.02	85.57	77.98	80.58	76.21	75.78	77.38	90.12
100	104.89	96.96	101.61	93.84	96.5	92.03	91.59	93.23	106.27

Figure 9. Maximum 24-hour rainfall values (mm) in different return periods for sub-basins.

As shown in Figure 9, the Darabad basin exhibits the highest maximum 24-hour rainfall values across all return periods, with 41.36 mm for the 2-year return period, 51.88 mm for the 5-year return period, and 106.27 mm for the 100-year return period. In contrast, sub-basin F demonstrates the lowest rainfall values, with 31.72 mm for the 2-year return period, 41.57 mm for the 5-year return period, and 92.03 mm for the 100-year return period.

Notably, the differences between sub-basins become more pronounced as the return period increases. For example, the rainfall for the 100-year return period in sub-basin C reaches 101.61 mm, while in sub-basin E, it is only 93.84 mm. These variations highlight the spatial heterogeneity of extreme rainfall events across the study area, which can be attributed to differences in elevation, topography, and climatic conditions.

This analysis provides valuable insights into the potential impacts of extreme rainfall events in different parts of the Darabad basin, which is crucial for designing appropriate flood management strategies and hydraulic infrastructure.

3.3. *Investigating Flooding Potential in the Basin*

After the initial estimation of stormwater volume and peak discharge with the created model, the SCS-CN model was calibrated, and the final model was prepared. The maximum instantaneous discharge statistics of Qalak station were used to calibrate the model. The parameters of soil maintenance coefficient and initial loss were considered as parameters of calibration or optimization of the model with 20% variance. The initial storage amount and the maximum storage capacity depend on various factors such as soil type, vegetation type, permeability, pit storage, and initial soil moisture. After calibrating the model, the peak discharge and flood volume in the sub-basins were calculated.

After preparing the SCS-CN rainfall–runoff model and calibrating it in the study area, the values of the maximum specific instantaneous discharge with different return periods were calculated for each sub-basin. The results obtained were used to classify flooding potential and prepare the corresponding layer, as one of the criteria for site selection of LID-BMPs. The maximum specific instantaneous discharge of each sub-basin is shown in Figure 10. After preparing the SCS-CN rainfall–runoff model and calibrating it for the study area, the values of the maximum specific instantaneous discharge for different return periods were calculated for each sub-basin. These results were utilized to classify flooding potential and prepare the corresponding layer as a criterion for the site selection of LID-BMPs. The maximum specific instantaneous discharge values for each sub-basin are shown in Figure 10.

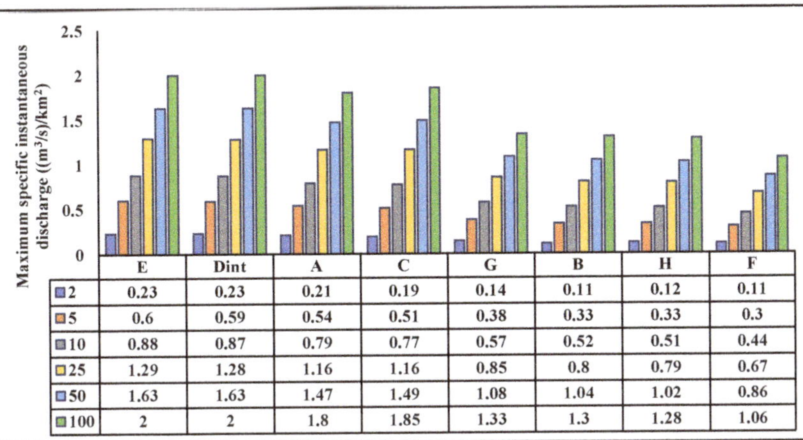

	E	Dint	A	C	G	B	H	F
2	0.23	0.23	0.21	0.19	0.14	0.11	0.12	0.11
5	0.6	0.59	0.54	0.51	0.38	0.33	0.33	0.3
10	0.88	0.87	0.79	0.77	0.57	0.52	0.51	0.44
25	1.29	1.28	1.16	1.16	0.85	0.8	0.79	0.67
50	1.63	1.63	1.47	1.49	1.08	1.04	1.02	0.86
100	2	2	1.8	1.85	1.33	1.3	1.28	1.06

Figure 10. Maximum specific instantaneous discharge with different return periods for each sub-basin.

As shown in Figure 10, sub-basin E exhibits the highest maximum specific discharge values across all return periods, with 0.23 $m^3/s/km^2$ for the 2-year return period, 0.6 $m^3/s/km^2$ for the 5-year return period, and 2 $m^3/s/km^2$ for the 100-year return period.

On the other hand, sub-basins F and B show the lowest values, with sub-basin F recording only 0.11 $m^3/s/km^2$ for the 2-year return period and 1.06 $m^3/s/km^2$ for the 100-year return period. The differences in maximum specific discharge between the sub-basins are influenced by factors such as land use, soil type, and topographic conditions. For instance, sub-basin A, with a maximum specific discharge of 1.8 $m^3/s/km^2$ for the 100-year return period, reflects moderate flooding potential, while sub-basin G, with a value of 1.33 $m^3/s/km^2$ for the same return period, shows relatively lower flooding potential. These findings indicate significant spatial variations in flooding potential within the study area. The classification of flooding potential based on these results plays a critical role in the strategic placement of LID-BMPs to mitigate flood risks effectively.

3.4. The Relationship Between Land Use/Land Cover Change and the Increase in Peak Discharge and Flood Volume

For investigating the relationship between land use change and the increase in peak discharge and flood volume in the study area, the years of 2000 and 2022 were selected as index years. The curve number values were calculated for each of the years, and according to the rainfall values at the time of concentration and the SCS-CN calibrated model, the peak discharge and flood volume values were estimated for the years. As it can be seen in Figures 11 and 12, the amounts of changes in peak discharge and flood volume are the highest in the 2-year return period, and as the return period increases, the amounts of changes are reduced. The reason is that, as the return period becomes longer, the peak discharge and, naturally, the destructive power of the flood increase, for instance, in a rainfall with a return period of 100 years, which is extremely powerful, and the land use parameter is less effective on the peak discharge and flood volume. Only urban sub-basins that were affected by land use changes are shown in this graph. According to the graph, it can be seen that the lowest amount of change in peak discharge and flood volume per land use change belongs to sub-basin B, because there was a slight change in land use. The highest amounts of changes in stormwater volume and peak discharge per land use change occurred in sub-basin F because it has been affected more by transformations caused by urbanization than other sub-basins.

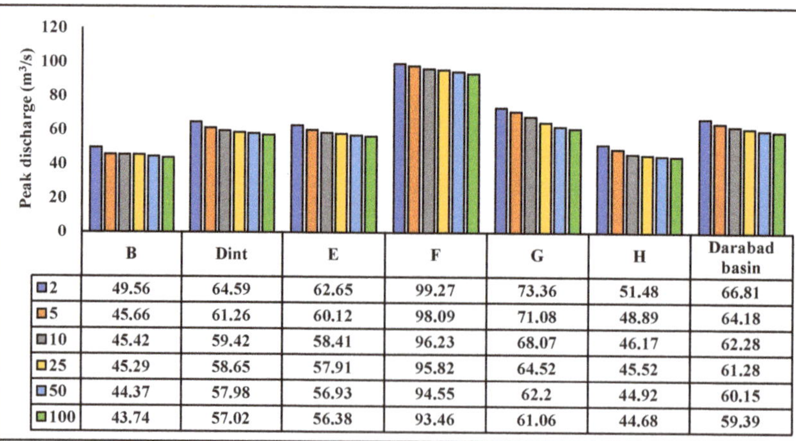

	B	Dint	E	F	G	H	Darabad basin
2	49.56	64.59	62.65	99.27	73.36	51.48	66.81
5	45.66	61.26	60.12	98.09	71.08	48.89	64.18
10	45.42	59.42	58.41	96.23	68.07	46.17	62.28
25	45.29	58.65	57.91	95.82	64.52	45.52	61.28
50	44.37	57.98	56.93	94.55	62.2	44.92	60.15
100	43.74	57.02	56.38	93.46	61.06	44.68	59.39

Figure 11. The relationship between LULC change and increase in peak discharge of urban sub-basins in different return periods.

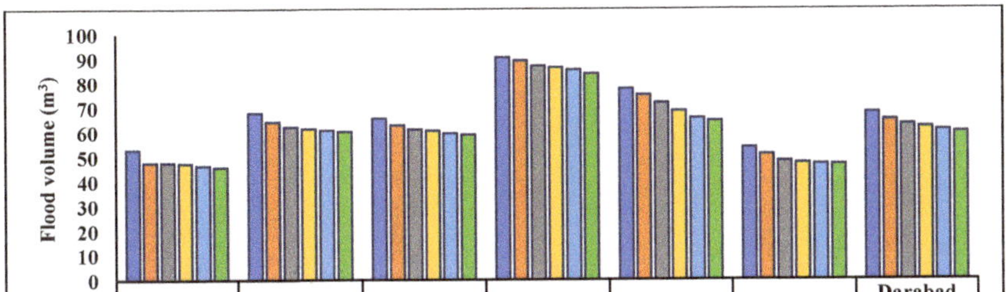

	B	Dint	E	F	G	H	Darabad basin
2	52.99	68.02	65.82	90.61	77.74	53.9	68.18
5	47.84	64.32	63.11	89.36	75.16	51.2	65.16
10	47.78	62.37	61.37	87.09	72.18	48.5	63.21
25	47.55	61.6	60.96	86.54	68.77	47.7	62.18
50	46.64	60.88	59.77	85.49	65.82	47.17	60.96
100	45.97	60.42	59.22	83.84	64.65	47	60.18

Figure 12. The relationship between LULC change and increase in flood volume of urban sub-basins in different return periods.

3.5. Site Selection of Optimal Low-Impact Developments

The fuzzy maps of site selection criteria are shown in Figure 13. Map values are between 0 and 1, where 0 represents the lowest value and 1 represents the highest value. The criteria map shows that the high values are different in the study area. For example, in the slope map, the areas with high values are mostly in the central and southern regions of the studied region. In the standard map of rainfall, the northern regions have high values compared to other regions. By combining these maps, it becomes evident that different criteria emphasize distinct areas of the study region, underlining the importance of multi-criteria decision-making approaches in site selection processes. This integrated analysis allows for the identification of optimal locations that balance the varying influences of slope, rainfall, and other relevant factors.

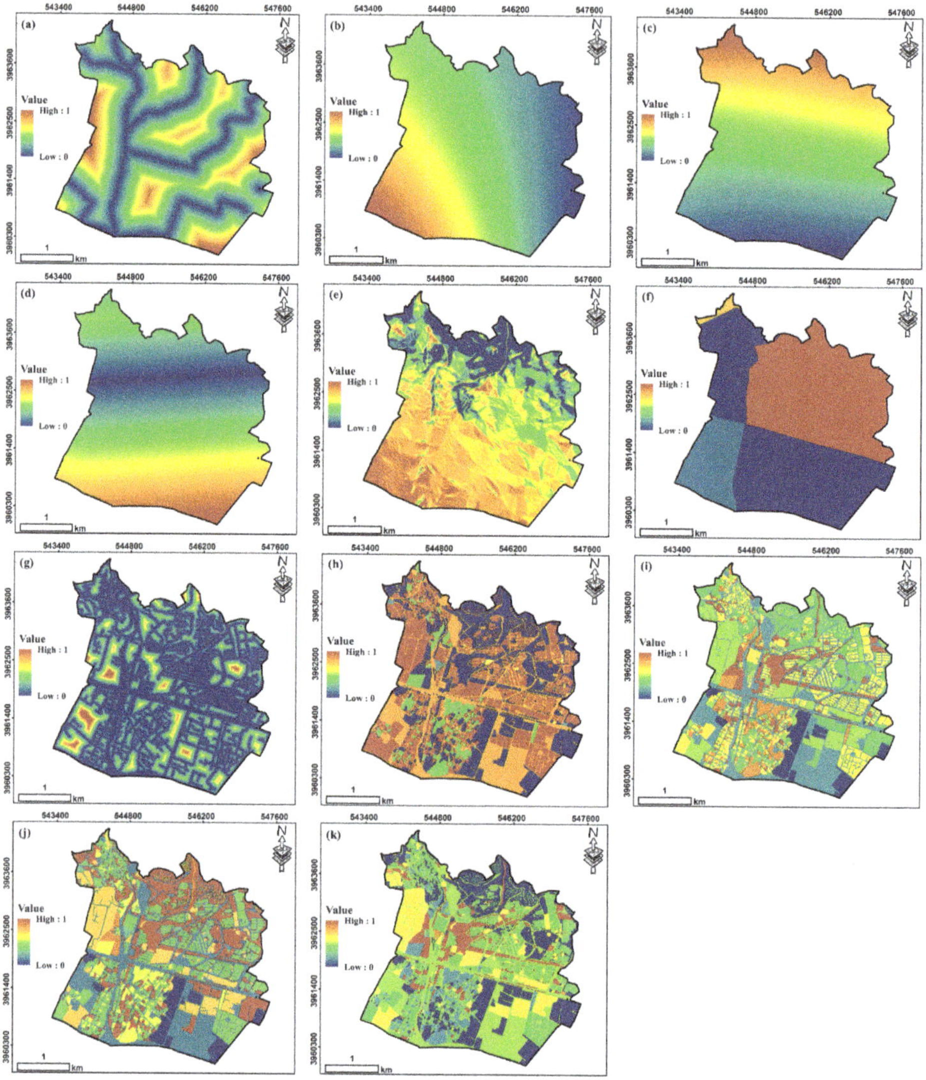

Figure 13. Fuzzy maps of site selection criteria: (**a**) stream, (**b**) groundwater, (**c**) rainfall, (**d**) distance from fault line, (**e**) slope, (**f**) flood potential, (**g**) distance from street, (**h**) green roof, (**i**) bioretention basin, (**j**) grass swale, and (**k**) porous pavement.

After determining the appropriate areas for each LID-BMP, their location raster map was divided into five classes to ascertain the best class that represents the best areas for each LID-BMP. After extracting the best areas for each LID-BMP, these areas were put together. In the areas that were found to be suitable for two or three LID-BMP types, according to the characteristics of the LID-BMPs and the priority of the land use for each LID-BMP, these area was assigned to one LID-BMP model. In Figure 14, the final map of the suitable locations for all four LID-BMPs of bioretention basin, green roof, grass swale, and porous pavement is shown together in the urban part of the Darabad basin. Finally, 119.85 ha for grass swale, 67.77 ha for bioretention basin, 37.31 ha for green roof, and

18.84 ha for porous pavement are proposed in the urban part of Darabad basin. The ratio of the area of LID-BMPs to the area of each sub-basin can be seen in Table 10. In sub-basins A, B, and C, which are non-urban, there is no effect on the pick discharge and stormwater volume, so in the continuation of the research, the numbers related to these three basins have been avoided.

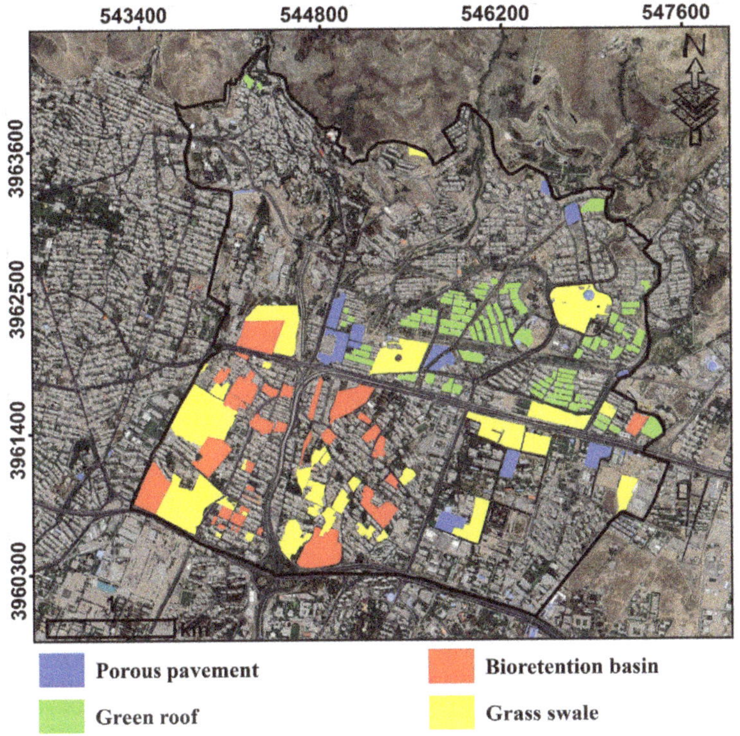

Figure 14. Final map of the suitable locations for all four LID-BMPs.

Table 10. The ratio of the area of LID-BMPs to the total area of each sub-basin.

Sub-Basin	Area of LID-BMP (ha)	Total Area (ha)	Ratio
Dint	27.60	323.98	8.51%
E	49.72	398.53	12.47%
F	63.23	377.67	16.74%
G	56.43	140.03	40.29%
H	15.04	118.82	12.65

After determining appropriate sites for each LID-BMP, it is time to model the SCS-CN model assuming the use of LID-BMPs in the sub-basins. The effectiveness of each LID-BMP model is calculated according to Table 2. As mentioned earlier, after applying LID-BMPs in each sub-basin, the CN values change for them. In Table 11, the CN values of the sub-basins after the use of LID-BMPs are discussed.

Table 11. Values of the CN in the sub-basins after applying LID-BMPs.

Sub-Basin	A	B	C	Dint	E	F	G	H	Darabad Basin
CN values	94.54	94.93	94.46	95.81	95.66	87.62	90.33	93.11	93.72

3.6. *Simulation Results*

The results show that the greatest impact of LID-BMPs on the peak discharge in all return periods occurred in sub-basin G because the ratio of the area where LID-BMPs were applied to the total area of the sub-basin was higher in sub-basin G. The simulation results of the rainfall–runoff model showed that, if LID-BMPs are applied, the average percentage of peak discharge reduction in urban sub-basins in a return period of 25 years will be 28.82%. Figure 15 show the percentage of change in peak discharge with the application of LID-BMPs. It is also evident from the graphs that, under the condition of using LID-BMPs, the percentage of peak discharge changes in shorter return periods is higher and better than that in longer return periods. In other words, LID-BMPs perform better in shorter return periods.

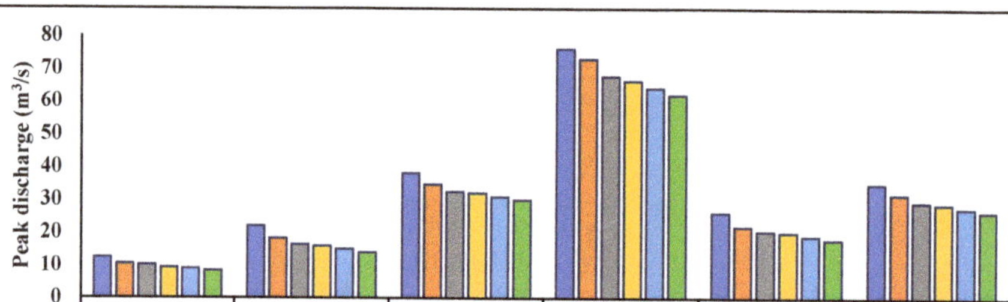

	Dint	E	F	G	H	Darabad basin
2	12.43	22	38.15	76.05	26.17	34.96
5	10.59	18.31	34.73	73.08	21.95	31.73
10	10.19	16.43	32.5	67.83	20.52	29.49
25	9.44	16.02	32.15	66.38	20.15	28.82
50	9.1	15.02	31	64.12	19.03	27.65
100	8.48	14.04	30	62.02	18	26.5

Figure 15. The effect of LID-BMPs on the peak discharge of urban sub-basins in different return periods.

As shown in Figure 16, if LID-BMPs are used, the average percentage of flood volume reduction in urban basins in the 25-year return period will be 32.72%. Again, a better performance of LID-BMPs in shorter return periods can be seen here.

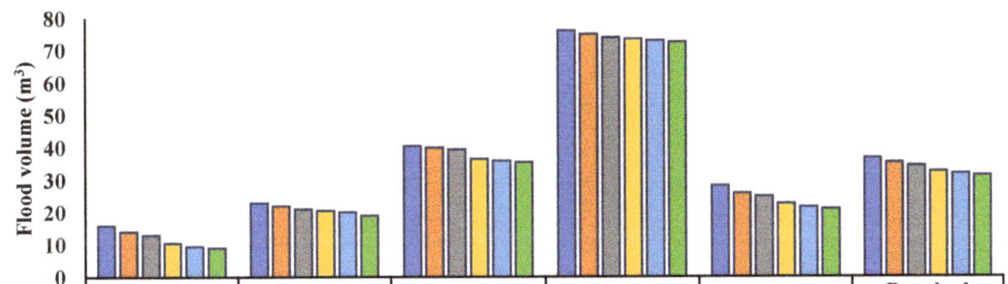

	Dint	E	F	G	H	Darabad basin
2	15.97	22.94	40.52	76.13	28.22	36.75
5	14.06	21.98	40.05	75.04	25.91	35.4
10	12.98	21.03	39.5	73.98	24.98	34.49
25	10.56	20.47	36.48	73.55	22.56	32.72
50	9.42	20	35.98	73.01	21.54	31.99
100	9.01	18.99	35.49	72.47	20.91	31.37

Figure 16. The effect of LID-BMPs on the flood volume of urban sub-basins in different return periods.

4. Discussion

The methodological approach has provided sufficient evidence that there is a relationship between temporal LULC change and changes in important hydrological parameters of stormwater volume and peak discharge. According to a study in Indiana, a 46% increase in urban land use can cause a 75% growth in stormwater volume [60]. Another survey conducted in Florida from 1920 to 1990 showed a 22–55% rise in stormwater volume due to land use changes from green cover to built-up [61]. A report from China illustrated that a 10% increase in built-up area led to an 11% increase in surface stormwater due to urbanization effects [62]. It is concluded that there is a connection between flow rate, as a measure of flood hazard intensity, the type of LULC, the intensity of rainfall, and return periods.

As a natural phenomenon, flooding cannot totally be avoided and leaves a significant impact on local livelihoods and socioeconomic development. In this research, after determining the level of vulnerability of each of the sub-basins against floods as a result of LULC change, their vulnerability in higher return periods was also estimated. The use of LID-BMPs was suggested as a management solution to control stormwater volume and peak discharge, and the selected models were located according to their functional characteristics and the spatial characteristics of the sub-basins. In addition, the impact of LID-BMPs in reducing flood potential was calculated by reducing the two parameters of stormwater volume and peak discharge for different return periods.

4.1. Urban LULC Change

Our results show a significant increase in urbanization in Tehran, particularly in the peri-urban areas. This trend has led to substantial changes in LULC, similar to the findings of other global studies. For instance, a study by Olivera and DeFee [63] observed a 181% increase in impervious areas in the Whiteoak Bayou watershed in Texas, USA, which led to a 77% increase in annual runoff. Similarly, Huq and Abdul-Aziz [64] found that urban expansion led to a marked increase in flood volume, supporting the conclusion that urbanization significantly impacts stormwater management.

In comparison to previous studies, such as the one by Moniruzzaman et al. [65] which reported a 32.92% increase in stormwater volume due to urbanization, our results for Tehran's Darabad basin show a notable shift from vegetative and barren lands to built-up areas, which has directly influenced both stormwater volume and peak discharge. This highlights the accelerated urbanization pattern in Tehran, which is more rapid than observed in many other cities. In particular, sub-basin F, being the most urbanized area in our study, exhibited the highest changes in flood volume, aligning with similar patterns reported globally where urban areas have been most affected by changes in LULC.

4.2. Effect of LULC Change on Urban Stormwater

The regional hydrological cycle and flood control have been significantly impacted by rapid urban development [66]. According to a prediction, if the expansion of built-up areas is not controlled, the risk of flooding is likely to rise in the future because of the increase in stormwater [67]. Tehran's geomorphological features include mountains in the north, a Piedmont zone in the center, and a desert in the south. These features result in several rivers and watercourses flowing through the city, which could potentially lead to flooding. Due to the lack of proper drainage systems, flash floods caused by overflowing rivers or extreme rainfall occur frequently in Tehran. The analysis of satellite images of the study area indicates the reduction in vegetation cover and bare land and their conversion to built-up area. The results of rainfall–runoff modeling showed that the lowest and highest changes in peak discharge and flood volume occurred in sub-basin B and F, respectively. This is because almost half of sub-basin B is located in the mountainous part of the area, while sub-basin F is located in the most urban part of the study area and, consequently, is more affected by developmental land use changes.

4.3. Site Selection of Low-Impact Developments

Determining the location of LID-BMPs is inherently a spatial decision and depends on various criteria such as topography, access, and environment. Therefore, in this study, a GIS-MCDA model was used to locate different LID-BMPs. The integration of MCDA and GIS creates synergy by combining spatial data analysis with user preferences, enabling structured and creative decision making [68–70]. The accuracy and efficiency of GIS-MCDA models depend on the considered effective criteria, the weight determined for the criteria, and the efficiency of aggregation models [71,72]. By reviewing past studies and considering experts' opinion, effective site selection criteria and the weight of each criterion were calculated using the AHP method. Among the advantages of this method is the simultaneous consideration of quantitative and qualitative criteria, the possibility of checking the consistency of the results, and easy implementation.

4.4. Limitation and Uncertainties

One of the limitations of this study is the lack of access to free satellite images with high spatial resolution, which can cause the impact of urban infill on flooding to be underestimated or overestimated [73]. Another argument inferred that, in order to accurately represent the unique sub-catchment responses to storm events, detailed topological and geographic information is required rather than simply classifying the impervious land cover [74].

The lack of knowledge about the spatial vulnerability of floods can lead to a poor understanding of its management. Also, the lack of meteorological and rain gauge stations in the study area were other limitations of this study in preparing relevant data, and therefore, rainfall–runoff modeling was performed using the statistics of the nearest stations, which increased the uncertainty.

5. Conclusions

The results of this study assessed the impact of land use changes and the application of low-impact development best management practices (LID-BMPs) on stormwater management in an urban basin located in Northern Tehran. The findings showed that, between 2000 and 2022, green and barren land uses decreased from 357 and 262 hectares to 327 and 128 hectares, respectively, while urban land use increased by 13.02%. These changes highlight the urgent need for effective management strategies to address the challenges of urbanization. The study evaluated four types of LID-BMPs, including bioretention basins, green roofs, grass swales, and porous pavements. Site selection analysis revealed that land use is the most important criterion for choosing bioretention basins and green roofs, while slope and distance from streets are the most decisive factors for grass swales and porous pavements. The results indicated that the study area has the highest potential for grass swales, with approximately 120 hectares of urban land identified as suitable for their implementation. Changes in peak discharge and flood volume before and after the application of LID-BMPs demonstrated that these practices significantly reduce peak discharge and flood volume, particularly for shorter return periods, highlighting their effectiveness in managing frequent, low-intensity storm events. Based on these findings, it is recommended that urban planners prioritize the implementation of grass swales in the identified areas and integrate different types of LID-BMPs into urban designs based on the specific characteristics of each area. Additionally, due to the positive impact of these practices on reducing short-term floods, they should be used to manage frequent storm events. Finally, future research should focus on integrating high-resolution satellite imagery and detailed topographic data for the more accurate modeling of LULC changes and hydrological impacts. Expanding meteorological monitoring networks, examining the combined effects of urbanization and climate change, and employing advanced AI-based modeling techniques could significantly enhance flood risk assessments. Additionally, incorporating community participation and policy integration will help develop more effective and sustainable urban flood mitigation solutions.

Author Contributions: Conceptualization, S.R. and B.M.; methodology, S.R., B.M. and F.M.; software, S.R. and F.M.; validation, S.R. and F.M.; formal analysis, S.R. and F.M.; investigation, S.R.; resources, S.R. and F.M.; data curation, S.R.; writing—original draft preparation, S.R., F.M. and B.M.; writing—review and editing, B.M. and J.J.A.; supervision, B.M. and J.J.A. All authors have read and agreed to the published version of the manuscript.

Funding: This research received no external funding.

Data Availability Statement: The raw data supporting the conclusions of this article will be made available by the authors on request.

Conflicts of Interest: The authors declare no conflicts of interest.

References

1. Huang, C.-L.; Hsu, N.-S.; Liu, H.-J.; Huang, Y.-H. Optimization of low impact development layout designs for megacity flood mitigation. *J. Hydrol.* **2018**, *564*, 542–558. [CrossRef]
2. Pour, S.H.; Abd Wahab, A.K.; Shahid, S.; Asaduzzaman, M.; Dewan, A. Low impact development techniques to mitigate the impacts of climate-change-induced urban floods: Current trends, issues and challenges. *Sustain. Cities Soc.* **2020**, *62*, 102373. [CrossRef]

3. Shahid, S.; Wang, X.-J.; Harun, S.B.; Shamsudin, S.B.; Ismail, T.; Minhans, A. Climate variability and changes in the major cities of Bangladesh: Observations, possible impacts and adaptation. *Reg. Environ. Chang.* **2016**, *16*, 459–471. [CrossRef]
4. Quichimbo-Miguitama, F.; Matamoros, D.; Jiménez, L.; Quichimbo-Miguitama, P. Influence of Low-Impact Development in Flood Control: A Case Study of the Febres Cordero Stormwater System of Guayaquil (Ecuador). *Sustainability* **2022**, *14*, 7109. [CrossRef]
5. Sohn, W.; Kim, J.-H.; Li, M.-H.; Brown, R. The influence of climate on the effectiveness of low impact development: A systematic review. *J. Environ. Manag.* **2019**, *236*, 365–379. [CrossRef]
6. Cooper, R.T. Projection of future precipitation extremes across the Bangkok Metropolitan Region. *Heliyon* **2019**, *5*, e01678. [CrossRef] [PubMed]
7. Liu, J.; Niyogi, D. Meta-analysis of urbanization impact on rainfall modification. *Sci. Rep.* **2019**, *9*, 7301. [CrossRef]
8. Bartholomew, I.; Nienow, P.; Sole, A.; Mair, D.; Cowton, T.; Palmer, S.; Wadham, J. Supraglacial forcing of subglacial drainage in the ablation zone of the Greenland ice sheet. *Geophys. Res. Lett.* **2011**, *38*, L08502. [CrossRef]
9. Qin, H.-P.; Li, Z.-X.; Fu, G. The effects of low impact development on urban flooding under different rainfall characteristics. *J. Environ. Manag.* **2013**, *129*, 577–585. [CrossRef] [PubMed]
10. Zhu, Z.; Chen, Z.; Chen, X.; Yu, G. An assessment of the hydrologic effectiveness of low impact development (LID) practices for managing runoff with different objectives. *J. Environ. Manag.* **2019**, *231*, 504–514. [CrossRef]
11. Fan, G.; Lin, R.; Wei, Z.; Xiao, Y.; Shangguan, H.; Song, Y. Effects of low impact development on the stormwater runoff and pollution control. *Sci. Total Environ.* **2022**, *805*, 150404. [CrossRef]
12. Huang, J.J.; Xiao, M.; Li, Y.; Yan, R.; Zhang, Q.; Sun, Y.; Zhao, T. The optimization of Low Impact Development placement considering life cycle cost using Genetic Algorithm. *J. Environ. Manag.* **2022**, *309*, 114700. [CrossRef] [PubMed]
13. Bae, C.; Lee, D.K. Effects of low-impact development practices for flood events at the catchment scale in a highly developed urban area. *Int. J. Disaster Risk Reduct.* **2020**, *44*, 101412. [CrossRef]
14. Hager, J.; Hu, G.; Hewage, K.; Sadiq, R. Performance of low-impact development best management practices: A critical review. *Environ. Rev.* **2019**, *27*, 17–42. [CrossRef]
15. Zhang, X.; Jia, H. Low impact development planning through a comprehensive optimization framework: Current gaps and future perspectives. *Resour. Conserv. Recycl.* **2023**, *190*, 106861. [CrossRef]
16. Naeini, A.M.; Tabesh, M.; Soltaninia, S. Modeling the effect of land use change to design a suitable low impact development (LID) system to control surface water pollutants. *Sci. Total Environ.* **2024**, *932*, 172756. [CrossRef] [PubMed]
17. Yaraghifard, M.; Shokouhibidhendi, M. Futuristic research of resilience of water resources with scenario planning approach based on case study: Zayandeh Rood watershed. *Water Soil Manag. Model.* **2024**, *5*, 26–37.
18. Chen, P.-Y.; Tung, C.-P.; Li, Y.-H. Low impact development planning and adaptation decision-making under climate change for a community against pluvial flooding. *Water* **2017**, *9*, 756. [CrossRef]
19. Houdeshel, C.D.; Pomeroy, C.A.; Hair, L.; Moeller, J. Cost-estimating tools for low-impact development best management practices: Challenges, limitations, and implications. *J. Irrig. Drain. Eng.* **2011**, *137*, 183–189. [CrossRef]
20. Zhang, K.; Chui, T.F.M. A comprehensive review of spatial allocation of LID-BMP-GI practices: Strategies and optimization tools. *Sci. Total Environ.* **2018**, *621*, 915–929. [CrossRef] [PubMed]
21. Jia, H.; Yao, H.; Yu, S.L. Advances in LID BMPs research and practice for urban runoff control in China. *Front. Environ. Sci. Eng.* **2013**, *7*, 709–720. [CrossRef]
22. Zhou, Q.; Lai, Z.; Blohm, A. Optimising the combination strategies for pipe and infiltration-based low impact development measures using a multiobjective evolution approach. *J. Flood Risk Manag.* **2019**, *12*, e12457. [CrossRef]
23. Dietz, M.E. Low impact development practices: A review of current research and recommendations for future directions. *Water Air Soil Pollut.* **2007**, *186*, 351–363. [CrossRef]
24. Todeschini, S.; Papiri, S.; Ciaponi, C. Performance of stormwater detention tanks for urban drainage systems in northern Italy. *J. Environ. Manag.* **2012**, *101*, 33–45. [CrossRef] [PubMed]
25. Bai, Y.; Zhao, N.; Zhang, R.; Zeng, X. Storm water management of low impact development in urban areas based on SWMM. *Water* **2018**, *11*, 33. [CrossRef]
26. Bibi, T.S.; Kara, K.G. Evaluation of climate change, urbanization, and low-impact development practices on urban flooding. *Heliyon* **2023**, *9*, e12955. [CrossRef] [PubMed]
27. Wu, Y.; She, D.; Xia, J.; Song, J.; Xiao, T.; Zhou, Y. The quantitative assessment of impact of pumping capacity and LID on urban flood susceptibility based on machine learning. *J. Hydrol.* **2023**, *617*, 129116. [CrossRef]
28. Alfredo, K.; Montalto, F.; Goldstein, A. Observed and modeled performances of prototype green roof test plots subjected to simulated low-and high-intensity precipitations in a laboratory experiment. *J. Hydrol. Eng.* **2010**, *15*, 444–457. [CrossRef]
29. Wu, J.; Chen, Y.; Yang, R.; Zhao, Y. Exploring the optimal cost-benefit solution for a low impact development layout by zoning, as well as considering the inundation duration and inundation depth. *Sustainability* **2020**, *12*, 4990. [CrossRef]

30. Seo, M.; Jaber, F.; Srinivasan, R. Evaluating various low-impact development scenarios for optimal design criteria development. *Water* **2017**, *9*, 270. [CrossRef]
31. Zhan, W.; Chui, T.F.M. Evaluating the life cycle net benefit of low impact development in a city. *Urban For. Urban Green.* **2016**, *20*, 295–304. [CrossRef]
32. Khosravi, H.; Abrishami, M.; Mehrian, M.R.; Chamberlain, B. The positive impact of transit-oriented-development characteristics on Metro Station usage: A case study of Tehran's metro stations and TOD index calculation. *Cities* **2024**, *148*, 104840. [CrossRef]
33. Recanatesi, F.; Petroselli, A.; Ripa, M.; Leone, A.; Ripa, M.N. Assessment of stormwater runoff management practices and BMPs under soil sealing. *J. Environ. Manag.* **2017**, *201*, 6–18. [CrossRef]
34. Salaudeen, A.; Shahid, S.; Ismail, A.; Adeogun, B.K.; Ajibike, M.A.; Bello, A.-A.D.; Salau, O.B. Adaptation measures under the impacts of climate and land-use/land-cover changes using HSPF model simulation: Application to Gongola river basin, Nigeria. *Sci. Total Environ.* **2023**, *858*, 159874. [CrossRef] [PubMed]
35. Shorabeh, S.N.; Kakroodi, A.; Firozjaei, M.K.; Minaei, F.; Homaee, M. Impact assessment modeling of climatic conditions on spatial-temporal changes in surface biophysical properties driven by urban physical expansion using satellite images. *Sustain. Cities Soc.* **2022**, *80*, 103757. [CrossRef]
36. Nadizadeh Shorabeh, S.; Hamzeh, S.; Zanganeh Shahraki, S.; Firozjaei, M.K.; Jokar Arsanjani, J. Modelling the intensity of surface urban heat island and predicting the emerging patterns: Landsat multi-temporal images and Tehran as case study. *Int. J. Remote Sens.* **2020**, *41*, 7400–7426. [CrossRef]
37. Miralha, L.; Muenich, R.L.; Schaffer-Smith, D.; Myint, S.W. Spatiotemporal land use change and environmental degradation surrounding CAFOs in Michigan and North Carolina. *Sci. Total Environ.* **2021**, *800*, 149391. [CrossRef]
38. Ahmadi, M.; Ghamary Asl, M. Monitoring urban growth in Google Earth Engine from 1991 to 2021 and predicting in 2041 using CA-MARKOV and geometry: Case study—Tehran. *Arab. J. Geosci.* **2023**, *16*, 107. [CrossRef]
39. Mohimi, A.; Esmaeily, A. Spatiotemporal analysis of urban sprawl using a multi-technique approach and remote sensing satellite imagery from 1990 to 2020: Kerman/Iran. *Environ. Dev. Sustain.* **2024**, *26*, 18033–18068. [CrossRef]
40. Naikoo, M.W.; Rihan, M.; Ishtiaque, M. Analyses of land use land cover (LULC) change and built-up expansion in the suburb of a metropolitan city: Spatio-temporal analysis of Delhi NCR using landsat datasets. *J. Urban Manag.* **2020**, *9*, 347–359. [CrossRef]
41. Devianti; Jayanti, D.S.; Mulia, I.; Sitorus, A.; Sartika, T.D. Model of surface runoff estimation on oil palm plantation with or without biopore infiltration hole using SCS-CN and ANN methods. *IOP Conf. Ser. Earth Environ. Sci.* **2019**, *365*, 012065. [CrossRef]
42. Walega, A.; Amatya, D.M.; Caldwell, P.; Marion, D.; Panda, S. Assessment of storm direct runoff and peak flow rates using improved SCS-CN models for selected forested watersheds in the Southeastern United States. *J. Hydrol. Reg. Stud.* **2020**, *27*, 100645. [CrossRef]
43. Verma, S.; Singh, P.; Mishra, S.K.; Singh, V.; Singh, V.; Singh, A. Activation soil moisture accounting (ASMA) for runoff estimation using soil conservation service curve number (SCS-CN) method. *J. Hydrol.* **2020**, *589*, 125114. [CrossRef]
44. McCuen, R.H.; Wong, S.L.; Rawls, W.J. Estimating urban time of concentration. *J. Hydraul. Eng.* **1984**, *110*, 887–904. [CrossRef]
45. Sultan, D.; Tsunekawa, A.; Tsubo, M.; Haregeweyn, N.; Adgo, E.; Meshesha, D.T.; Fenta, A.A.; Ebabu, K.; Berihun, M.L.; Setargie, T.A. Evaluation of lag time and time of concentration estimation methods in small tropical watersheds in Ethiopia. *J. Hydrol. Reg. Stud.* **2022**, *40*, 101025. [CrossRef]
46. Biantoro, A.W.; Wahyudi, S.; Ni'am, M.F.; Mahardika, A. Time of Concentration Estimated Using Some Methods and Application in The Ciliwung Watershed, Jakarta. *Int. J. Adv. Res. Eng. Innov.* **2022**, *4*, 17–25.
47. Tan, X.; Mai, Q.; Chen, G.; Liu, B.; Wang, Z.; Lai, C.; Chen, X. Intensity-duration-frequency curves in the Guangdong-Hong Kong-Macao Greater Bay Area inferred from the Bayesian hierarchical model. *J. Hydrol. Reg. Stud.* **2023**, *46*, 101327. [CrossRef]
48. Koutsoyiannis, D.; Kozonis, D.; Manetas, A. A mathematical framework for studying rainfall intensity-duration-frequency relationships. *J. Hydrol.* **1998**, *206*, 118–135. [CrossRef]
49. Ghahraman, B.; Abkhezr, H. Improvement in intensity-duration-frequency relationships of rainfall in Iran. *JWSS-Isfahan Univ. Technol.* **2004**, *8*, 1–14.
50. Al-Ghobari, H.; Dewidar, A.; Alataway, A. Estimation of surface water runoff for a semi-arid area using RS and GIS-based SCS-CN method. *Water* **2020**, *12*, 1924. [CrossRef]
51. Jia, H.; Yao, H.; Tang, Y.; Yu, S.L.; Zhen, J.X.; Lu, Y. Development of a multi-criteria index ranking system for urban runoff best management practices (BMPs) selection. *Environ. Monit. Assess.* **2013**, *185*, 7915–7933. [CrossRef]
52. Jia, H.; Yao, H.; Tang, Y.; Shaw, L.Y.; Field, R.; Tafuri, A.N. LID-BMPs planning for urban runoff control and the case study in China. *J. Environ. Manag.* **2015**, *149*, 65–76. [CrossRef] [PubMed]
53. Shorabeh, S.N.; Firozjaei, M.K.; Nematollahi, O.; Firozjaei, H.K.; Jelokhani-Niaraki, M. A risk-based multi-criteria spatial decision analysis for solar power plant site selection in different climates: A case study in Iran. *Renew. Energy* **2019**, *143*, 958–973. [CrossRef]

54. Qureshi, S.; Shorabeh, S.N.; Samany, N.N.; Minaei, F.; Homaee, M.; Nickravesh, F.; Firozjaei, M.K.; Arsanjani, J.J. A new integrated approach for municipal landfill siting based on urban physical growth prediction: A case study mashhad metropolis in Iran. *Remote Sens.* **2021**, *13*, 949. [CrossRef]
55. Khosravian, J.; Qureshi, S.; Rostamzadeh, S.; Moradi, B.; Derakhshesh, P.; Yousefi, S.; Jamali, K.; Ahmadi, R.; Nickravesh, F. Evaluating the feasibility of constructing shopping centers on urban vacant land through a spatial multi-criteria decision-making model. *Front. Sustain. Cities* **2024**, *6*, 1373331. [CrossRef]
56. Saadat Foomani, M.; Malekmohammadi, B. Site selection of sustainable urban drainage systems using fuzzy logic and multi-criteria decision-making. *Water Environ. J.* **2020**, *34*, 584–599. [CrossRef]
57. Altuwaijri, H.A.; Alotaibi, M.H.; Almudlaj, A.M.; Almalki, F.M. Predicting urban growth of Arriyadh city, capital of the Kingdom of Saudi Arabia, using Markov cellular automata in TerrSet geospatial system. *Arab. J. Geosci.* **2019**, *12*, 135. [CrossRef]
58. Shorabeh, S.N.; Argany, M.; Rabiei, J.; Firozjaei, H.K.; Nematollahi, O. Potential assessment of multi-renewable energy farms establishment using spatial multi-criteria decision analysis: A case study and mapping in Iran. *J. Clean. Prod.* **2021**, *295*, 126318. [CrossRef]
59. Zarei, M.; Mohseni, F.; Sohrabi, P. Correlates of spatial structure variability in Bushehr port-city: A comprehensive analysis using fuzzy cognitive mapping methodology. *J. Infrastruct. Policy Dev.* **2024**, *8*, 8789. [CrossRef]
60. Tang, Z.; Engel, B.; Pijanowski, B.C.; Lim, K. Forecasting land use change and its environmental impact at a watershed scale. *J. Environ. Manag.* **2005**, *76*, 35–45. [CrossRef] [PubMed]
61. Kim, Y.; Engel, B.A.; Lim, K.J.; Larson, V.; Duncan, B. Runoff impacts of land-use change in Indian River Lagoon watershed. *J. Hydrol. Eng.* **2002**, *7*, 245–251. [CrossRef]
62. Zhou, F.; Xu, Y.; Chen, Y.; Xu, C.-Y.; Gao, Y.; Du, J. Hydrological response to urbanization at different spatio-temporal scales simulated by coupling of CLUE-S and the SWAT model in the Yangtze River Delta region. *J. Hydrol.* **2013**, *485*, 113–125. [CrossRef]
63. Olivera, F.; DeFee, B.B. Urbanization and Its effect on runoff in the Whiteoak Bayou Watershed, Texas 1. *JAWRA J. Am. Water Resour. Assoc.* **2007**, *43*, 170–182. [CrossRef]
64. Huq, E.; Abdul-Aziz, O.I. Climate and land cover change impacts on stormwater runoff in large-scale coastal-urban environments. *Sci. Total Environ.* **2021**, *778*, 146017. [CrossRef]
65. Moniruzzaman, M.; Thakur, P.K.; Kumar, P.; Ashraful Alam, M.; Garg, V.; Rousta, I.; Olafsson, H. Decadal urban land use/land cover changes and its impact on surface runoff potential for the Dhaka City and surroundings using remote sensing. *Remote Sens.* **2020**, *13*, 83. [CrossRef]
66. Zhao, Y.; Xia, J.; Xu, Z.; Qiao, Y.; Shen, J.; Ye, C. Impact of urbanization on regional rainfall-runoff processes: Case study in Jinan City, China. *Remote Sens.* **2023**, *15*, 2383. [CrossRef]
67. Shrestha, B.B. Approach for analysis of land-cover changes and their impact on flooding regime. *Quaternary* **2019**, *2*, 27. [CrossRef]
68. Shorabeh, S.N.; Firozjaei, H.K.; Firozjaei, M.K.; Jelokhani-Niaraki, M.; Homaee, M.; Nematollahi, O. The site selection of wind energy power plant using GIS-multi-criteria evaluation from economic perspectives. *Renew. Sustain. Energy Rev.* **2022**, *168*, 112778. [CrossRef]
69. Gholamalizadeh, A.; Shorabeh, S.N.; Choubineh, K.; Karimi, A.; Ghahremani, L.; Firozjaei, M.K. Exploring the climatic conditions effect on spatial urban photovoltaic systems development using a spatial multi-criteria decision analysis: A multi-city analysis. *Sustain. Cities Soc.* **2024**, *116*, 105941. [CrossRef]
70. Ebrahimi Sirizi, M.; Taghavi Zirvani, E.; Esmailzadeh, A.; Khosravian, J.; Ahmadi, R.; Mijani, N.; Soltannia, R.; Jokar Arsanjani, J. A Scenario-Based Multi-Criteria Decision-Making Approach for Allocation of Pistachio Processing Facilities: A Case Study of Zarand, Iran. *Sustainability* **2023**, *15*, 15054. [CrossRef]
71. Shorabeh, S.N.; Varnaseri, A.; Firozjaei, M.K.; Nickravesh, F.; Samany, N.N. Spatial modeling of areas suitable for public libraries construction by integration of GIS and multi-attribute decision making: Case study Tehran, Iran. *Libr. Inf. Sci. Res.* **2020**, *42*, 101017. [CrossRef]
72. Boloorani, A.D.; Shorabeh, S.N.; Samany, N.N.; Mousivand, A.; Kazemi, Y.; Jaafarzadeh, N.; Zahedi, A.; Rabiei, J. Vulnerability mapping and risk analysis of sand and dust storms in Ahvaz, IRAN. *Environ. Pollut.* **2021**, *279*, 116859. [CrossRef] [PubMed]
73. Li, J.; Bortolot, Z.J. Quantifying the impacts of land cover change on catchment-scale urban flooding by classifying aerial images. *J. Clean. Prod.* **2022**, *344*, 130992. [CrossRef]
74. Miller, J.D.; Kim, H.; Kjeldsen, T.R.; Packman, J.; Grebby, S.; Dearden, R. Assessing the impact of urbanization on storm runoff in a peri-urban catchment using historical change in impervious cover. *J. Hydrol.* **2014**, *515*, 59–70. [CrossRef]

Disclaimer/Publisher's Note: The statements, opinions and data contained in all publications are solely those of the individual author(s) and contributor(s) and not of MDPI and/or the editor(s). MDPI and/or the editor(s) disclaim responsibility for any injury to people or property resulting from any ideas, methods, instructions or products referred to in the content.

land

MDPI

Article

A Systematic Literature Review of Water-Sensitive Urban Design and Flood Risk Management in Contexts of Strategic Urban Sustainability Planning

Tahia Tasnia [1] and Anna Growe [2,*]

1 Independent Researcher, 34125 Kassel, Germany; tahiatasnia@gmail.com
2 Faculty of Architecture, Urban Planning and Landscape Planning, Institute of Urban Developments, Kassel University, 34127 Kassel, Germany
* Correspondence: anna.growe@uni-kassel.de

Abstract: Despite various sustainable urban development frameworks, over the years, inadequate land use patterns and infrastructure have worsened existing problems related to climate disasters such as flooding, heavy precipitation and droughts. Based on a systematic PRISMA literature search and bibliographic analysis, we analyzed statistical data from 44 articles relevant to water-sensitive urban design (WSUD) and flood risk management (FRM) worldwide from 2013 to 2023. We focused on specific selection criteria that focused on settlement typologies and outcomes, indicators and planning approaches to analyze the impact of flooding on urban infrastructure of four different settlement types in 23 case studies, summarized into nine different approaches. The results show that WSUD and FRM have shared sustainability goals but differ in their focus and applicability depending on the settlement type and indicators. In the context of strategic planning for urban sustainability, it can be stated that WSUD has a much stronger focus on integration and future orientation than FRM. Therefore, WSUD seems better suited to be linked to strategic planning for urban sustainability than FRM. Finally, we propose to extend WSUD to "water-sensitive regional design (WSRD)". This broader framework would integrate regional hydrological, ecological and socioeconomic aspects to address water issues at a larger scale.

Keywords: water-sensitive urban design; flood management; systematic literature review; indicators; strategic planning; urban sustainability

1. Introduction

The world has recently experienced rapid urbanization and a substantial surge in urban populations. Approximately 54% of the population resides in urban areas, a figure projected to increase to nearly two-thirds by 2050 [1]. These trends reflect the end of profound urbanization that has unfolded over 7 decades, which may merely increase the effects of disaster risk, such as floods, heavy rain and droughts.

These challenges must be considered in spatial planning to create sustainable spatial structures. In the light of Sustainable Development Goal (SDG) 11, a particular focus of this article is on urban structures. As the creation of urban sustainability is a complex and long-term task, spatial planning must be able to pursue the objectives and content of planning strategically and across many (sub)projects and (sub)concepts. This is achieved through strategic planning. Strategic planning is a key instrument of urban development that defines long-term goals and identifies ways in which cities can meet the challenges of urbanization, climate change and social inequality. It enables an integrated perspective in which different interests, stakeholders and resource dimensions are harmonized [2]. The advantage lies in the ability to proactively manage complex dynamics and ensure sustainable development. However, strategic planning also harbours risks: it can become rigid due to bureaucratic processes, lead to inflexible visions and restrict actual flexibility in dealing with unexpected developments. There is also a risk that—over time—topics are neglected and challenges are not sufficiently interwoven.

The latter challenge is the focus of this article. The aim of this paper is to systematically analyze which topics and approaches should play a central role in strategic planning for urban sustainability in an important field of sustainable planning, namely, the preparation for water-based challenges caused by, e.g., floods, heavy rain and droughts. The focus of this article is on all factors of the urban water cycle, melding the functional aspects of water management and flood protection with the principles of urban design [3]. Water-sensitive urban design (WSUD) and flood risk management (FRM) are increasingly recognized as critical frameworks for sustainable urban development. Both concepts respond to the escalating pressures of urbanization, climate change, and extreme weather events, which exacerbate water-related challenges in cities. However, their approaches, objectives and implementation pathways often diverge, creating opportunities and tensions in the context of integrated spatial planning [4–8]. In this context, it is noteworthy that the discussion on WSUD and FRM in this article takes place from an abstract–analytical planning perspective. It should be noted that the perspective on WSUD and FRM from an aquatic ecology or engineering perspective can bring other topics into focus.

Despite their differences, WSUD and FRM share several thematic roots. Both concepts underscore the importance of adaptive and systems-based management, recognizing the dynamic interactions between human and natural systems [9,10]. They also emphasize resilience building and the need for participatory governance to achieve equitable and sustainable outcomes. Furthermore, both advocate for the preservation and restoration of natural hydrological processes to enhance ecosystem services and mitigate climate risks.

The subjects of WSUD and FRM are different but related, manifesting with questions of functionality and outcomes. This paper addresses the following research questions:

- What are the differences and similarities between the concepts of WSUD and FRM? Can these be identified in the literature?
- Have the topics of WSUD and FRM changed over the last 10 years?
- Do WSUD and FRM address various topics depending on different urban locations?

The first question serves as a framework question, while the other two address the spatial and temporal variations in the understanding of WSUD and FRM. To address the research questions, this study analyses various strands of research in the domain of water and flood management, providing a brief overview of the theoretical understanding of WSUD and FRM. Subsequently, we characterize the methodological approach in detail in the analysis of this study. The results of a systematic review of 44 articles sourced from English-language discourse on WSUD and FRM are presented and critically examined. Subsequently, we analyze 23 case studies and determine which approaches are used and their outcomes. To identify the relationship between these case studies, we run these research articles through VOSviewer and generate a bibliometric map to determine the connections among the keywords. Afterward, it is observed that the selected case studies represent various climatic and geographical locations. Below, we briefly discuss the possible outcomes of our questions.

First, the differences and similarities between WSUD and FRM are simple to identify. WSUD aims to integrate urban water management by imitating natural water cycles, also focusing on stormwater reuse, pollution reduction and sustainable urban development. However, FRM emphasizes risk mitigation through drainage systems, emergency responses and flood resilience. It mainly targets flood prevention, preparedness and post-flood recovery [11]. In summary, FRM addresses specific flood risks and disaster preparedness, whereas WSUD is more anticipatory and comprehensive. Each method's goals are to improve urban resilience and long-term sustainability and share techniques with local and more applicable approaches such as sponge cities, integrated urban water management (IUWM), low-impact development (LID) or green infrastructure (GI) [8,12,13].

Second, the past 10 years have seen the progressive evolution of WSUD and FRM to address the difficulties posed by climate change, with greater emphasis on resilience frameworks and new ideas combining ecological restoration and urban sustainability, such as sponge cities, IUWM, LID, GI, water-sensitive cities (WSCs) or blue–green infrastructure [11,12,14–17]. Additionally, FRM has been included in fresh local methodologies such as local knowledge (LK) and multi-criteria assessment (MCA) to support local indicators [18,19]. Furthermore, both strategies were originally applied to newly developed metropolitan regions; however, WSUD is currently being modified for suburban and brownfield areas, indicating a shift toward distributed urban resilience.

Third, WSUD and FRM address distinct priorities based on geography and climate across different land types. In large cities, as in Europe, Australia or Asia, WSUD emphasizes greenfield development, rainwater harvesting and stormwater management, while FRM addresses flood resilience through drainage improvements and emergency measures [20–22]. In tropical regions, such as Southeast Asia and Africa, mainly villages and suburban areas, FRM often focuses on seasonal flood control, resettlement and LK sharing, while WSUD develops public water access, drought prevention and water preservation [23,24]. In temperate climates such as Europe, WSUD solutions are often retrofitted to improve urban water quality and reduce runoff, with public green park development aligning with sustainability goals; however, FRM adopts local policy and indicators to reduce flood damage [25]. All these case studies also had different land types, climates, local indicators and regions, so they are discussed broadly in the Literature Review and Narrative Analysis sections.

Furthermore, WSUD is associated with strategic planning because it requires a comprehensive, long-term approach to integrating water management practices within urban development. Strategic planning in WSUD is essential for a structured framework for transitioning urban water systems toward sustainability [6]. It plays a fundamental role in urban sustainability by addressing complex challenges related to urban water management, ensuring more resilient and adaptable urban infrastructure. These can create multifunctional spaces that support urban flood resilience and green infrastructure [14,20,26–28].

Finally, theoretical and practical considerations for the application of solutions tailored to specific circumstances aim to maximize effectiveness and efficacy. To achieve these objectives, this study employs a multi-period difference-in-differences analysis, as well as PRISMA and bibliometric methods, to answer the research questions.

2. Literature Review: Discussion of WSUD and FRM

2.1. Research on the Basic Concepts of Urban Water Infrastructure

2.1.1. Concepts Based on Urban Water Planning

The role of WSUD in strategic planning is significant, as it aligns infrastructure development with long-term ecological health and the goal of climate resilience to acquire urban sustainability [5]. It preserves natural hydrological regimes, protecting urban streams and minimizing ecological impacts. Also, it can mitigate the environmental stress worsened by climate change and urbanization, improving water quality and reducing stormwater runoff [4,14,26,27]. However, strategic programs are transitioning from centralized to decentralized water management solutions. The development of decentralized water systems manifests the role of strategic planning in facilitating WSUD application. Decentralized solutions promote resilience and flexibility and allow urban areas to manage stormwater and freshwater resources effectively [5–8,14]. Therefore, these can prepare cities for future uncertainties and accommodate diverse land types, stimulating a comprehensive and flexible approach to urban development. Additionally, WSUD with strategic planning enhances resource efficiency and urban livability by integrating natural water cycles into urban planning, also employing practical measures such as permeable surfaces, water treatment, stormwater harvesting, conservation and preservation of fresh water, resulting in urban sustainability [28,29].

The concept of WSUD is adopted when the above issues are escalating due to several factors, such as the rapid pace of urbanization, urban flooding and climate change, which are prominent drivers of this paradigm shift [30,31]. It has the potential to reduce freshwater pollution for human use, decrease flooding potential, protect the ecosystem and improve urban life. It can seamlessly integrate with the urban landscape, mitigating detrimental impacts on nature, the ecosystem and built infrastructure regardless of the land type. Therefore, a framework is required that envelops the vision, scale, scope and methodology to include multi-value benefits by using WSUD and mitigate flooding [32]. This process also aims to retain water in the urban landscape through stormwater harvesting, reuse and infiltration into soil [30–32]. Considering flood management, other than WSUD, a few alternative approaches include sustainable urban drainage systems (SUDSs), low-impact development (LID), green infrastructure (GI) and sponge cities, depending on the governing jurisdiction and the practical implementation of integrated urban watershed management, which produces the same results to support the water systems and handle disasters [15,32,33].

Due to finite natural disasters and urban demands, many cities in the UK, Australia, Europe, and the sub-Saharan region yield valuable insights into their applicability. Moreover, they introduced various approaches promoting the management of natural and synthetic processes of water supply concerning groundwater, wastewater and stormwater within urban areas [12,13,31]. We now examine the potential of two new approaches, GI and SUDSs, which are an extension of the GI framework. These approaches advocate effectively addressing on-site and catchment urban surface water issues and promoting urban drainage and climatic adaptation [12,32].

The widespread complex IUWM and WSUD challenges create a paradigm-shifting context yet stimulate frequency and volumes to mitigate damage from flood events, both of which are consequences of increased surface area [15]. This situation is addressed through numerous approaches to urban water planning, particularly concerning urban drainage, which have emerged globally with a similar focus on distributed, environmentally friendly systems against disaster. These approaches encompass SUDSs, LID, GI, WSUD, sustainable water management (SWM) and sponge cities. Although the concepts may vary in scope and context, they share a common philosophy: rather than disregarding the natural hydrological cycle, they emphasize the "activation of natural process" [27]. Such an endeavor presents the prospect of reshaping this urbanization wave, transforming cities from centers of crisis to resilient cities of the future, which are potentially more sustainable, as a paradigm of "watercentric urbanism" [13,34,35]. To summarize, WSUD is an interdisciplinary approach that seeks to incorporate the urban water cycle into the design and planning of cities. It aims to enhance water quality, ensure efficient water use, and mitigate urban heat through practices such as green roofs, permeable pavements and constructed wetlands [36,37]. Central to WSUD is the principle of "living with water", emphasizing the integration of natural water systems into urban environments to promote ecological sustainability and human well-being.

2.1.2. Concepts Based on Flood Management

In contrast, FRM is primarily concerned with minimizing the risks associated with flooding. It involves a systematic process of hazard assessment, risk evaluation and the implementation of structural and non-structural measures to reduce the likelihood and impact of flood events [38,39]. Traditional FRM approaches have focused on engineered solutions, such as levees and dams, but there is an increasing emphasis on nature-based solutions and adaptive management strategies to enhance resilience [40].

Countries such as China are embracing an innovative approach known as the "sponge city program" [11,21,41,42]. Even regions such as the UK, which has historically had different climatic conditions, have seen a movement toward adopting similar methodologies. The concept of sponge cities revolves around philosophical principles such as harmonizing with nature and prioritizing the preservation of ecological functions within urban water systems [13,42,43]. This approach entails enhancing the quantity of urban green spaces, promoting multifunctionality within urban landscapes, improving habitat quality and incorporating biodiversity-friendly design principles into green and blue infrastructure, including urban rainwater management measures and direct participation during floods [22,27]. Similarly, China is advancing a novel approach known as the "blue–green city". A blue–green city strives to replicate a naturally oriented water cycle within urban settings by restoring or newly constructing water bodies with GI [17,22].

However, Flanders (Belgium), Hamburg (Germany), Copenhagen (Denmark), and Dordrecht (The Netherlands) are also adopting local flood risk approaches, which are effective and cost-efficient [44–47]. These regions were positively impacted after adopting FRM to address floods. The region of Lagos, Nigeria, is adopting FRM with the basic steps of drainage widening, relocation and resettlement schemes [23]. The concept of mainstreaming seeks advantageous synergies, aiming to enhance the resilience of vulnerable urban areas, systems and functions through adaptation integrated into the routine process of redeveloping and regenerating existing areas [14]. As highlighted, an area of current interest in hydrological modeling involves assessing the influence of land use changes on hydrological responses. These models serve as invaluable tools for predicting the impacts of land use alterations, including various development scenarios, on flow regimes [22].

2.1.3. Differences and Similarities of Discussed Approaches

It is vital to understand that the management of stormwater and water-related issues is crucial for urban development. We have familiarized ourselves with various approaches related to WSUD and FRM, as explained in Sections 2.1.1 and 2.1.2. The following text details the differences and similarities between the discussed approaches:

- The focus: WSUD combines a problem-solving method in urban planning where human development strives to replicate the natural water cycle. An SUDS is an engineered system for designing and managing stormwater, while LID is a process of managing and developing the urban area to achieve the least impact on the volume of stormwater runoff [35]. GI addresses the integration of environmental elements into cities [12,34]. The sponge city concept involves retaining water as much as possible while ensuring the quality of the available water [11,21,40,41]. IUWM encompasses the concept of design and the presentation of water in structures and other uses in the community [14]. However, FRM requires the detailed consideration of several factors of flooding and their effectiveness in corresponding to the needs of the communities. Moreover, MCA is a co-emerged method from FRM. MCA is particularly valuable when assessing the strategic significance of determining the relative advantages and disadvantages of social, economic and infrastructure vulnerability [18,48].

- Types of practices and technologies: The approaches in metropolitan areas may differ in the types of practices and technologies they incorporate, such as green roofs, rain gardens, permeable pavements, urban wetlands, and river and stream restoration [8]. FRM involves any impact on infrastructure to support emergency measures, for example, existing or proposed flood warning systems, sharing community-based knowledge, evacuation routes and response strategies and the assessment of any impacts of FRM.
- Geographic origin: Geographical locations such as the UK, Europe, Australia, Asia (mainly China), Africa and South America are mentioned broadly in the context of various water management approaches, case studies and innovative programs, such as sponge cities, blue–green cities and GI. However, MCA, FRM and LK are introduced in suburban and rural areas in Asia (India, Vietnam and Thailand), Africa and Europe (Belgium and Romania), that is, in tropical climates where seasonal flooding is common [23,24,30,48–50].
- Emphasis on water cycle: Some approaches, such as WSUD and SUDSs, place greater emphasis on the water cycle, including rainfall, runoff, drought and evaporation, while others, such as GI and LID, place greater emphasis on reducing runoff and improving water quality [32]. The long-term viability of options can be tested considering the impacts of climate change and development scenarios on existing flood behavior with FRM implemented.

2.2. Contributions of the Study

The differences between WSUD and FRM, however, are also significant and can pose challenges to their integration into strategic planning for urban sustainability. WSUD often operates at the scale of individual developments or urban precincts, focusing on long-term ecological outcomes and the co-benefits of green infrastructure. In contrast, FRM typically addresses larger catchment or regional scales, with a strong emphasis on risk quantification and immediate hazard reduction. These differing scales, goals and metrics of success can result in different priorities and implementation.

This study contributes to the literature by comparing the literature on SWM that addresses both urban and non-urban areas. Unlike the existing literature primally focused of urban-centric WSUD and FRM, this study also highlights that integrated, ecological and community-based non-engineering approaches are as effective as engineering solutions, especially in the suburban and rural contexts. Thus, this study identifies approaches that can be integrated in strategic urban planning more easily than others.

3. Materials and Methods

With the uncontrolled urban sprawl and increased frequency of disasters, increasing numbers of academic papers have been published. It is particularly important to evaluate the quality of such a large number of research papers and obtain valuable information. Below, we discuss our systematic and narrative review of data collection.

3.1. PRISMA Method

The PRISMA (Preferred Reporting Items for Systematic Reviews and Meta-Analyses) method is a structured framework widely used in academic research to improve the transparency, reliability and reproducibility of systematic reviews and meta-analyses. Its core objective is to standardize the process of identifying, selecting and synthesizing research findings across diverse studies, ensuring methodological rigor and reducing bias [51].

3.1.1. Data Collection of Water- and Flood-Related Issues

We conducted a qualitative review of all scientific articles written in English on WSUD, FRM or urban flooding measures indexed in Google Scholar, following the PRISMA guidelines. This produced a precise article collection, determining research gaps and systematic methods to analyze results and outcomes. Using Google Scholar's search engine, keywords such as WSUD, FRM and mitigation were searched for each year from 2013 to 2023 (last updated in December 2023). An advanced keyword search in Google Scholar resulted in more than 17,000 references related to the keywords. The search engine was used for every 2 years, such as 2013 to 2014 and 2017 to 2018, to ensure comprehensive coverage and avoid missing any relevant articles. Several results have been obtained by using this approach [52]. Only articles with WSUD or FRM in their titles or abstracts were considered to refine the search and ensure relevance to our research. Additionally, articles including terms such as IUWM, SUWM, SWM or WSC in their title or abstract, as well as keywords related to flood management, urban floods, stormwater, heavy rain, urban storms or extreme precipitation, were deemed acceptable. Keywords such as urban drainage, mitigation and SUD may or may not be present.

3.1.2. Data Collection for Case Studies

For case studies, a different set of search criteria was employed. Keywords such as FRM and four land types (cities, villages/rural areas, brown/infill development and suburban/small towns) were searched for each year from 2013 to 2023. Despite using specific search terms for each land type, many articles were obtained. To narrow down the focus of our research, we categorized the outcomes of case studies of specific land types into six categories: FRM, water reuse, water protection, water hygiene, urban water management (in combination with development and SWM),and water treatment management (in combination with development and SWM), as well as water treatment. These outcomes were selected based on their relevance to water-related problems and solutions, particularly focusing on floods.

3.1.3. Additional Methods of Specific Data Collection

After selecting relevant articles based on the specified outcomes, 120 articles remained. To further refine our article selection and ensure specificity, we introduced another criterion: the journal ranking. We checked the ranking of the selected articles by using the Scimago Journal & Country Rank search engine. Only journal articles categorized in the quartile charts, Q1 and Q2, were considered. This additional step ensured that the selected articles met high-quality publication standards. After applying this criterion, we were left with a final selection of 45 articles, including 23 case studies and a journal article based on the data collection method.

3.2. Bibliometric Method

This study aimed to analyze the scope of research on selected materials by using a bibliometric review and data mapping process. The mapping process was conducted by using a tool named VOSviewer 1.6.20. VOSviewer is a software tool designed for constructing and visualizing bibliometric networks. It is widely used for exploring large-scale research data, including citation networks, co-authorship relations and keyword co-occurrence patterns. The tool supports the analysis of bibliographic databases such as Web of Science, Scopus and PubMed.

A total of 44 relevant articles (excluding the journal article referring to the systematic literature review) generated a bibliometric map for quantitative analysis based on the most used keywords (mentioned at least three times) and the links between them. This review can provide a reference point for further research, which is related to selected materials and how they vary and are connected in their heterogeneity. For the case studies, the generated map represents how different land types and their specific methods are related. However, it also expresses the correspondence between the most crucial factors.

In VOSviewer, the distance between two journals in the visualization indicates the relatedness of the journals regarding co-citation links. Generally, the closer the two journals are to each other, the stronger their applicability is. Lines also represent the strongest co-citation links between journals. The indicator shows the current publications from blue (old) to purple (mixed period) to yellow (new). The size of the circles represents the frequency of appearance of the keywords. The distance between the two circles indicates their correlation.

4. Discussion of Empirical and Narrative Analysis

The motivation behind creating a bibliometric map is to explore the bibliographic data related to WSUD and FRM, which are frequently discussed. To achieve this, we visualized the connections between keywords in the overall literature to understand the appropriate planning processes for various situations and their outcomes.

4.1. Analysis of Empirical Results

4.1.1. Descriptive Statistics of Systematic Literature Review

The colored lines in Figure 1 help illustrate the relationships among these topics within the same realm. A word cloud, or the size of the circle, shows the frequencies of the keywords occurring more than three times. It is clearly indicated that "risk" was the most frequent word, followed by "WSUD", "opportunity", "case study" and "FRM" (shown in black dotted box). It provides key areas of research, and this visualization helps understand the evolution of keywords, approach and indicator use over time, providing insights into trends and shifts in research focuses. Keywords highlighted in yellow are from recent publications (in this analysis, from 2020), and keywords highlighted in blue are from older publications (in this analysis, from 2015/2016). The focus of the scientific discussion in the text corpus of this analysis has shifted from a focus on risks and flood risk management to an emphasis on opportunities and solutions.

Figure 2a,b show these network maps of the trend topics according to the keywords used from 2013 to 2023. The color legend in the figure does not show the years 2013 and 2014. These years are omitted because the analysis only considered articles with at least three keyword links. In 2013 and 2014, no articles could be identified that met this criterion. Based on the identified papers, we could summarize how WSUD, SUWM, GI, FRM and sponge cities are connected (red dotted circle) and major relatedness to keywords such as opportunity, risk, water management and case study, which are shown in blue, green and yellow (in black dotted box). These represent the need to reduce flood damage by using these three or four most crucial objectives to implement an approach based on the keyword selection.

4.1.2. Case Study Analysis

As indicated in Figures 3 and 4, three themes of case studies based on four land types are presented. Figure 3 maps adjoined bibliometric maps based on suburban, village and brownfield, and Figure 4 addresses the map of core cities. These maps enable the individual understanding of specific case studies and the most used keywords.

Indeed, we have only one brownfield case study, three village case studies and five suburban case studies. In VOSviewer, even with each set of land types selected separately and the minimal repetition criterion of two keyword sets, very few appeared to produce a compelling map. Thus, in Figure 3, we combined three different land types and set the threshold to the least repetition of three phrases. Consequently, two significant clusters of keywords appeared, each bordered by a different colored curved line.

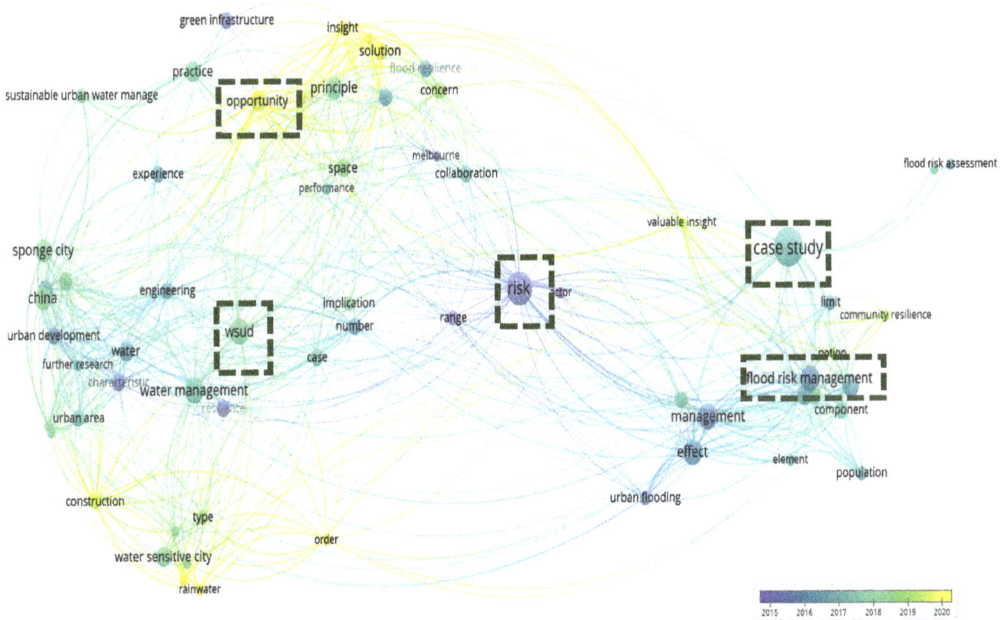

Figure 1. Bibliometric analysis of keyword links among 44 selected articles (excluding the journal referring to the systematic literature review).

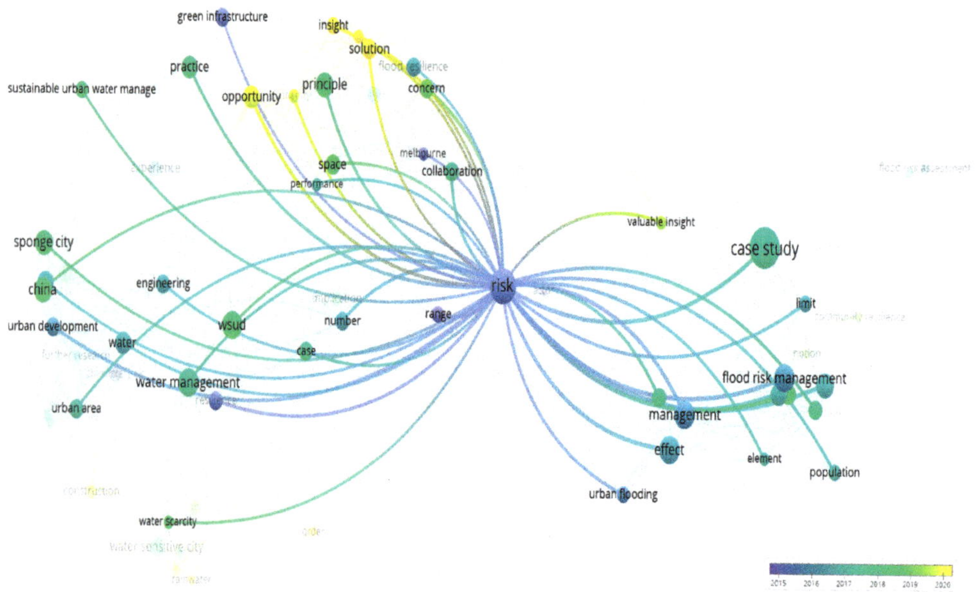

(**a**)

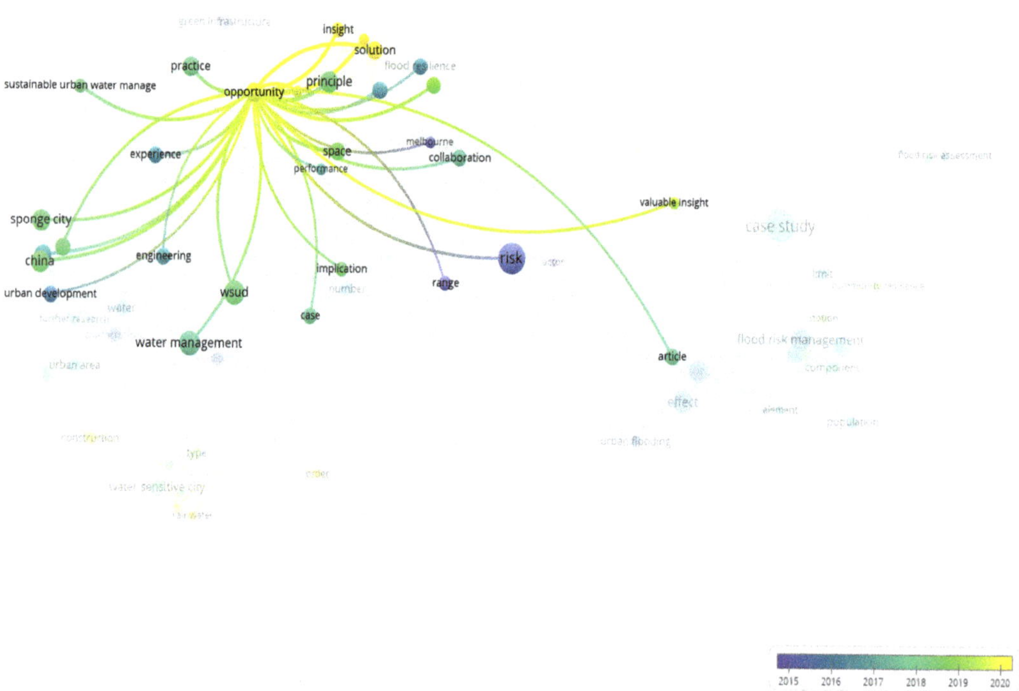

(**b**)

Figure 2. (**a**) Keyword "risk" in the second panel; (**b**) keyword "opportunity" in the second panel.

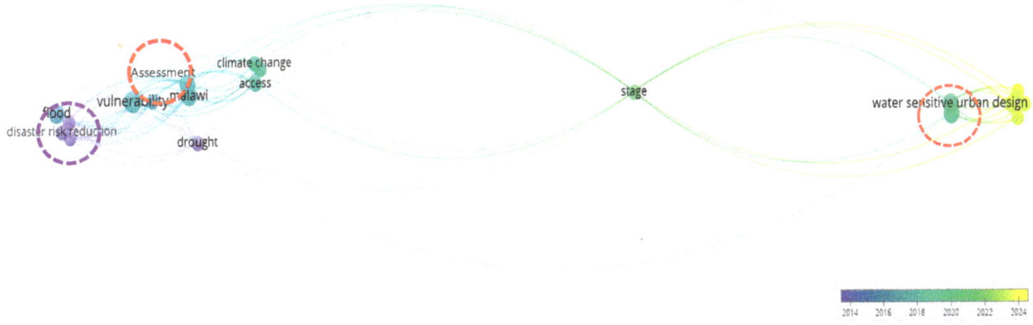

Figure 3. Bibliometric analysis of keyword links among five suburban, three village and one brownfield case study.

Figure 4 addresses the most frequent issues in 14 core cities' case studies based on various climates and regions.

The links among keywords in the chosen case studies are shown in Figure 5 which illustrate how connected activities and techniques have changed over time despite various indicators and land types. Six clusters of the selected 23 case studies were identified by using a bibliometric approach. The top six clusters, displayed from dark blue (old) to green to yellow (new), represent the research topics of management, flood control, risk, disaster reduction, case studies and assessment in Figure 5 (in black dotted box).

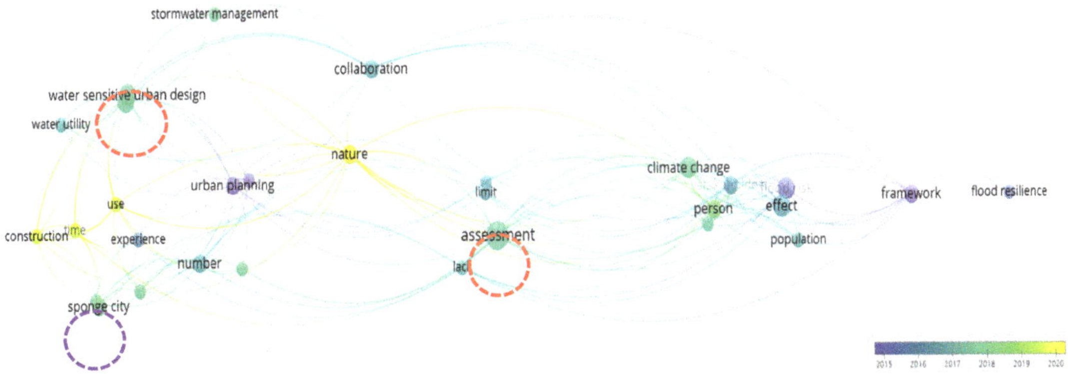

Figure 4. Bibliographic analysis of keyword links among studies of 14 core cities.

Additionally, we made a comparison with the previously provided land-based individual and adjoined map to determine which land types are associated with any particular keywords. For instance, when selecting the term "WSUD", Figures 3–5 display it with a red dotted circle corresponding to each type of land. The term "assessment" is also visible in all three figures.

The term "disaster reduction" in a dark dotted purple circle only appears in Figure 3, which represents the underdeveloped land types. In Figure 4, the word "sponge city" only surfaced in the main cities. So while the academic literature discusses core cities with various (urban) planning-related concepts (WSUD as a planning concept as well as sponge city), the focus in the other settlement types is on discussing WSUD with disaster risk reduction and vulnerability.

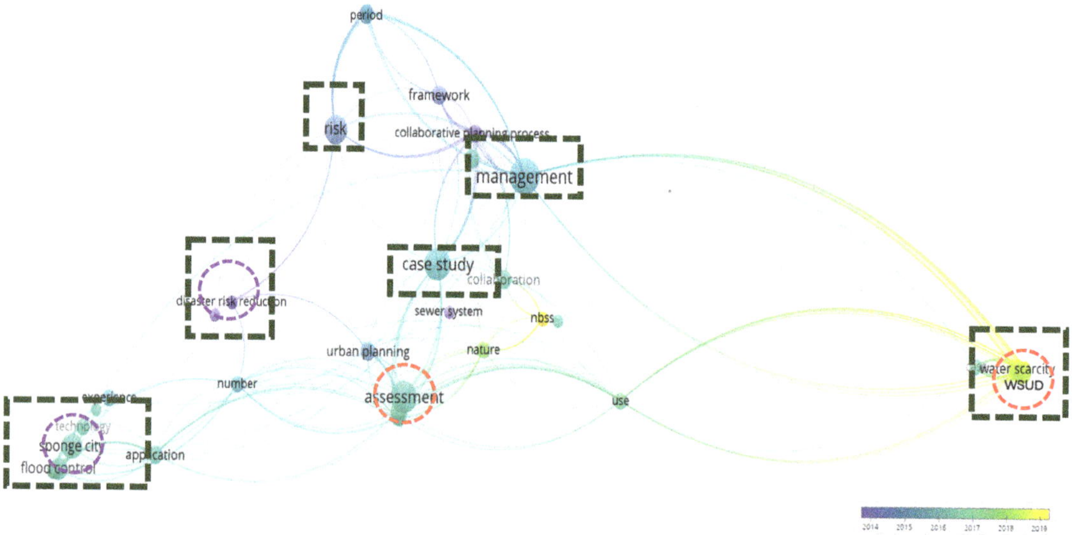

Figure 5. Bibliometric analysis of links among keywords of 23 case studies.

4.2. Analysis of Narrative Results

Based on all the literature and specific land type-based bibliographic maps, we can identify the major indicators playing a vital role in flood- and water-related problems and approaches. To provide a comprehensive overview of international research activity on water and flood management, the following chapter summarizes the most discussed land types and regions and a heterogeneity analysis of the selected publications. Moreover, this chapter represents case studies, areas of investigation and research findings, with the latter considering different limitations and opportunities of the research field and ways to apply empirical research.

4.2.1. Most Discussed Land Types and Regions

Certain key details are essential to identifying and localizing fine-grain urban morphology in our case studies and investigate their typologies. Understanding these factors is crucial to identifying and characterizing fine-grain urban morphology and typologies within different geographic contexts. Figure 6 highlights the countries with different colors to represent four land types. With the PRISMA method and our specific keywords, we produced 24 case studies based on four land types worldwide, which (together with the selection process) are discussed in Section 3.1. The process was specific, so the collection of the case studies was not random.

Table 1 presents the findings to provide an overview of the regions and climates most frequently discussed in FRM and water-related urban design. This table provides an overview of the regions, climates, built environments and land types frequently discussed in FRM and water-related urban design literature.

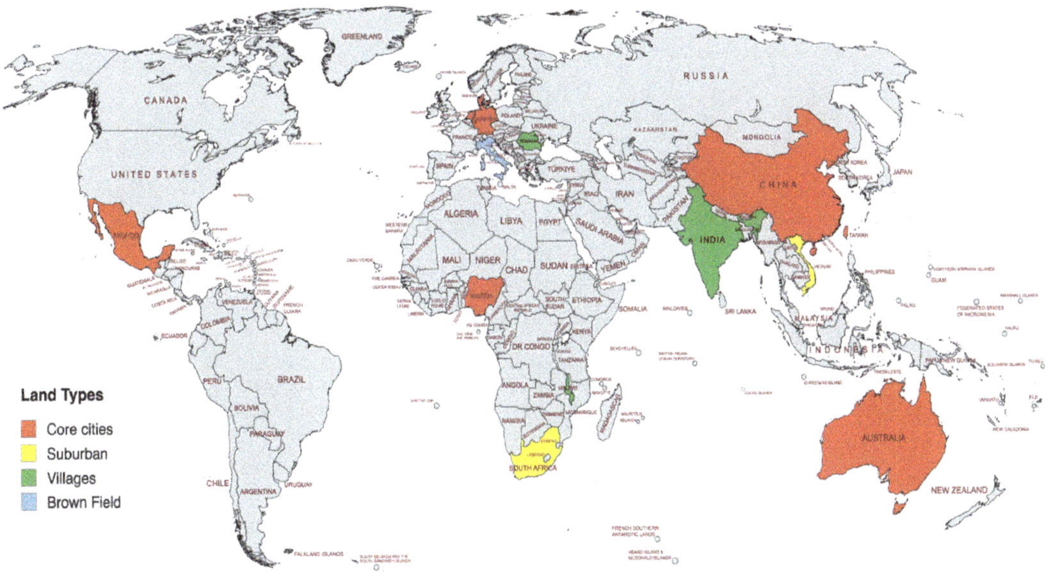

Figure 6. Countries in which the case studies from the articles identified in the systematic literature analysis are located.

4.2.2. Heterogeneity Analysis

We synthesized the results from the review of current knowledge on water services, land types and climates provided by WSUD and FRM (Table 1). In our analysis of 23 research articles on flood management from diverse case studies, we focused on different research strands and identified various actors, indicators, tools, planning processes, approaches, outcomes and limitations associated with WSUD and FRM in Table 2.

Table 1. Overview of various land types, climates and discussed approaches.

Land Type	Region	Climate	Approaches
City [11,21,23,25,36,41,42,45–47,52–56]	South Asia (China) [11,21,41,42,52,53], Vietnam [25] and Taiwan [54]	Marine monsoon humid subtropical	Resilience adaptability, spatial analyses, sponge city, rainwater storage infiltration system and WSUD
	Australia [36,55]	Temperate oceanic climate	SWM, WSUD and green open spaces
	Europe (Copenhagen [45], Hamburg [46] and Dordrecht [47])	Oceanic climate	Flood resilience, system analysis and improvement in sewer system
	South America (Mexico) [56]	Humid subtropical climate	WSUD
	Africa (Nigeria) [23].	Tropical wet and savanna climate	FRM
Suburban [18,44,49,57,58]	Southeast Asia (Vietnam [49] and Thailand [57]) Africa (Malawi) [18] and Western Cape province [58]	High humidity and hot temperatures, tropical subtropical climate	Anticipation and incorporation of ecological process, WSUD, local engagement and SWM
	Europe (Belgium) [44]	Temperate maritime climate	FRM locally
Villages [19,48,50]	South Asia (India) [19]	Tropical monsoon rainforest	Flood risk assessment
	Africa (Malawi) [48]	Subtropical climate	Sharing local knowledge
	Europe (Romania) [50]	Continental climate	Multi-criteria method
Brownfields [59]	Europe (Italy) [59]	Humid subtropical continental climate	WSUD

Table 2. Overview of approaches, drivers and outcomes.

Approaches	Drivers	Planning, Design and Management	Outcome
Water-sensitive urban design (WSUD) [45,56–59]; land types: city, suburban and brownfield	1. In Copenhagen, stormwater runoff in sewer catchment. 2. In Italy, Trento, heat waves, extreme climatic conditions, water pollution and inaccessible green public areas. 3. In Mexico and Thailand, climatic changes in local hydrological cycles, increased ground and surface water levels and destruction of riverbanks. 4. Western Cape George, heat wave, drought and restrictions on water use based on income.	1. In Copenhagen, WSUD focuses on land use planning and sewer systems divided into 60% for the city and 40% for the sewer system to prevent runoff in the sewer catchment. 2. WSUD focuses on turning brownfields into green public areas and recreational spaces in Italy, Trento. 3. In Mexico, WSUD focuses on the water cycle and utility infrastructure, and in Chiang Mai, it focuses on determining the source of water pollution. 4. In George, WSUD focuses on water preservation from rainwater, water reuse and green roof installation.	1. Copenhagen introduced specific solutions for each land type, stormwater treatment and land use reduction to 60%. 2. Green public areas, water management and regeneration of brownfields. 3. Water management, water protection and hygiene, and flood risk management (FRM) in Mexico and Thailand. 4. Water reuse, management and accessibility for the public.
Sponge city [11,21,41,42]; land type: city	Urban flooding, increased water pollution, drainage blockage and unplanned land expansion.	Data on existing drainage, flood hazard mapping, stormwater retention, brownfield redevelopment and increased green public areas.	Land use planning, mitigation and enhancement of social and ecological sectors. Encouraging Chinese government to broaden sponge city ideas.
Flood risk assessment [17,35,48]; land types: city, suburban and villages	Drainage management, inadequate sewer systems, seasonal flooding, lack of urban planning strategies and topography of coastal basin or riverside.	Rainwater storage infiltration, control of water pollution, adjustment of demands and supply of water, relocation of settlements of floodplains and wetlands and local need-based approaches.	Water protection, urban water redevelopment and mitigation.
Stormwater management [47,53]; land type: city	Urban flooding and narrow drainage.	Stormwater harvesting, improving drainage systems, non-engineering measures and institutional process of different phases of the transition.	Active participation, strengthening the construction, FRM, water protection with urbanization and multi-dimensional approaches.
Flood resilience [46,49] and urban resilience [54]; land types: city and suburban	Seasonal flooding, sea-level rise, climate change, continuous urban growth and increasing pluvial flooding on localized flood response.	Visual tools for resilience strengthen vulnerability assessment, capacities and practice action regarding FRM; identify current measures; anticipate and accommodate the community sharing of flood information; and incorporate ecological processes.	Urban resilience, flood risk management by creating robustness, adaptability and transformability—for the specific context of flooding in cities.
Multi-criteria assessment [19]; land type: suburban	Seasonal chronic flooding, geological position and riverside unplanned urban areas.	Adopt flood measures, considering social, economic and infrastructure vulnerability; identify regions frequently affected by floods.	Flood risk reduction, economic stability, social and environmental benefits, and community sharing.

Table 2. *Cont.*

Approaches	Drivers	Planning, Design and Management	Outcome
Spatial analysis approach (SAA) [24]; land type: city	Climate change, unplanned expansion of built areas, and social and economic vulnerability.	SAA is applied to assess future urban flood damage, construction of structural measures and drainage and sharing flood information efficiently.	This approach reduces flood damage by 29% and prepares to implement WSUD in the near future.
Green infrastructure [36,55]; land type: city	Unplanned urbanization, filling wetland and natural water bodies, to achieve flood resilience and ecosystem services to improve the environment.	Avoid urban settlements in floodplains; place green open spaces alongside, with multifunctional attributes such as providing active transport corridors, recreational areas, amenities and habitat.	Flood management and urban resilience achieved by increasing green open spaces.
Local knowledge (LK) and community services [48,57]; land type: suburban and villages	Natural disasters such as earthquakes, droughts, strong winds, seasonal floods, lack of a government, no economic support and negation of urban planning.	Engaging communities in communities, sharing LK, categorizing local indicators and using indigenous material to build.	LK to be included in FRM for disaster risk reduction.

5. Findings

Advancing with our research, we conducted an in-depth analysis of the specific challenges discussed in the case studies and proposed corresponding solutions. Based on our evaluation of 23 case studies, we identified approximately nine different approaches to addressing water and flood-related challenges. Below, we discuss all the case studies to understand their uniqueness.

5.1. Overview of City Case Studies

In South Asia, Kaohsiung City, Taiwan, has strengthened the resilience of urban retailers toward flood risk. It emphasizes the short-term objectives and highlights the concept of "living with risk" rather than solely focusing on hazard prevention [54]. Another city of South Asia, Hanoi, Vietnam, has adopted a spatial analysis approach, combining both structural and non-structural flood mitigation methods, such as assessing tangible direct flood damage, drainage enlargement, early warning systems, etc. [24]. In Beijing and the Yangtze River Delta, China, these regions have adopted the "Sponge City" concept, which is a holistic approach to managing stormwater. This concept integrates stormwater harvesting, flood hazard mapping, non-engineering measures (e.g., urban greening) and urban redevelopment. As a multi-disciplinary strategy, it integrates various approaches, such as stormwater harvesting, flood hazard mapping, non-engineering measures based on resilience and urban redevelopment. The adaptive management aspect is evident in how the system responds to flooding challenges, climate change, and rapid urbanization [11,21,41,42,53].

Similarly, in the context of Europe and its temperate climate, the case study in Hamburg, Germany, showed a framework developed to assess the flood resilience of cities, emphasizing their robustness, adaptability and transformability. The primary objective of this framework is to facilitate adaptive management through a multi-disciplinary approach. This enables Hamburg to adapt to evolving circumstances and to ensure the sustainable management of flood risks, reflecting a comprehensive perspective [46]. The Dordrecht project in the Netherlands applies a multi-layer safety approach by integrating protection, prevention, preparedness and emergency relief strategies. This case addresses flood risk from various angles, from infrastructure development to community engagement from a holistic perspective. It combines multi-disciplinary aspects, such as engineering, policy and creating a more resilient urban environment. The approach also allows for adaptive management, as it can be modified in response to new data and evolving flood risks. It enhances national flood policy and promotes regional flood resilience, focusing on urban redevelopment and stormwater management [47]. Denmark provides an excellent case study of a holistic approach to FRM and urban resilience. The Danish strategy is multi-disciplinary in nature and involves the retrofitting of WSUD in order to reduce stormwater runoff and limit combined sewer overflows to 20 per year. The project's goals of disconnecting 60% of unaffected surfaces from the sewer system, managing runoff through green spaces and permeable surfaces and collaborating with stakeholders highlight its adaptive management and flexibility [45].

In Melbourne, Australia, the management of stormwater through the implementation of green infrastructure practices was examined, thereby elucidating the role of agencies and institutional processes in the context of sustainability transitions. This exemplifies the principles of adaptive management and flexibility. This case emphasizes the importance of a multilevel and multi-dimensional understanding of change processes, integrating environmental, social and institutional aspects of water management, and reflects the holistic and multi-disciplinary approach to transition toward more sustainable stormwater systems [36,55].

Finally, the challenges of WSCs are addressed in Guadalajara, Mexico, where the WSUD approach is proposed to analyze water cycle dynamics and water utility infrastructure [56]. The review of flood management in Lagos, Nigeria, highlights the challenges of rapid urbanization, poor drainage infrastructure and floodplain development, exacerbated by climate risks. Current strategies of FRM focus on controlling floods through drainage expansion but face challenges in mitigating long-term impacts on vulnerable populations and infrastructure [23]. WSUD and FRM are both highly applicable to cities.

5.2. Overview of Suburban Case Studies

The case studies focus on FRM in suburban areas across different regions and climates. The first case study set in the Vietnamese Mekong Delta aimed to develop localized flood response capacity to promote urban flood resilience. The actors included seasonal flooding and basin-scale management, which emphasizes floodwater retention. This suburban case emphasizes the need for adaptive and flexible management, using both engineering solutions and traditional ecological knowledge. It reflects a holistic approach, as it combines flood risk management with social development, climate adaptation and community engagement. The approach focused on accommodating flooding, incorporating ecological processes and revealing flood dynamics to the public [49].

The second case study, located in Chiang Mai, Thailand, aimed to address water quality issues by using WSUD. The approach involved research design collaboration in WSUD, focusing on addressing non-point source pollution. The outcomes included flood management, water hygiene and water protection measures [57]. The third case study, conducted in George Municipality in South Africa, focused on mitigating drought scenarios through WSUD activities. Actors included the increased occurrence of heat waves and droughts leading to water demand management. The outcome suggests the short-term effects of reducing water consumption through WSUD activities [58]. Both these case studies adapted WSUD as a preventive approach.

The fourth case study, located in Flanders, Belgium, focused on FRM in a temperate maritime climate. Actors included dense river networks and urbanization, making landers sensitive to flood hazards. This approach was risk-based, comprising flood hazards, vulnerability and damage calculations. The outcome included building trust, raising awareness, educating, reaching an agreement and motivating actions to manage flood risks effectively [44]. The final case study, set in Karonga, Malawi, addressed disaster risk reduction in the absence of a local government. Actors included the absence of a local government, which negatively affects urban planning and disaster risk reduction. The planning process involved engaging communities in vulnerability assessments and data collection for disaster events [18]. Thus, WSUD also applied in suburban areas, while FRM and local adaptive programs were more acceptable for suburban land types.

5.3. Overview of Village Case Studies

This research study outlines three case studies focusing on FRM in villages or rural areas across different regions and climates. In the first case study, located in Assam, India, the aim was to mitigate the disaster risk posed by chronic flooding in the Kopili River Basin. This study employs multi-criteria analysis to assess the social, economic and agricultural climate characterized by warm summers [19].

The second case study, situated in Marginea village, Romania, aimed to examine the vulnerability of the population and buildings to floods in a temperate continental climate. Vulnerability is assessed based on proximity to the river, the presence of hydro-engineering works, building materials and conditions, and demographic and economic factors. This study highlights the importance of delineating floodplain areas to ensure economic and population security [50].

Finally, the third case study conducted in southern Malawi explores LK across the FRM cycle. It identifies various categories of local indicators and strategies devised by vulnerable populations living in hazard-prone areas. In this case, the absence of local government support for the integration of both local knowledge and scientific flood risk assessments is notable. The involvement of the community in vulnerability assessments and data collection lends support to the concept of adaptive management and flexibility, as well as holistic approaches to the reduction in disaster risk in rural areas.

This study emphasizes the significance of incorporating LK into FRM to enhance community resilience [48]. The FRM and LK approaches are broadly accepted in rural areas; however, the application of WSUD on these land types is limited.

5.4. Overview of Brownfield Case Study

The case study focuses on brownfield regeneration by WSUD in Trento, Italy, within a marine west coast climate characterized by warm summers. The actors involved the lack of green spaces and facilities in certain neighborhoods, extreme climatic events in the northern part of the city and the presence of heat waves and urban flooding. The study aimed to enhance public green areas to achieve a more balanced distribution throughout the city, providing equal recreational opportunities for all citizens. The regeneration of brownfield into green spaces in Trento aligns with a holistic approach to urban development. Regenerating these brownfields located near the city center can offer new recreational opportunities for citizens, particularly those residing in areas with low availability of urban parks and proximity to natural areas. Overall, the case study illustrates how brownfield regeneration can address environmental challenges, such as urban flooding and heat waves, by enhancing urban green spaces and improving residents' overall quality of life [59]. This WSUD approach addresses urban flooding, heatwaves and quality-of-life improvements—

demonstrating the integration of multi-disciplinary solutions and adaptive management in response to urban environmental challenges. Thus, WSUD can regenerate greenfield and flood prevention even on abandoned land.

6. Results

The data analysis was conducted within the framework outlined in the introduction, focusing on key aspects such as climatic conditions, land types, social and economic status and approaches to water and flood management, which are elaborately analyzed above. After an extensive review of the literature, we identified the following key findings and limitations to answer our research questions.

6.1. Overlap and Similarities of WSUD and FRM and Limitations of Urban Water Management

WSUD and FRM share several goals but differ in focus and application. Both concepts promote sustainable water management and share new features with various approaches, such as IUWM, SWM, MCA, GI and sponge cities, using traditional methods. Additionally, this research study highlighted that the planning approaches of water management in existing urban areas are fully functioning but have limitations regarding implementation in suburban and rural areas [1–3,8,11–19].

WSUD is a very comprehensive approach to urban planning and design that integrates water management into urban environments. It addresses several aspects of water management at the same time (e.g., stormwater management, water efficiency and conservation, but also integrated water cycle management). However, the particular challenge for strategic urban planning is not to take up different approaches in parallel but to develop urban design with water in mind. This is particularly challenging, as large urban infrastructure systems are usually trapped in existing structures due to institutional inertia and persistent socio-technical regulations. Connecting multiple perspectives on water in the city (from a design perspective to an ecological perspective to an engineering perspective) beyond established paths is not easy. Against this background, the question arises as to how WSUD is discussed in the academic context, which other approaches and (sub)spaces are associated with the approach and how the perception of WSUD has changed over time (from highlighting risks to highlighting opportunities).

FRM primarily focuses on reducing flood damage, integrating local approaches such as MCA or LK to mitigate flood damage. While both approaches recognize the importance of environmental sustainability, WSUD is more applicable in urban spaces, prioritizing water-sensitive planning, whereas FRM is applicable to urban and non-urban areas through floodplain management and vulnerability assessments. Moreover, both approaches must consider the natural water cycle.

This research study adds to the literature by demonstrating the need for context-specific application of water management strategies and thus the call for the development of broader and more evenly applicable frameworks for managing flood and urban water systems.

6.2. The Evolution of WSUD and FRM over the Past Decade

Over the past decade, WSUD and FRM have addressed emerging environmental challenges and urbanization trends. The Kaohsiung and Beijing case studies focus on enhancing urban resilience to flood risks exacerbated by climate change and urbanization. Each city faces challenges from extreme weather events, such as heavy rainfall and sea-level rise, which overwhelm existing drainage systems. Both studies emphasize the importance of adaptive practices—Kaohsiung highlights resilience among local retailers, while Beijing advocates for the sponge city concept to restore natural water retention. Additionally, both approaches prioritize community engagement and integrated land use planning to mitigate flood risks and improve overall urban livability [11,21,41,42,53].

Suburban sites such as the "Mekong Delta, Vietnam", where the district is more rural, are even now attempting to implement WSUD in the same area, using localized solutions for better results [49]. These examples emphasize that water-related conflict resolution through non-engineering methods such as ecologically and community-based seasonal flooding management is as crucial as hardware design implementation. Stakeholder sensitization, public education and even community and social efforts appear to be important actors in FRM within the basin economy in these regions. It is preferable as an approach encouraging sharing community knowledge based on floods and adapting livelihoods with these seasonal flooding issues as a natural method rather than addressing flood damage reduction. These ideas reflect a shift toward adapting water risks in livelihood rather than solely preventing hazards.

Brownfield regeneration efforts in Trento, Italy, present opportunities for environmental and social renewal. By converting underdeveloped areas into public green spaces, cities can mitigate urban flooding, manage extreme heat and enhance residents' quality of life. These findings align with global trends toward climate-resilient urban parks and the importance of integrating sustainable urban design with community well-being [57].

The Kopili River Basin in Assam, India, and Marginea village in Romania focus on assessing flood vulnerability within rural communities affected by distinct climatic conditions. Each study employs a multi-criteria analysis to evaluate various dimensions of vulnerability, including social, economic and infrastructural factors, to inform FRM strategies. In Assam, the assessment emphasizes the frequent and chronic flooding during the monsoon season, while in Romania, the focus is on the physical vulnerability of buildings and population demographics. Ultimately, both case studies aim to identify high-risk areas to implement targeted mitigation measures and enhance community resilience against flooding [19,50]. The case study in southern Malawi demonstrates the effectiveness of "LK" and local practices to manage flood risks. These communities have successfully incorporated floodplain mapping, agricultural adaptations, and social vulnerability assessments, showing that low-cost, high-impact solutions can build resilience in resource-limited areas [18,48].

The evolution of WSUD and FRM in the past decade has shifted toward integrated approaches emphasizing sustainability, community engagement and resilience against climate impacts. Increasingly, LK is recognized as vital for effective FRM. Additionally, reliance on data-driven decision making to assess vulnerabilities and inform indicators is growing. Overall, the focus has moved from merely preventing flooding to adapting to water risks through nature-based solutions.

6.3. Contextual Application of WSUD and FRM in Different Urban Locations

The implementation of WSUD and FRM varies significantly among urban, suburban and rural settings. In cities, advanced solutions such as SUDS and WSUD practices are often employed to enhance living quality and mitigate climate risks through multifunctional urban design. However, rural and suburban areas benefit more from localized, context-sensitive strategies combining traditional practices with ecological management, as seen in the Mekong Delta, Vietnam, and Marginea, Romania [24,50].

Brownfield regeneration through WSUD practices, including in European cities such as those in Italy, provides insights into the transformation of neglected sites into flood-resilient urban spaces through the integration of public green areas with sustainable drainage systems [58].

Moreover, the limitations of applying urban water management strategies in suburban and rural areas highlight the importance of creating solutions for local contexts. For instance, the localized strategies used in Southeast Asia and parts of Africa demonstrate how ecological processes and public engagement can support water management where conventional urban approaches have limited applicability.

To transcend individual cities and projects, we propose expanding WSUD into "water-sensitive regional design (WSRD)". This broader framework would integrate regional hydrological, ecological and socioeconomic aspects to address water challenges at a landscape scale. WSRD acknowledges that water management must not only respond to individual urban needs but also align with the broader regional water cycle.

Lastly, the need for improved communication among stakeholders—municipalities, engineers, developers and residents—presents a challenge and an opportunity. Effective collaboration supported by continuous technical training and community engagement is essential to the long-term maintenance and scaling of innovative water management practices. Adopting these insights enables practitioners to develop flexible, resilient water management frameworks adaptive to climate change and aligned with local needs.

7. Conclusions

To conclude, the above analysis indicates that factors such as land types, climates, social and economic status, and unplanned urban and drainage systems affect the adoption and effectiveness of WSUD and FRM to reduce the effects of climatic disasters on urban infrastructure at a global level. This study highlights critical gaps and opportunities in urban, suburban, rural and brownfield water management systems. While cities have advanced water management strategies, such as WSUD, SWM, GI and IUWM, these approaches face challenges in suburban and rural areas, largely due to limitations in funding, regulation and local applicability. The findings suggest that water management strategies must be more adaptive to the socioeconomic conditions and ecological contexts of non-urban areas, blending traditional gray infrastructure with GI.

Strategic planning plays a critical role in ensuring that water management strategies like WSUD are integrated into urban development with a focus on long-term sustainability. It provides the structure necessary for implementing adaptive solutions that consider local contexts, ensuring that urban water management is resilient to future challenges such as climate change, population growth and evolving environmental conditions. By setting clear goals and fostering collaboration among various stakeholders, strategic planning helps optimize water-sensitive approaches while aligning them with broader urban sustainability objectives.

Although the spatial scale of FRM (when considering catchment and hazard areas) usually covers larger areas than WSUD, which has so far mostly taken place at the neighborhood level, WSUD offers more starting points for integration into strategic urban planning. Although larger scales are often associated with more abstract and strategic planning and less with tangible planning projects, this is not the case for FRM.

While crucial to reducing flood hazards, FRM's traditional focus on risk assessment and engineering measures can limit its integration with broader urban development objectives. The focus of flood risk management on specific flood events and risk metrics is often siloed from other urban considerations, which can lead to fragmented policy implementation [10,39,60]. Although the recent shift towards nature-based solutions has improved the compatibility of FRM with sustainability goals, the focus remains on hazards rather than integration.

In contrast, WSUD is closely aligned with the principles of sustainable development and promotes multifunctional landscapes that provide ecosystem services, improve the quality of life in the city and contribute to climate change adaptation [37]. The focus of WSUD on the integration of natural hydrological processes into urban environments ensures that water management becomes a core element of urban design, promoting synergies among ecological, social and economic objectives [60]. This is in line with the holistic and future-oriented nature of strategic planning.

To summarize, in the context of strategic planning for urban sustainability, the following relationships can be identified for WSUD and FRM: WSUD has a much stronger focus on integration and future orientation than FRM. Thus, WSUD seems better suited to be linked to strategic planning for urban sustainability than FRM. Among the similarities, the following can be found:

- WSUD and strategic planning for urban sustainability follow a holistic perspective: WSUD promotes resource-efficient and adaptive water management that integrates natural cycles and strengthens the resilience of urban systems to climate change and extreme weather events. Strategic planning approaches take up these goals by developing a long-term vision for sustainable cities that integrate social, ecological and economic dimensions.
- Both approaches emphasize the need for multi-disciplinary collaboration among actors from different institutions: WSUD strategies require coordination among engineering, ecology, urban planning and politics, while strategic planning provides a platform for integrating such sectoral perspectives.
- Both approaches are linked by adaptive management and flexibility. The case studies in this paper show that WSUD requires a flexible framework responding successfully to changing environmental and climate conditions. Strategic planning approaches provide a framework for the flexibility needed and can use WSUD as a model for iterative planning processes.

These case studies reflect the value of holistic approaches, multi-disciplinary strategies and adaptive management in urban flood risk management, making them strong examples of the principles outlined above. Each of these cities or regions employs a combination of infrastructure development, environmental strategies, community engagement and flexible management systems to address the complex challenges posed by flooding and climate change. In the meanwhile, context-specific strategies are essential, with rural and suburban areas benefiting from LK and ecological practices that balance the cost of infrastructure investments. Successful case studies, such as in the Mekong Delta and Chiang Mai, demonstrate the importance of community engagement and non-engineering solutions in FRM. Similarly, brownfield regeneration, as seen in Trento, illustrates how neglected urban spaces can be renewed to address environmental challenges while improving the quality of life.

This research study highlights the need for improved stakeholder communication and collaboration, as detachment among municipalities, planners and local communities obstructs effective water and flood management. This consideration signifies close and comprehensive interaction among experts in urban planning and water management in various fields. Communities of practice (CoPs) play a vital role in advancing WSUD and FRM in contexts of strategic urban sustainability planning by fostering collaboration, knowledge exchange and capacity building among stakeholders [61]. CoPs bring together practitioners, researchers, policymakers and community members to co-create solutions that address complex water-related challenges in urban environments. Through shared learning and collective problem solving, CoPs enhance the integration of social, ecological and technical perspectives, promoting innovative and context-sensitive approaches [62]. Furthermore, CoPs support the dissemination of best practices and the adaptation of WSUD principles to diverse local conditions, ensuring resilience and sustainability in urban water systems. As WSUD and FRM—due to their different foci—can be addressed in different CoPs, opening the interaction between different CoPs is essential to addressing the multifaceted nature of urban water challenges [63].

Finally, this study recommends expanding WSUD into WSRD, which combines ecological, hydrological and socioeconomic dimensions on a broader scale. This framework can better address water management challenges across diverse landscapes, ensuring resilience to climate risks while meeting regional needs. The application of these insights enables practitioners to develop adaptive and sustainable water management systems that facilitate the integration of urban and rural areas, extending beyond the boundaries of the city [64]. By presenting successful case studies, this research study provides insights for developing resilient, context-sensitive urban planning. These insights can inspire strategic planning toward sustainable, climate-adaptive cities, promoting effective water management across various land types.

In summary, WSUD and strategic planning for urban sustainability are linked as complementary approaches and offer potential for re-scaling processes [64]. While WSUD offers practical solutions for urban water management, strategic planning provides the framework for embedding these solutions in a broader sustainability concept. Strategic urban planning must extend beyond individual projects to address water challenges at regional and landscape scales. Expanding WSUD into "Water-Sensitive Regional Design" (WSRD) would integrate hydrological, ecological and socioeconomic factors, enabling more cohesive planning strategies. Strategic planning also requires improved communication among municipalities, engineers and residents to ensure the effective implementation and scaling of innovative water management practices. The results of this paper emphasize the critical role of strategic urban planning in fostering adaptive and resilient water-sensitive frameworks.

Author Contributions: Conceptualization, T.T. and A.G.; methodology, A.G.; software, T.T.; validation, T.T.; formal analysis, T.T.; investigation, T.T.; data curation, T.T.; writing—original draft preparation, T.T.; writing—review and editing, T.T. and A.G.; visualization, T.T.; supervision, A.G. All authors have read and agreed to the published version of the manuscript.

Funding: This research received no external funding.

Data Availability Statement: No new data were created or analyzed in this study. Data sharing is not applicable to this article.

Conflicts of Interest: The authors declare no conflict of interest.

References

1. Palazzo, E. From Water Sensitive to Floodable: Defining *Adaptive Urban Design* for Water Resilient Cities. *J. Urban Des.* **2019**, *24*, 137–157. [CrossRef]
2. Hersperger, A.M.; Oliveira, E.; Pagliarin, S.; Palka, G.; Verburg, P.; Bolliger, J.; Grădinaru, S. Urban Land-Use Change: The Role of Strategic Spatial Planning. *Glob. Environ. Chang.* **2018**, *51*, 32–42. [CrossRef]
3. Vercruysse, K.; Dawson, D.A.; Wright, N. Interoperability: A Conceptual Framework to Bridge the Gap between Multifunctional and Multisystem Urban Flood Management. *J. Flood Risk Manag.* **2019**, *12*, e12535. [CrossRef]
4. Lerer, S.; Arnbjerg-Nielsen, K.; Mikkelsen, P. A Mapping of Tools for Informing Water Sensitive Urban Design Planning Decisions—Questions, Aspects and Context Sensitivity. *Water* **2015**, *7*, 993–1012. [CrossRef]

5. Van Der Steen, P.; Howe, C. Managing Water in the City of the Future; Strategic Planning and Science. *Rev. Environ. Sci. Biotechnol.* **2009**, *8*, 115–120. [CrossRef]
6. Ferguson, B.C.; Frantzeskaki, N.; Brown, R.R. A Strategic Program for Transitioning to a Water Sensitive City. *Landsc. Urban Plan.* **2013**, *117*, 32–45. [CrossRef]
7. Nicolini, E.; Mamì, A. Circular Water Management in Public Space—Experimental Feasibility Studies in Different Urban Contexts. *Sustainability* **2023**, *15*, 12025. [CrossRef]
8. Khodadad, M.; Aguilar-Barajas, I.; Khan, A.Z. Green Infrastructure for Urban Flood Resilience: A Review of Recent Literature on Bibliometrics, Methodologies, and Typologies. *Water* **2023**, *15*, 523. [CrossRef]
9. Fletcher, T.D.; Shuster, W.; Hunt, W.F.; Ashley, R.; Butler, D.; Arthur, S.; Trowsdale, S.; Barraud, S.; Semadeni-Davies, A.; Bertrand-Krajewski, J.-L.; et al. SUDS, LID, BMPs, WSUD and More—The Evolution and Application of Terminology Surrounding Urban Drainage. *Urban Water J.* **2015**, *12*, 525–542. [CrossRef]
10. Wheater, H.; Evans, E. Land Use, Water Management and Future Flood Risk. *Land Use Policy* **2009**, *26*, S251–S264. [CrossRef]
11. Chan, F.K.S.; Griffiths, J.A.; Higgitt, D.; Xu, S.; Zhu, F.; Tang, Y.-T.; Xu, Y.; Thorne, C.R. "Sponge City" in China—A Breakthrough of Planning and Flood Risk Management in the Urban Context. *Land Use Policy* **2018**, *76*, 772–778. [CrossRef]
12. Mguni, P.; Herslund, L.; Jensen, M.B. Sustainable Urban Drainage Systems: Examining the Potential for Green Infrastructure-Based Stormwater Management for Sub-Saharan Cities. *Nat. Hazards* **2016**, *82*, 241–257. [CrossRef]
13. Ellis, J.B. Sustainable Surface Water Management and Green Infrastructure in UK Urban Catchment Planning. *J. Environ. Plan. Manag.* **2013**, *56*, 24–41. [CrossRef]
14. Bach, P.M.; Rauch, W.; Mikkelsen, P.S.; McCarthy, D.T.; Deletic, A. A Critical Review of Integrated Urban Water Modelling—Urban Drainage and Beyond. *Environ. Model. Softw.* **2014**, *54*, 88–107. [CrossRef]
15. Kirshen, P.; Aytur, S.; Hecht, J.; Walker, A.; Burdick, D.; Jones, S.; Fennessey, N.; Bourdeau, R.; Mather, L. Integrated Urban Water Management Applied to Adaptation to Climate Change. *Urban Clim.* **2018**, *24*, 247–263. [CrossRef]
16. Bichai, F.; Cabrera Flamini, A. The Water-Sensitive City: Implications of an Urban Water Management Paradigm and Its Globalization. *WIREs Water* **2018**, *5*, e1276. [CrossRef]
17. Tang, Y.; Chan, F.K.S.; O'Donnell, E.C.; Griffiths, J.; Lau, L.; Higgitt, D.L.; Thorne, C.R. Aligning Ancient and Modern Approaches to Sustainable Urban Water Management in China: Ningbo as a "Blue-Green City" in the "Sponge City" Campaign. *J. Flood Risk Manag.* **2018**, *11*, e12451. [CrossRef]
18. Šakić Trogrlić, R.; Wright, G.; Duncan, M.; Van Den Homberg, M.; Adeloye, A.; Mwale, F.; Mwafulirwa, J. Characterising Local Knowledge across the Flood Risk Management Cycle: A Case Study of Southern Malawi. *Sustainability* **2019**, *11*, 1681. [CrossRef]
19. Sharma, S.V.S.; Roy, P.S.; Chakravarthi, V.; Rao, G.S. Flood Risk Assessment Using Multi-Criteria Analysis: A Case Study from Kopili River Basin, Assam, India. *Geomat. Nat. Hazards Risk* **2018**, *9*, 79–93.
20. Leigh, N.G.; Lee, H. Sustainable and Resilient Urban Water Systems: The Role of Decentralization and Planning. *Sustainability* **2019**, *11*, 918. [CrossRef]
21. Zhang, S.; Li, Y.; Ma, M.; Song, T.; Song, R. Storm Water Management and Flood Control in Sponge City Construction of Beijing. *Water* **2018**, *10*, 1040. [CrossRef]
22. Akhter, M.; Hewa, G. The Use of PCSWMM for Assessing the Impacts of Land Use Changes on Hydrological Responses and Performance of WSUD in Managing the Impacts at Myponga Catchment, South Australia. *Water* **2016**, *8*, 511. [CrossRef]
23. Nkwunonwo, U.C.; Whitworth, M.; Baily, B. Review Article: A Review and Critical Analysis of the Efforts towards Urban Flood Risk Management in the Lagos Region of Nigeria. *Nat. Hazards Earth Syst. Sci.* **2016**, *16*, 349–369. [CrossRef]
24. Kefi, M.; Mishra, B.; Kumar, P.; Masago, Y.; Fukushi, K. Assessment of Tangible Direct Flood Damage Using a Spatial Analysis Approach under the Effects of Climate Change: Case Study in an Urban Watershed in Hanoi, Vietnam. *Int. J. Geo-Inf.* **2018**, *7*, 29. [CrossRef]
25. Oral, H.V.; Carvalho, P.; Gajewska, M.; Ursino, N.; Masi, F.; Hullebusch, E.D.V.; Kazak, J.K.; Exposito, A.; Cipolletta, G.; Andersen, T.R.; et al. A Review of Nature-Based Solutions for Urban Water Management in European Circular Cities: A Critical Assessment Based on Case Studies and Literature. *Blue-Green Syst.* **2020**, *2*, 112–136. [CrossRef]
26. Schuetze, T.; Chelleri, L. Integrating Decentralized Rainwater Management in Urban Planning and Design: Flood Resilient and Sustainable Water Management Using the Example of Coastal Cities in The Netherlands and Taiwan. *Water* **2013**, *5*, 593–616. [CrossRef]
27. Kuller, M.; Bach, P.M.; Ramirez-Lovering, D.; Deletic, A. Framing Water Sensitive Urban Design as Part of the Urban Form: A Critical Review of Tools for Best Planning Practice. *Environ. Model. Softw.* **2017**, *96*, 265–282. [CrossRef]
28. Larsen, T.A.; Hoffmann, S.; Lüthi, C.; Truffer, B.; Maurer, M. Emerging Solutions to the Water Challenges of an Urbanizing World. *Science* **2016**, *352*, 928–933. [CrossRef]

29. Makropoulos, C.K.; Natsis, K.; Liu, S.; Mittas, K.; Butler, D. Decision Support for Sustainable Option Selection in Integrated Urban Water Management. *Environ. Model. Softw.* **2008**, *23*, 1448–1460. [CrossRef]
30. Rashetnia, S.; Sharma, A.K.; Ladson, A.R.; Browne, D.; Yaghoubi, E. A Scoping Review on Water Sensitive Urban Design Aims and Achievements. *Urban Water J.* **2022**, *19*, 453–467. [CrossRef]
31. Rodriguez, C.N.A.S.; Ashley, R.; Gersonius, B.; Rijke, J.; Pathirana, A.; Zevenbergen, C. Incorporation and Application of Resilience in the Context of Water-sensitive Urban Design: Linking European and Australian Perspectives. *WIREs Water* **2014**, *1*, 173–186. [CrossRef]
32. Wu, W.; Jamali, B.; Zhang, K.; Marshall, L.; Deletic, A. Water Sensitive Urban Design (WSUD) Spatial Prioritisation through Global Sensitivity Analysis for Effective Urban Pluvial Flood Mitigation. *Water Res.* **2023**, *235*, 119888. [CrossRef]
33. Siekmann, T.; Siekmann, M. Resilient Urban Drainage—Options of an Optimized Area-Management. *Urban Water J.* **2015**, *12*, 44–51. [CrossRef]
34. Lennon, M.; Scott, M.; O'Neill, E. Urban Design and Adapting to Flood Risk: The Role of Green Infrastructure. *J. Urban Des.* **2014**, *19*, 745–758. [CrossRef]
35. Bai, Y.; Zhao, N.; Zhang, R.; Zeng, X. Storm Water Management of Low Impact Development in Urban Areas Based on SWMM. *Water* **2018**, *11*, 33. [CrossRef]
36. Brown, R.R.; Farrelly, M.A.; Loorbach, D.A. Actors Working the Institutions in Sustainability Transitions: The Case of Melbourne's Stormwater Management. *Glob. Environ. Chang.* **2013**, *23*, 701–718. [CrossRef]
37. Wong, T.H.F.; Brown, R.R. The Water Sensitive City: Principles for Practice. *Water Sci. Technol.* **2009**, *60*, 673–682. [CrossRef] [PubMed]
38. Apel, H.; Aronica, G.T.; Kreibich, H.; Thieken, A.H. Flood Risk Analyses—How Detailed Do We Need to Be? *Nat. Hazards* **2009**, *49*, 79–98. [CrossRef]
39. Schanze, J. Flood risk management—A basic framework. In *Flood Risk Management: Hazards, Vulnerability and Mitigation Measures*; Schanze, J., Zeman, E., Marsalek, J., Eds.; NATO Science Series; Springer Netherlands: Dordrecht, The Netherlands, 2006; Volume 67, pp. 1–20.
40. Raška, P.; Bezak, N.; Ferreira, C.S.S.; Kalantari, Z.; Banasik, K.; Bertola, M.; Bourke, M.; Cerdà, A.; Davids, P.; Madruga De Brito, M.; et al. Identifying Barriers for Nature-Based Solutions in Flood Risk Management: An Interdisciplinary Overview Using Expert Community Approach. *J. Environ. Manag.* **2022**, *310*, 114725. [CrossRef] [PubMed]
41. Wang, H.; Mei, C.; Liu, J.; Shao, W. A New Strategy for Integrated Urban Water Management in China: Sponge City. *Sci. China Technol. Sci.* **2018**, *61*, 317–329. [CrossRef]
42. Zevenbergen, C.; Fu, D.; Pathirana, A. Transitioning to Sponge Cities: Challenges and Opportunities to Address Urban Water Problems in China. *Water* **2018**, *10*, 1230. [CrossRef]
43. Rubinato, M.; Nichols, A.; Peng, Y.; Zhang, J.; Lashford, C.; Cai, Y.; Lin, P.; Tait, S. Urban and River Flooding: Comparison of Flood Risk Management Approaches in the UK and China and an Assessment of Future Knowledge Needs. *Water Sci. Eng.* **2019**, *12*, 274–283. [CrossRef]
44. Kellens, W.; Vanneuville, W.; Verfaillie, E.; Meire, E.; Deckers, P.; De Maeyer, P. Flood Risk Management in Flanders: Past Developments and Future Challenges. *Water Resour Manag.* **2013**, *27*, 3585–3606. [CrossRef]
45. Fryd, O.; Backhaus, A.; Birch, H.; Fratini, C.F.; Ingvertsen, S.T.; Jeppesen, J.; Panduro, T.E.; Roldin, M.; Jensen, M.B. Water Sensitive Urban Design Retrofits in Copenhagen—40% to the Sewer, 60% to the City. *Water Sci. Technol.* **2013**, *67*, 1945–1952. [CrossRef]
46. Restemeyer, B.; Woltjer, J.; Van Den Brink, M. A Strategy-Based Framework for Assessing the Flood Resilience of Cities—A Hamburg Case Study. *Plan. Theory Pract.* **2015**, *16*, 45–62. [CrossRef]
47. Van Herk, S.; Zevenbergen, C.; Gersonius, B.; Waals, H.; Kelder, E. Process Design and Management for Integrated Flood Risk Management: Exploring the Multi-Layer Safety Approach for Dordrecht, The Netherlands. *J. Water Clim. Chang.* **2014**, *5*, 100–115. [CrossRef]
48. Manda, M.Z. Where There Is No Local Government: Addressing Disaster Risk Reduction in a Small Town in Malawi. *Environ. Urban.* **2014**, *26*, 586–599. [CrossRef]
49. Liao, K.-H.; Le, T.A.; Nguyen, K.V. Urban Design Principles for Flood Resilience: Learning from the Ecological Wisdom of Living with Floods in the Vietnamese Mekong Delta. *Landsc. Urban Plan.* **2016**, *155*, 69–78. [CrossRef]
50. Romanescu, G.; Hapciuc, O.E.; Minea, I.; Iosub, M. Flood Vulnerability Assessment in the Mountain–Plateau Transition Zone: A Case Study of Marginea Village (Romania). *J. Flood Risk Manag.* **2018**, *11*, S502–S513. [CrossRef]
51. Page, M.J.; McKenzie, J.E.; Bossuyt, P.M.; Boutron, I.; Hoffmann, T.C.; Mulrow, C.D.; Shamseer, L.; Tetzlaff, J.M.; Akl, E.A.; Brennan, S.E.; et al. The PRISMA 2020 Statement: An Updated Guideline for Reporting Systematic Reviews. *BMJ* **2021**, *372*, n71. [CrossRef] [PubMed]
52. Xiao, Y.; Watson, M. Guidance on Conducting a Systematic Literature Review. *J. Plan. Educ. Res.* **2019**, *39*, 93–112. [CrossRef]

53. Li, G.-F.; Xiang, X.-Y.; Tong, Y.-Y.; Wang, H.-M. Impact Assessment of Urbanization on Flood Risk in the Yangtze River Delta. *Stoch. Environ. Res. Risk Assess.* **2013**, *27*, 1683–1693. [CrossRef]
54. Ling, T.-Y.; Chiang, Y.-C. Strengthening the Resilience of Urban Retailers towards Flood Risks—A Case Study in the Riverbank Region of Kaohsiung City. *Int. J. Disaster Risk Reduct.* **2018**, *27*, 541–555. [CrossRef]
55. Schuch, G.; Serrao-Neumann, S.; Morgan, E.; Low Choy, D. Water in the City: Green Open Spaces, Land Use Planning and Flood Management—An Australian Case Study. *Land Use Policy* **2017**, *63*, 539–550. [CrossRef]
56. Gleason, J.A.; Casiano Flores, C. Challenges of Water Sensitive Cities in Mexico: The Case of the Metropolitan Area of Guadalajara. *Water* **2021**, *13*, 601. [CrossRef]
57. Wanitchayapaisit, C.; Charoenlertthanakit, N.; Surinseng, V.; Yaipimol, E.; Rinchumphu, D.; Suppakittpaisarn, P. Enhancing Water-Sensitive Urban Design in Chiang Mai through a Research–Design Collaboration. *Sustainability* **2023**, *15*, 16127. [CrossRef]
58. Lottering, N.; Du Plessis, D.; Donaldson, R. Coping with Drought: The Experience of Water Sensitive Urban Design (WSUD) in the George Municipality. *Water SA* **2014**, *41*, 1. [CrossRef]
59. Cortinovis, C.; Geneletti, D. Mapping and Assessing Ecosystem Services to Support Urban Planning: A Case Study on Brownfield Regeneration in Trento, Italy. *One Ecosyst.* **2018**, *3*, e25477. [CrossRef]
60. Brown, R.R.; Keath, N.; Wong, T.H.F. Urban Water Management in Cities: Historical, Current and Future Regimes. *Water Sci. Technol.* **2009**, *59*, 847–855. [CrossRef]
61. Fulgenzi, A.; Brouwer, S.; Baker, K.; Frijns, J. Communities of practice at the center of circular water solutions. *WIREs Water* **2020**, *7*, e1450. [CrossRef]
62. King, P.; Martin-Ortega, J.; Armstrong, J.; Ferré, M.; Bark, R.H. Mainstreaming nature-based solutions: What role do Communities of Practice play in delivering a paradigm shift? *Environ. Sci. Policy* **2023**, *144*, 53–63. [CrossRef]
63. Matsumoto, Y.; Kasamatsu, H.; Sakakibara, M. Challenges in Forming Transdisciplinary Communities of Practice for Solving Environmental Problems in Developing Countries. *World Futures* **2022**, *78*, 546–565. [CrossRef]
64. Yan, S.; Growe, A. Regional Planning, Land-Use Management, and Governance in German Metropolitan Regions—The Case of Rhine–Neckar Metropolitan Region. *Land* **2022**, *11*, 2088. [CrossRef]

Disclaimer/Publisher's Note: The statements, opinions and data contained in all publications are solely those of the individual author(s) and contributor(s) and not of MDPI and/or the editor(s). MDPI and/or the editor(s) disclaim responsibility for any injury to people or property resulting from any ideas, methods, instructions or products referred to in the content.

land

MDPI

Article

Evolution and Mechanism of Cooperative Innovation Networks Based on Strategic Emerging Industries: Evidence from Northeast China

Xiaodong Zhou †, Tong Liu †, Peng Zhang *, Xujia Zhang * and Nanchen Chu

College of Geography Science, Harbin Normal University, Harbin 150025, China; zhouxd0926@163.com (X.Z.); 15645280886@163.com (T.L.); chunanchen_1992@163.com (N.C.)

* Correspondence: zhangp575@nenu.edu.cn (P.Z.); hsdzhangxj@163.com (X.Z.)

† These authors contributed equally to this work.

Abstract: The rapid flow of resources and participation of diverse entities in the information age drive the formation of innovation networks, yet how these networks evolve in strategic emerging industries—particularly in transitioning regions like Northeast China—remains underexplored. This study investigates the spatial formation and evolution of cooperative innovation networks in strategic emerging industries, as well as the underlying mechanisms driving their development and transformation. Focusing on China's three northeastern provinces—a representative old industrial base undergoing transformation—we construct and analyze inter- and intra-city cooperative innovation networks from 2009 to 2023 using invention patent data. Employing social network analysis and spatial econometric models, we reveal that while the external network exhibits limited expansion with Beijing as a dominant hub, the internal network demonstrates increasing cohesion, transitioning to a multi-core structure. Factors such as urban innovation capacity, economic proximity, and institutional proximity play key roles in shaping these networks. These findings offer policymakers empirical evidence to strategically optimize cross-regional innovation ecosystems in resource-dependent regions, supporting urban sustainability and fostering innovation-driven development.

Keywords: strategic emerging industries; cooperative innovation network; negative binomial gravity model; three northeastern provinces

1. Introduction

Innovation is widely recognized as the primary driver of economic development, particularly within the domain of strategic emerging industries. These industries, defined by their high-tech nature, substantial growth potential, and technological advantages [1], include sectors such as information technology, new energy, and biomedicine. Their strategic importance lies in their ability to drive global economic transformation, enhance industrial upgrading, and strengthen national competitiveness [2]. The primary essence of strategic emerging industries lies in their strategic significance, which is reflected in several aspects. These industries are typically green, low-carbon, and environmentally friendly, aligning with the long-term and sustainable development goals of cities. Moreover, they are closely integrated with government strategies for urban development. The technologies employed in these industries often possess strategic breakthrough potential, relying on innovation and technology-driven development.

However, to fully realize this potential, the achievement of high-quality development in these industries depends on innovation networks based on collaborative relationships. These networks, originally proposed by Freeman [3], facilitate technological synergy and resource sharing among firms, academic and research institutions, and government agencies, thereby accelerating technological breakthroughs and industrialization.

With the advancement of strategic emerging industries, the development of innovation activities increasingly depends on inter-organizational and cross-regional network collaboration. The transition from "local space" to "mobility space" and "cyberspace" has significantly reshaped the dynamics of innovation networks [4,5], driving the reconfiguration of city-centered innovation within cyberspace [6]. Globalization, informatization, and networking have facilitated the strengthening of inter-city connections, disrupting the traditional hierarchical urban organization model and giving rise to a networked urban structure. The innovative capacity of cities is not only dependent on internal innovation resources but is also reflected in the increased connectivity of external networks. These synergistic networks enhance the complementarity and advancement of technology and knowledge [7].

To construct an effective city innovation network, it is essential to quantify the innovation linkages between cities. However, the cooperation process involves a significant amount of tacit knowledge transfer and diffusion, which is difficult to measure directly. Consequently, scholars often use joint patent and collaborative research paper data to construct technology innovation networks and scientific knowledge networks, respectively. Patent collaboration data reflect the cooperation between cities in technological development [8], while co-authored papers between research institutions indicate the distribution and flow of academic resources [9]. These data provide a foundation for quantifying the structural characteristics and evolution of urban innovation networks.

Research on innovation networks, both domestically and internationally, primarily focuses on three key areas. Firstly, case studies of different types of strategic emerging industries are conducted. Domestic research predominantly centers on innovation networks in sectors such as electronic information [10] and biomedicine [11], investigating their formation mechanisms and characteristics. In contrast, international scholars tend to focus more on areas such as artificial intelligence [12,13] and biotechnology [14], examining cross-border collaboration, technology diffusion, and the flow of innovation resources within the global value chain. However, most of these studies are industry-specific, making it challenging to comprehensively capture the characteristics of innovation networks in the context of regional cooperation. As a critical region for advancing strategic emerging industries and facilitating economic restructuring in China, the three northeastern provinces possess unique regional characteristics and development needs in the construction of inter-city and trans-regional innovation networks. Consequently, their innovation network development patterns and evolutionary mechanisms warrant in-depth exploration.

Secondly, research on the spatio-temporal evolution of innovation networks in strategic emerging industries has gained significant attention. The evolution of innovation networks has become a key topic for scholars in economic geography [15], primarily focusing on the changes in topological structure, spatial patterns, and linkages among innovation actors [16]. However, the evolution of innovation networks within strategic emerging industries, particularly the spatial characteristics of inter-city and cross-regional networks, remains an area requiring further investigation.

Thirdly, research on the factors influencing the formation and evolution of innovation networks has focused on both internal and external determinants. Internal factors include the composition of innovation agents and their collaborative linkages, while external factors involve the innovation environment and policy frameworks. In recent years, scholars, notably Boschma [17], have investigated how various proximities, such as geographical, cognitive, and organizational proximities, facilitate or restrict knowledge flow and innovation diffusion, influencing the construction and evolution of innovation networks [18]. The interaction of these multidimensional proximities has been identified as a critical factor in fostering innovation cooperation. However, the impact of proximity varies across different stages of development, spatial scales, and industrial contexts [19]. Further research is needed to understand how different proximities shape the evolution of innovation networks and cooperation in strategic emerging industries, particularly in the three northeastern provinces.

In terms of research methods for innovation networks, Jaffe introduced social network analysis (SNA), which has become the primary method for studying inter-regional innovation linkages [20]. SNA is used to examine the characteristics, topological relationships, and structural evolution of innovation networks. Regarding influencing factors, commonly employed methods include the negative binomial regression model [21] and QAP regression analysis [22], which are used to explore the endogeneity of urban innovation capacity and its spatial spillover effects.

Based on the above discussion, the key issues this study aims to address are as follows: (1) What are the spatial and structural formations and evolutions of the cross-regional cooperative innovation networks in strategic emerging industries in the three northeastern provinces? (2) What are the spatial and structural formations and evolutions of the inter-city cooperative innovation networks in strategic emerging industries in the three northeastern provinces? (3) What role does multidimensional proximity play in shaping the evolution of these networks?

This study examines the spatial and structural evolution of innovation networks in strategic emerging industries across China's three northeastern provinces, leveraging patent data from 2009 to 2023. By employing ArcGIS10.8 and social network analysis (SNA), this research traces the development of cross-regional and inter-city innovation networks. Additionally, a negative binomial gravity regression model is applied to investigate the underlying mechanisms of network evolution through the lens of multidimensional proximity. This study aims to systematically examine how innovation strategies can drive sustainable industrial transformation in Northeast China, with a focus on optimizing cross-regional and inter-city innovation networks to inform evidence-based policies for revitalizing the region's economy through technology-led development pathways.

The main contributions of this study are as follows: (1) Focusing on the three traditional industrial bases in Northeast China, this study provides an in-depth analysis of the evolution mechanism and driving factors of urban innovation networks within the context of the region's ongoing economic transformation. This region, once heavily reliant on traditional industries, is undergoing a profound shift toward strategic emerging industries. Understanding the mechanisms behind innovation network evolution in this context is critical for policymakers, as it offers insights into how these regions can navigate their transition from industrial decline to economic revitalization. This study addresses a significant gap in the literature by providing an empirical model specifically tailored to the needs of declining industrial regions, an area that has received limited attention in existing research. (2) This study broadens the scope by examining the entire spectrum of strategic emerging industries, rather than focusing on single industries. It explores the role of cross-industry synergies in the evolution of regional innovation networks, addressing a significant gap in previous studies that have overlooked inter-industry connections. (3) This study introduces a novel "inter-regional and intra-city" framework, analyzing both external collaborations and internal synergies within urban innovation networks. This multi-scale approach provides new insights into how innovation-driven development unfolds in transitioning regions, offering a more comprehensive view compared to traditional studies focused solely on intra-regional interactions.

2. Material and Methods

2.1. Study Area

The three northeastern provinces of China—Liaoning, Jilin, and Heilongjiang—encompassing 36 prefecture-level administrative units (see Figure 1), are critical old industrial bases and hold a key position in the national regional innovation strategy. With a combined resident population of 85.1241 million in 2023 (6.04% of China's total), this region forms a cohesive geographic and socioeconomic entity characterized by shared historical trajectories, industrial structures, and policy frameworks. As the core target of China's "Northeast Revitalization" strategic initiatives since 2003, these provinces face systemic challenges distinct from other regions. However, on a national scale, the northeast region lags behind economically developed areas, such as Beijing–Tianjin–Hebei, the Yangtze River Delta, and the Pearl River Delta, particularly in terms of economic development and scientific and technological innovation capacity. Consequently, there is a significant gap in the development of strategic emerging industries [23,24]. This region faces challenges in industrial upgrading and innovation-driven development, but at the same time, it offers significant opportunities for exploring new models of innovation cooperation.

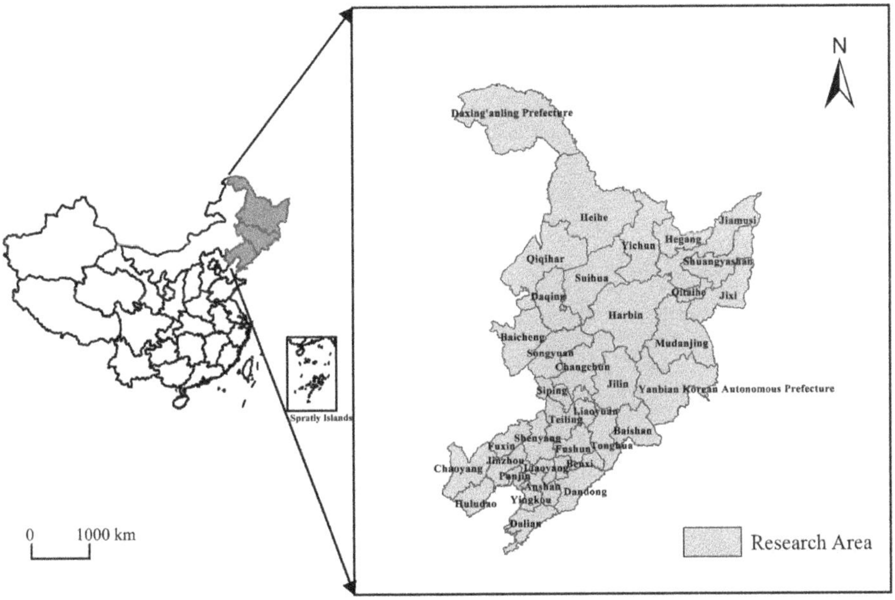

Figure 1. Study area.

The innovation environment in the three northeastern provinces exhibits distinct regional characteristics. On one hand, state-owned enterprises (SOEs) dominate the innovation landscape, with a well-established industrial system but limited innovation vitality. On the other hand, the secondary industry accounts for a high proportion of the economic structure, with traditional manufacturing sectors largely driving regional economic development, while high-tech industries remain relatively underdeveloped [25]. Moreover, compared to the southern coastal regions, the northeast has a lower degree of marketization, a relatively rigid institutional environment, and a less developed private economy. In terms of inter-city and cross-regional cooperation, the model in this region relies heavily on policy support, with relatively weak market-driven cooperation dynamics [26]. These factors contribute to an uneven distribution of internal resources and limited synergistic efficiency within the innovation networks, reflecting certain imbalances. As such, the three northeastern provinces provide a typical and representative case for study.

2.2. Data Sources and Data Processing

Patents, as a product of technological innovation, serve as an ideal indicator of innovation cooperation between cities, with cooperative patents representing their primary form [27]. The data on cooperative patents are sourced from the Wisdom Sprout (PatSnap) Global Patent Database (https://www.zhihuiya.com/, accessed on 28 May 2024), utilizing the SEIC classification of the 2018 version of Strategic Emerging Industries to identify nine industry categories. The patent right holders considered in this study include companies, institutions, and government entities. Since China officially introduced the "strategic emerging industries" strategy in 2009, the research period for this study spans from 2009 to 2023, with a total of 389,355 authorized invention patents being screened for analysis.

Pre-processing of patent data: (1) Data Screening: Patents filed individually were removed, and those involving individuals (where addresses could not be confirmed), as well as patents from Hong Kong, Macao, Taiwan, and foreign enterprises, were excluded. This ensures that this study focuses solely on the cooperative innovation network within the three northeastern provinces, excluding any transnational external influences. As a result, 26,913 valid collaborative patents were retained. (2) Multiple cooperation records: Patents with three or more co-applicants were recorded as multiple cooperation records using pairwise combinations. (3) Geographic information matching: Using the "Tianyancha" website and manual searching, each applicant was matched with their respective city. (4) Constructing an undirected weighted matrix: Based on the joint applications of invention patents between cities, an undirected weighted matrix was established with cities as nodes, patent cooperation relationships as edges, and the number of cooperations as weights. This matrix was used to construct a city cooperation and innovation network in the three northeastern provinces.

This study divides the research period into three time intervals—2009–2013, 2014–2018, and 2019–2023—using a five-year boundary to systematically analyze the evolution of the urban innovation network for strategic emerging industries in the three northeastern provinces.

2.3. Method and Model Selection

2.3.1. Social Network Analysis

Social network analysis is a crucial method for studying the interrelationships within networks, revealing both the structural characteristics of the overall network and individual nodes [11]. It is widely used in innovation network studies. We selected network size, connection strength, and small-world indicators to examine the overall characteristics of the cooperative innovation network among northeastern cities. Additionally, the positions of city nodes in the network were assessed based on their centrality.

2.3.2. Geographical Information System (GIS) Spatial Visualization Method

In the spatial analysis, cooperation between innovation entities was converted into cooperation between cities based on the locations of the applicants. The XY to Line function in GIS was then used to derive the spatial pattern map of the innovation network [28].

2.3.3. Negative Binomial Gravity Model

The gravity model theory posits that interactions between entities are primarily influenced by the entities' own attributes and the spatial distance between them. Scherngell [29] proposed a negative binomial regression method based on the gravity model to analyze the mechanisms via which various factors impact inter-regional innovation cooperation [30]. In this study, we adopted the model proposed by these researchers and used it to examine the influencing factors of the internal innovation network in the three northeastern provinces across the three time periods and to analyze its evolutionary mechanisms. The specific model was set up as follows:

$$\begin{aligned} C_{ij} &= \alpha + \beta_1 PAT_i + \beta_2 PAT_j + \beta_3 Geo_{ij} + \beta_4 Bor_{ij} + \beta_5 Eco_{ij} + \beta_6 Ind_{ij} \\ &+ \beta_7 Ins_{ij} + \beta_8 Tec_{ij} + \beta_9 Soc_{ij} \cdot \text{gov} + \beta_{10} Soc_{ij} \cdot \text{edu} + \varepsilon_{ij} \end{aligned} \quad (1)$$

where C_{ij} is the intensity of cooperation between city i and city j; α is the constant term; β is the estimated parameter; and ε_{ij} is the random error term. The meanings and data sources of the variables are detailed in Table 1, details of the specific data are provided in the Supplementary Material.

Table 1. Explanations of the independent variables in the negative binomial gravity regression model.

Variable	Explanation	Data Sources and Processing
PAT_i	Total number of patent applications for inventions in city i	The number of patent applications for each time period is a five-year average.
PAT_j	Total number of patent applications for inventions in city j	The number of patent applications for each time period is a five-year average.
Geo_{ij}	Normalized spatial distance between city i and city j. Together with Bor_{ij}, this characterizes the geographic proximity between cities.	$Geo_{ij} = 1 - \ln\left(\frac{d_{ij}}{\max d_{ij}}\right)$ (2) d_{ij} denotes the geographic distance between node i and node j, calculated by city latitude and longitude, and $\max d_{ij}$ is the maximum distance in the sample.
Bor_{ij}	Administrative boundary proximity between city i and city j. Together with Geo_{ij}, this characterizes the geographic proximity between cities.	A dummy variable that takes the value of 1 if the two cities share an administrative boundary
Eco_{ij}	Degree of similarity in the level of economic development between city i and city j. This characterizes the economic proximity between cities.	Expressed as inter-city GDP ratio.
Ind_{ij}	Degree of similarity in the level of industrial structure between city i and city j. This characterizes the industrial proximity between cities.	Expressed as the ratio of tertiary value added to GDP between cities [31].
Ins_{ij}	Degree of institutional difference between city i and city j. This characterizes institutional proximity between cities.	Cities in the same province are assigned a value of 1, while different provinces are assigned 0.

Table 1. *Cont.*

Variable	Explanation	Data Sources and Processing
Tec_{ij}	Degree of similarity in technological knowledge between city i and city j. This characterizes technological proximity between cities.	$$Tec_{ij} = \frac{\sum_{m=1}^{8} PAT_{imt} \times PAT_{jmt}}{\sqrt{\sum_{m=1}^{8} PAT^{2}_{imt} \times PAT^{2}_{jmt}}} \quad (3)$$ Drawing on Jaffe's technological proximity measure formula [32], PAT_{imt} and PAT_{jmt} are the number of patent applications for cities i and j in year t in category m-th IPC classification number. The values range from 0 to 1, with larger values indicating greater technological proximity. The number of patents for each period is calculated using the sum of patent numbers over five-year intervals.
$Soc_{ij}{\cdot}gov$	The degree of similarity in governmental support between city i and city j, with $Soc_{ij}{\cdot}edu$ characterizing social proximity between cities.	Expressed as the ratio of financial expenditures on science and technology between cities.
$Soc_{ij}{\cdot}edu$	The extent to which city i and city j have similar levels of education, with $Soc_{ij}{\cdot}gov$ characterizing social proximity between cities.	Expressed as the ratio of financial expenditures on education between cities.

Note: (1) Eco_{ij}, Ind_{ij}, $Soc_{ij}{\cdot}gov$, and $Soc_{ij}{\cdot}edu$ are calculated as the ratios between the two cities, with the maximum value as the denominator and the minimum value as the numerator, resulting in a range of (0, 1). (2) The data for PAT_i, PAT_j, and Tec_{ij} were sourced from the Shanghai Intellectual Property Information Service Platform. (3) Eco_{ij}, Ind_{ij}, $Soc_{ij}{\cdot}gov$, and $Soc_{ij}{\cdot}edu$ data were obtained from the "China City Statistical Yearbook" and city statistical bulletins, with a five-year average taken for each period. (4) Yanbian Korean Autonomous Prefecture and Daxing'anling Prefecture were excluded from the analysis due to data unavailability.

3. Results

3.1. Evolution Characteristics of the External Inter-City Cooperative Innovation Network in the Three Northeastern Provinces

3.1.1. Evolution Characteristics of the Topological Structure of the External Inter-City Innovation Network

(1) Network expansion and loose density. With the development of strategic emerging industries in the three northeastern provinces, external connections have become increasingly close. The external innovation network in the three northeastern provinces has experienced significant growth over the years, as indicated by the increase in both the number of participating cities and collaborative ties (see Table 2). Specifically, the node count surged by 64% (149 cities in 2009–2013→245 cities in 2019–2023), while the number of collaborative ties expanded by 165% (293→775 relationships). However, despite this expansion, the network density declined slightly by 2.4% (from 0.02657 to 0.02593). This paradox, characterized by "growth without densification", reflects a limited integration of newly added cities into the core hubs of Northeast China. Consequently, the external network has developed an "umbrella-like" structure, where newly joined cities contribute only sporadically to the overall connectivity, leading to a dilution in network density.

(2) Increasingly close innovation ties, lacking small-world characteristics. As shown in Table 2, the average weighted degree centrality has increased by 267% (from 34.161 to 125.478), reflecting a deepening of bilateral collaborations across cities. Small-world characteristics typically manifest as a high clustering coefficient (tight local connections) and a low characteristic path length (efficient global connections) [33]. These features allow nodes in small-world networks to reach other nodes with few steps, although they are relatively rarely directly connected. Generally, a network is considered to have small-world

properties if the clustering coefficient is greater than 0.1 and the average path length is less than 10 [34]. However, the clustering coefficient remains near zero (<0.002), which is well below the small-world threshold of 0.1, indicating that local clusters are weak and fail to form dense, tightly knit groups. The average path length has decreased slightly (from 2.797 to 2.669), suggesting improved regional connectivity. These observations suggest that the network has expanded broadly but lacks the high level of local interconnectedness characteristic of small-world networks, which would typically facilitate more efficient information flow and collaboration.

Table 2. Overall characteristics of external cooperative innovation networks in the three northeastern provinces.

	Indicator	2009–2013	2014–2018	2019–2023
Network size	Number of network nodes	149	186	245
	Number of network relationships	293	453	775
	Network density	0.02657	0.02633	0.02593
Contact strength	Average degree	3.933	4.871	6.327
	Average weighted degree	34.161	68.022	125.478
Small-world characteristics	Average path length	2.797	2.778	2.669
	Average clustering coefficient	0	0	0.002

(3) External cooperation dominated by provincial capitals and municipalities, with evident spatial polarization. Beijing maintained its position as the dominant super-core hub throughout the entire period from 2009 to 2023 (Table 3). It consistently ranked first across all centrality metrics—degree centrality, weighted degree centrality, and betweenness centrality—underscoring its central role in the innovation network. Additionally, the influence of provincial capitals and municipalities has grown over time, with these cities comprising 80% of the top-10-ranked cities by 2019–2023, compared to 60% in 2009–2013. Among these emerging subcores, Nanjing saw a significant rise in weighted centrality, advancing from 4th to 1st place by 2019–2023, while Shenzhen also improved its degree centrality rank, moving from 6th to 5th. In contrast, non-administrative hubs like Suzhou gradually lost their prominence, dropping out of the top-10 rankings by the end of the study period. This trend highlights the increasing dominance of administrative centers in the network, reinforcing a core–periphery structure, in which cities like Beijing continue to anchor the network, while non-administrative cities in regions like the Beijing–Tianjin–Hebei, Yangtze River Delta, and Greater Bay Area struggle to maintain their position.

Table 3. Top-10 cities by centrality in the Northeast China-nationwide inter-city cooperative innovation network.

Indicator	Degree Centrality			Weighted Degree Centrality			Betweenness Centrality		
Ranking	2009–2013	2014–2018	2019–2023	2009–2013	2014–2018	2019–2023	2009–2013	2014–2018	2019–2023
1	Beijing *	Beijing *	Beijing *	Beijing *	Beijing *	Beijing *	Beijing *	Beijing *	Beijing *
2	Shanghai *	Tianjin *	Shanghai *	Shanghai *	Shanghai *	Nanjing *	Shanghai *	Nanjing *	Xi'an *
3	Tianjin *	Hangzhou *	Tianjin *	Yiyang	Nanjing *	Shenzhen	Tianjin *	Shanghai *	Tianjin *
4	Nanjing *	Nanjing *	Xi'an *	Nanjing *	Shenzhen	Shanghai *	Nanjing *	Tianjin *	Shanghai *
5	Wuxi	Shanghai *	Nanjing *	Tianjin *	Tianjin *	Xi'an *	Xi'an *	Hangzhou *	Shenzhen
6	Shenzhen	Shenzhen	Shenzhen	Suzhou	Qingdao	Ningbo	Wuxi	Shenzhen	Nanjing *
7	Chengdu *	Suzhou	Wuhan *	Shenzhen	Xuchang	Chongqing *	Wuhan *	Suzhou	Wuhan *
8	Qingdao	Shijiazhuang *	Hangzhou *	Chengdu *	Suzhou	Tianjin *	Guangzhou *	Shijiazhuang *	Hangzhou *
9	Xi'an *	Kunming *	Qingdao	Qingdao	Wuhan *	Hangzhou *	Chengdu *	Kunming *	Qingdao
10	Kunming *	Chengdu *	Guangzhou *	Guangzhou *	Guangzhou *	Chengdu *	Shenzhen	Urumqi *	Guangzhou *

* Provincial capitals and municipal cities.

3.1.2. The Spatial Evolution Characteristics of External Inter-City Innovation Network Structures

(1) The spatial distribution has gradually become more concentrated over the years, with significant clustering in the eastern and southern regions. Cities in the three northeastern provinces have increasingly engaged in innovation cooperation with cities nationwide, and the spatial distribution of external cooperation cities has expanded each year. From 2009 to 2013, the primary clusters were concentrated in the Beijing–Tianjin–Hebei region and the Yangtze River Delta. From 2014 to 2018, the clustering extended further south, focusing on the Pearl River Delta. Between 2019 and 2023, the distribution expanded toward the Chengdu–Chongqing city cluster and the Wuhan metropolitan area in the central and western regions. Many cities in the central and western regions have ended their isolation, becoming new members of the innovation network. The network structure has evolved toward greater complexity, ultimately displaying a pattern of "dense in the southeast, sparse in the northwest" (see Figure 2).

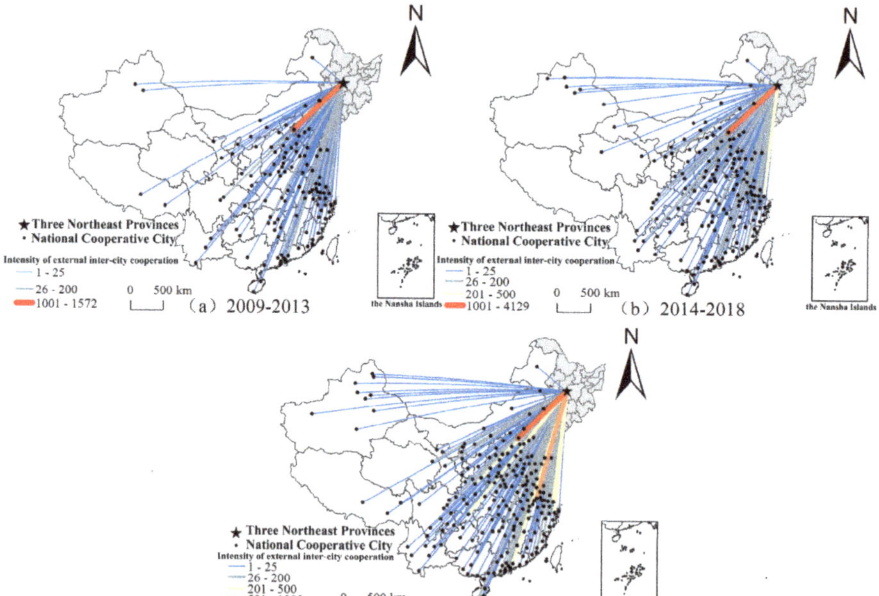

Figure 2. Spatial distribution of external cooperation and innovation links in the three northeastern provinces.

(2) In terms of network cooperation intensity, Beijing stands out as a prominent engine of innovation, with the influence of municipalities, provincial capitals, and sub-provincial cities steadily increasing. Most central nodes and key cooperative relationships are concentrated on the eastern side of the "Hu Huanyong Line" [35]. Using the natural breaks method, external innovation cooperation links are categorized into three levels: high intensity (frequency [501, 8664]), medium intensity (frequency [26, 500]), and low intensity (frequency [1, 25]). In the high-intensity category, the proportion of cooperation involving Beijing increased from 61.8% to 65.27% before decreasing to 57.61%. Despite the decline in cooperation intensity later on, the polarization effect remains evident. Medium-intensity cooperative cities are concentrated among municipalities, provincial capitals, and sub-provincial cities, such as Shanghai, Nanjing, and Shenzhen. These cities are gradually

becoming secondary hubs, with their cooperation intensity and network dominance increasing considerably. Low-intensity cooperation is widespread across the country but generally weakens, especially in regions such as Tibet, Qinghai, and Gansu, where participation is limited, and the cooperation network is fragile.

3.2. The Evolution Characteristics of the Inter-City Cooperative Innovation Network Within the Three Northeastern Provinces

3.2.1. Evolution Characteristics of the Topological Structure of the Inter-City Cooperative Innovation Network

(1) The network scale has gradually expanded and become more compact. From 2009 to 2013, the city cooperation rate was 83%, with the Heihe, Yichun, Qitaihe, Liaoyuan, and Daxing'anling area not participating (Table 4). Between 2014 and 2018, the cooperative innovation network continued to deepen, with only the Yichun and Daxing'anling area remaining isolated. From 2019 to 2023, all cities in the three northeastern provinces engaged in cooperative innovation, with increasingly frequent connections, witnessing a 20% growth in participating cities (30→36) and a 124% surge in collaborative ties (74→166), accompanied by a 55% increase in density, from 0.17011 to 0.26349. In the topological structure diagram (see Figure 3), Shenyang, Dalian, Changchun, and Harbin emerged as the main cooperative cities. Shenyang, Dalian, Changchun, and Harbin emerged as the main cooperative cities. This reflects enhanced peripheral integration through anchor hubs like the Shenyang–Dalian–Changchun–Harbin cluster, which mediated 68% of new connections post-2014. The intra-regional cooperative innovation network in Northeast China demonstrated marked structural transformations between 2009 and 2023, characterized by synergistic scaling, emergent small-world efficiency, and dynamic core–periphery reconfiguration.

Table 4. Overall characteristics of intra-regional cooperative innovation networks.

	Indicator	2009–2013	2014–2018	2019–2023
Network size	Number of network nodes	30	34	36
	Number of network relationships	74	107	166
	Network density	0.17011	0.19073	0.26349
Contact strength	Average degree	4.933	6.294	9.222
	Average weighted degree	40.067	65.235	198.611
Small-world characteristics	Average path length	2.103	1.95	1.768
	Average clustering coefficient	0.618	0.669	0.66

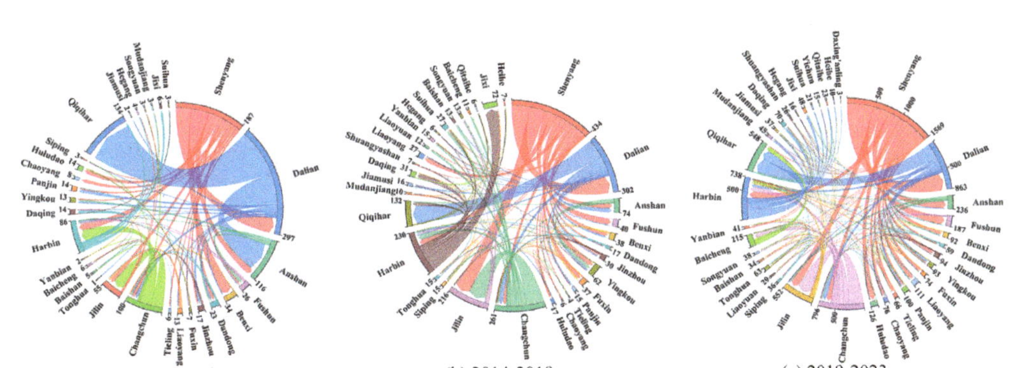

(a) 2009-2013 (b) 2014-2018 (c) 2019-2023

Figure 3. Topology of intra-regional cooperative innovation networks.

(2) The intensity of cooperation has increased, exhibiting small-world characteristics. The clustering coefficient, a critical indicator of small-world properties, remained above 0.618, which is over six-times the typical threshold of 0.1 for small-world networks. This reflects the emergence of tight-knit local clusters within the network. Additionally, the average path length decreased by 16%, from 2.103 to 1.768, crossing the small-world efficiency threshold of <2.0, which suggests a more efficient global connection within the network. Furthermore, the average weighted degree, which measures the intensity of connections, increased by 396%, from 40.067 to 198.611, indicating deeper and more frequent collaborative interactions, particularly in strategic emerging industries. These changes highlight the network's transition toward greater efficiency, with enhanced local specialization and accelerated knowledge transfer across clusters (see Table 4).

(3) City centers exhibit volatility, with a core–periphery phenomenon. As seen in Table 5, while the core cities, particularly those within the Shenyang–Dalian–Changchun–Harbin anchor cluster, continued to dominate the network, secondary hubs showed increased volatility. Harbin, for example, displaced Dalian as the leading city in degree centrality by 2019–2023, reflecting its growing importance in cross-provincial technology transfers, which led to a 214% increase in weighted degree centrality. Resource-based cities such as Daqing and Anshan retained their influence, while Baicheng gained prominence through agricultural innovation partnerships, increasing its number of collaborative ties by 17. Despite these shifts, a significant number of cities (12 out of 36) remained in the periphery, with low connectivity. This persistent inequality in connectivity suggests that the core–periphery structure of the network continues to be shaped by anchor-driven catch-up dynamics, with core cities serving as gatekeepers for the integration of peripheral cities, greatly impacting regional cooperation and the economic transformation of the northeast.

Table 5. Top-10 cities by centrality in the intra-regional cooperative innovation network of the three northeastern provinces.

Indicator	Degree Centrality			Weighted Degree Centrality			Betweenness Centrality		
Ranking	2009–2013	2014–2018	2019–2023	2009–2013	2014–2018	2019–2023	2009–2013	2014–2018	2019–2023
1	Shenyang	Shenyang	Harbin	Dalian	Shenyang	Shenyang	Harbin	Harbin	Harbin
2	Dalian	Changchun	Shenyang	Shenyang	Dalian	Dalian	Shenyang	Changchun	Jilin
3	Harbin	Harbin	Dalian	Qiqihar	Changchun	Changchun	Changchun	Shenyang	Changchun
4	Changchun	Dalian	Changchun	Anshan	Harbin	Harbin	Dalian	Dalian	Shenyang
5	Jilin	Jilin	Jilin	Changchun	Jilin	Jilin	Baishan	Jilin	Dalian
6	Fushun	Jinzhou	Anshan	Harbin	Qiqihar	Qiqihar	Jilin	Jinzhou	Anshan
7	Anshan	Qiqihar	Jinzhou	Jilin	Anshan	Anshan	Daqing	Jiamusi	Baicheng
8	Dandong	Jiamusi	Baicheng	Benxi	Jixi	Baicheng	Dandong	Qiqihar	Daqing
9	Liaoyang	Anshan	Chaoyang	Fushun	Yingkou	Fushun	Fushun	Qitaihe	Dandong
10	Daqing	Yingkou	Fushun	Dandong	Fushun	Huludao	Hegang	Daqing	Fuxin

3.2.2. The Spatial Evolution Characteristics of the Inter-City Innovation Network Structures

(1) The cooperative innovation network among cities in the three northeastern provinces is primarily focused on intra-city cooperation, supplemented by inter-city collaboration. Over time, the intensity of both intra- and inter-city cooperation has increased (see Figure 4). Cities, such as Shenyang, Dalian, Changchun, and Harbin, possess strong industrial clusters and research institutions, while resource-based cities such as Daqing and Anshan have specialized industrial parks, forming tight-knit internal cooperation networks. For example, Shenyang leads in high-end manufacturing and innovative materials, while Anshan leverages local resources and special funds to create industrial parks for the development of new types of materials and energy batteries. However, inter-city cooperation remains relatively weak, limiting the overall regional innovation capacity and the pace of industrial upgrading. To promote the economic development and industrial transformation of the northeast, it is necessary to strengthen inter-city cooperation, establish cooperative

policy frameworks, and provide financial support to facilitate the flow of technology, talent, and capital.

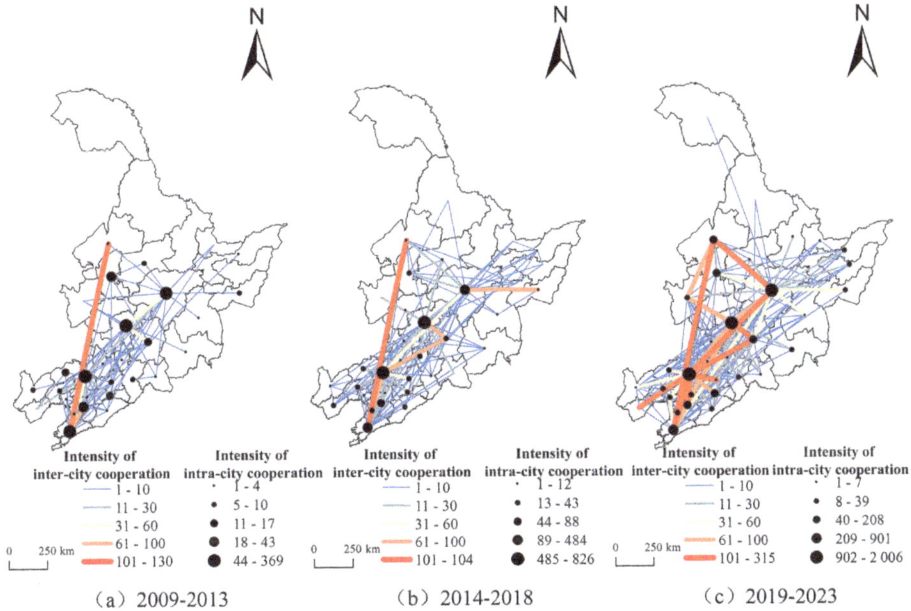

Figure 4. Spatial distribution of collaborative innovation linkages within regions.

(2) The cooperative innovation network has gradually evolved into a polygonal spatial pattern. From 2009 to 2013, Shenyang and Dalian served as the core, forming a center–periphery structure, with the cooperation intensity between Dalian and Qiqihar being the highest. This was primarily due to Qiqihar First Heavy Machinery Group establishing a branch in Dalian, resulting in 124 cooperative patents and the maintenance of close innovation links. From 2014 to 2018, the network shifted to a multi-core structure, with strengthened connections among Shenyang, Dalian, Changchun, and Harbin, exhibiting a hub-and-spoke formation. Between 2019 and 2023, the network further developed into a highly complex polygonal structure. The core positions of Shenyang, Dalian, Changchun, and Harbin were consolidated, with their radiating influence on surrounding cities enhanced. The inclusion of peripheral regions made the network more comprehensive.

3.3. *Analysis of the Mechanisms Influencing the Urban Cooperative Innovation Network*

3.3.1. Model Test

On the basis of the above research into the characteristics of cooperative innovation networks, we next explore the evolution mechanism of cooperative innovation networks from the perspective of proximity.

The explanatory variable "strength of cooperative ties" was a count variable with a variance significantly greater than its mean, indicating overdispersion. In such cases, Poisson regression is inappropriate due to its assumption of equidispersion. The negative binomial regression model was selected as it accounts for overdispersion by introducing an additional parameter for variance. This model was preferred over other alternatives, such as zero-inflated models, as it effectively addresses overdispersion without unnecessary complexity. Additionally, prior to the regression analysis, we conducted a collinearity test, and the variance inflation factors (VIFs) for all variables were found to be less than 5, indicating no multicollinearity issues. Thus, the negative binomial regression model is a suitable and efficient choice for analyzing the data. The regression results are shown in Table 6.

Table 6. Negative binomial gravity regression model results.

Variable	2009–2013	2014–2018	2019–2023
PATi	0.776 ***	0.891 ***	0.880 ***
	(0.071)	(0.059)	(0.058)
PATj	0.774 ***	0.889 ***	0.879 ***
	(0.072)	(0.059)	(0.058)
Geoij	−0.042	−0.151	0.354 **
	(0.261)	(0.196)	(0.174)
Borij	0.318	0.759 **	1.183 ***
	(0.348)	(0.302)	(0.240)
Ecoij	−1.358 **	−1.494 **	−1.630 ***
	(0.634)	(0.610)	(0.504)
Indij	1.420 **	2.260 **	−0.630
	(0.712)	(0.955)	(0.885)
Insij	1.027 ***	2.174 ***	0.925 ***
	(0.329)	(0.261)	(0.207)
Tecij	9.375 ***	−0.156	10.744 ***
	(3.012)	(1.472)	(1.848)
$Soc_{ij} \cdot gov$	0.135	−2.330 ***	0.561 *
	(0.578)	(0.485)	(0.333)
Socij·*edu*	−1.039	1.720 ***	1.323 **
	(0.805)	(0.626)	(0.546)
_cons	−18.748 ***	−13.530 ***	−22.027 ***
	(2.614)	(1.611)	(1.965)
N	812	1056	1122
Alpha	3.047	2.975	3.657
	(0.402)	(0.317)	(0.081)
Prob > chi^2	0.0000	0.0000	0.0000
Log likelihood	−599.29705	−865.75015	−1471.6152
Pseudo R^2	0.2232	0.2307	0.1685

Note: * $p < 0.1$, ** $p < 0.05$, *** $p < 0.01$, values in parentheses are standard errors.

3.3.2. Empirical Research Results

(1) The innovation capacity of cities, represented by PAT_i and PAT_j, remained significant at $p < 0.01$, with a positive regression coefficient. This aligns with the recombinant innovation theory, where cities with higher patent stocks exhibit stronger abilities to absorb and recombine cross-regional knowledge [36]. The three northeastern provinces' "14th Five-Year Plan for Digital Economy" (released in 2021) has accelerated this process through cloud-based innovation platforms. The three northeastern provinces of China have established a strategic emerging industry collaboration network centered on technological innovation through policy-driven initiatives and industrial transformation, significantly contributing to the region's sustainable development.

(2) The impact of geographical proximity on collaborative innovation networks has undergone significant changes over time. The spatial distance had a significantly positive regression coefficient between 2019 and 2023, and boundary proximity was significantly positive in both 2014–2018 and 2019–2023. This indicates that in the early stages, innovation activities primarily relied on the flow of technology and resources rather than geographical

proximity. As the industry matured, the importance of geographical proximity increased, reducing collaboration costs and promoting inter-city cooperation and technological innovation [37]. This finding refutes the "death of geographical proximity" hypothesis [38].

(3) In terms of economic proximity, the impact coefficient of internal innovation linkages was significantly negative across all three periods. This indicates that substantial economic disparities between cities can foster innovation cooperation [39]. Notably, complementary cooperative relationships have emerged among economic-center cities and resource-based cities. Resource-based cities provide raw materials and resources, while economic centers offer technological and knowledge support. Together, they drive balanced regional economic development and industrial upgrading.

(4) In terms of industrial proximity, this significantly promoted inter-city cooperation between 2009 and 2013 and between 2014 and 2018 but had a negative and insignificant effect between 2019 and 2023. This indicates that in the early stages, cities with similar industrial structures found it easier to collaborate. However, in the later stages, local protectionism and competition factors limited the flow of innovation elements, thereby inhibiting regional innovation linkages. Consequently, in recent periods, cities have sought cross-industry cooperation to achieve technological breakthroughs [40].

(5) In terms of institutional proximity, this has consistently and significantly positively impacted inter-city cooperation. Similar policy environments and institutional frameworks reduce cooperation risks and barriers, facilitating rapid collaboration. This strengthens regional economic integration and technological synergy, driving economic development and technological innovation [41].

(6) In terms of technological proximity, this significantly promoted inter-city cooperation between 2009 and 2013 and between 2019 and 2023, while having an insignificant negative impact between 2014 and 2018. In the early stages, the role of technological similarity in reducing communication costs and facilitating knowledge exchange is consistent with Boschma's proximity framework, which highlights technological proximity as a critical driver of collaborative innovation [17]. In the mid-term, the shift toward technological diversification aligns with Frenken et al., who argue that regions often engage in "related variety" to avoid lock-in effects and explore new innovation pathways [42]. Balland et al. further emphasize that excessive technological proximity can stifle novelty, necessitating strategic differentiation [43]. In the later stages, policy guidance enhanced technological sharing and knowledge transfer between cities, strengthening cooperation and driving technological innovation and economic transformation [44].

(7) In terms of social proximity, proximity in scientific expenditure and educational expenditure had different impacts on collaborative innovation networks. Initially, scientific expenditure proximity had a significantly negative effect, indicating that large differences in government support between cities promoted the flow of innovation elements from economic center cities to resource-based cities. This "compensatory diffusion" is consistent with Breschi and Lissoni, who argue that knowledge spillovers often follow gradients of resource inequality [45]. In the later stages, this turned positive, suggesting that inter-city cooperation relied increasingly on government funding support [46]. Educational expenditure proximity initially had an insignificant impact on the cooperation network, but as educational disparities decreased over time, it promoted inter-city innovation cooperation, strengthened knowledge flow and technology exchange, and provided a solid foundation of human and knowledge resources. Glaeser et al. argue that homogenized educational investments reduce cognitive barriers, enabling cities to share tacit knowledge and co-develop technologies [47].

4. Discussion

4.1. Analysis of Influencing Mechanisms in Urban Cooperative Innovation Networks in the Three Provinces of Northeast China

The aforementioned analysis explains the evolutionary mechanisms of the innovation network in the three northeastern provinces from the perspectives of urban innovation capacity and multidimensional proximity. Urban innovation capacity, economic proximity, and institutional proximity have consistently and significantly influenced the evolution of the innovation network. In contrast, geographical, industrial, technological, and social proximity have shown variations in their effects on network evolution. Together, these factors have jointly promoted the evolution of the innovation network in the northeastern provinces (see Figure 5).

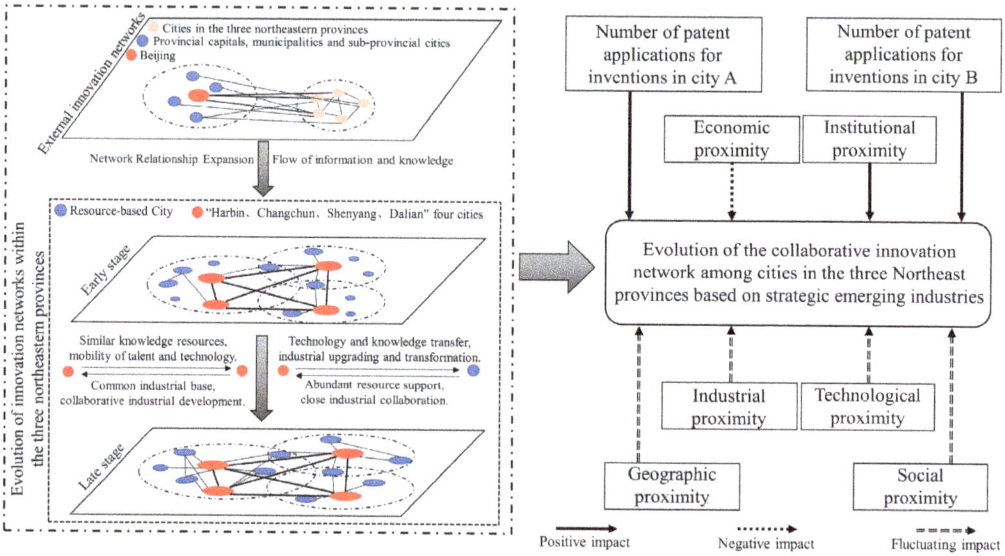

Figure 5. The mechanism of influence in urban cooperative innovation networks in the northeastern provinces of China.

Additionally, the external innovation network of the three northeastern provinces had an important impact on the internal network. The external network, characterized by strong innovation ties but low density, has enhanced the technological level and innovation capacity of enterprises and research institutions within the internal network through the knowledge spillover effect. The interaction between the external and internal networks has not only facilitated the sharing of technology and resources but also strengthened the competitiveness of the internal network. Over time, the internal network has continuously absorbed the advanced achievements of the external network, thereby improving its own innovation capacity and market competitiveness.

4.2. Recommendations

4.2.1. Institutionalize Cross-Regional Collaboration Mechanisms

A Northeast Regional Innovation Coordination Platform should be established to integrate digital resources across municipalities within and beyond the three northeastern provinces (Liaoning, Jilin, and Heilongjiang). This platform would facilitate efficient resource sharing and information exchange among various entities. Additionally, a dedicated

Interprovincial Innovation Fund should be created to support collaborative R&D projects between the northeastern provinces and major economic hubs, such as Beijing, Shanghai, and Tianjin. To further strengthen regional ties, an annual Northeast Strategic Industries Summit should be initiated, focusing on fostering partnerships in key sectors, such as advanced manufacturing, clean energy, and smart agriculture.

4.2.2. Optimize Spatial Development Patterns

A formal quad-city coordination framework should be established among Harbin, Changchun, Shenyang, and Dalian. This framework would include joint spatial planning committees for industrial zoning, cross-jurisdictional infrastructure investment protocols, and specialized transition funds to support resource-depleted cities like Fuxin and Daqing. To enhance connectivity, a high-frequency inter-city rail network should be expanded, and unified logistics data platforms should be developed to integrate supply chains. Furthermore, innovation voucher programs could be implemented to link R&D institutions in core and peripheral cities, promoting knowledge transfer and collaborative innovation.

4.2.3. Strengthen Regional Innovation Network Through Multidimensional Proximity

To advance urban innovation strategies and sustainable development, prioritize multidimensional proximity (institutional, technological, industrial) in Northeast China's regional planning. Anchor growth in the Harbin–Changchun–Shenyang–Dalian metropolitan cluster, fostering cross-city innovation corridors and green industrial hubs. Optimize sustainable industrial clusters (e.g., clean energy, smart manufacturing) with shared infrastructure to reduce ecological footprints. Establish a unified factor market to accelerate flows of talent, capital, and green technologies, supported by aligned regulatory frameworks and carbon-neutral incentives. Strengthen institutional proximity through joint R&D platforms and circular economy policies. Introduce flexible talent-sharing mechanisms, prioritizing skills in low-carbon sectors, while expanding cross-regional public transit to enhance connectivity.

4.3. Limitations and Prospects

4.3.1. Limitations

While the analysis of jointly filed patents (n = 26,913) effectively reveals formal innovation collaborations between cities, this methodological focus necessitated the exclusion of 389,355 individual patents (representing 6.91% of total patents) from the original dataset. These excluded patents, while not indicative of inter-organizational collaboration, may contain valuable insights into intra-organizational innovation patterns and solo inventor activities. Future investigations could employ alternative methodologies, such as citation network analysis or inventor mobility tracking, to extract latent collaboration patterns from this substantial body of excluded data.

4.3.2. Prospects

(1) Multi-Actor Collaboration Dynamics: Subsequent studies should systematically examine the triadic relationships between enterprise–university–research institute consortia, particularly analyzing how funding mechanisms influence collaboration patterns in advanced manufacturing and new energy sectors. Comparative analysis of collaboration efficiency metrics across different institutional combinations would yield actionable policy insights.

(2) Industry-Specific Network Topologies: The nine strategic emerging industries (including new-generation information technology and high-end equipment manufacturing) require discrete network analyses using multiplex network modeling. This approach would enable the identification of industry-specific knowledge flow channels, cross-sectoral technology convergence points, and critical path dependencies within regional innovation value chains.

5. Conclusions

This study is based on the PatSnap patent database to construct and map strategic emerging industry innovation networks among cities within the three northeastern provinces of China from 2009 to 2023. It employs GIS 10.8 spatial visualization and Gephi10.1-based topological analysis to characterize spatio-temporal evolution patterns and network structural dynamics, while applying a negative binomial gravity regression model to quantify influencing factors and elucidate the spatial interaction mechanisms of innovation networks. The main conclusions are as follows:

(1) The external innovation network of the three northeastern provinces has been expanding in scale, but its density has been decreasing year by year. The newly joined cities do not have close cooperation with the northeastern provinces, leading to a dilution of the network. We believe that this characteristic of the external collaborative innovation network does not meet the current economic development needs of the northeastern provinces. Simply expanding the network's scale without increasing its density is unlikely to effectively enhance the economic competitiveness of the region.

(2) The external innovation network of the three northeastern provinces is evolving into a pattern of "dense in the southeast and sparse in the northwest", resulting in an uneven spatial distribution. Cooperative relationships are concentrated on the eastern side of the "Hu Huanyong Line", with Beijing serving as the hub and municipalities directly under the central government, provincial capitals, and sub-provincial cities acting as the main cooperation nodes.

(3) The internal innovation network of the three northeastern provinces has become increasingly compact over the years, exhibiting small-world characteristics and forming a "core-periphery" structure. The four major cities of Harbin, Changchun, Shenyang, and Dalian remain at the core. The secondary core resource-based cities have experienced fluctuations; however, they continue to be crucial nodes for industrial development and coordinated transformation. The innovation network has evolved from a simple core–periphery structure to a multi-core hub-and-spoke network, eventually forming an extensive polygonal network structure. From an internal perspective, the development pattern of the cooperative innovation network in the northeastern region aligns with general trends. However, it is noteworthy that the intensity of cooperative innovation between cities is still relatively weak compared with the more developed regions in Southern China. While focusing on optimizing the internal structure, it is essential to enhance the intensity of cooperative innovation to drive the sustained and rapid development of strategic emerging industries.

(4) In the impact mechanism on cooperative innovation networks, urban innovation capacity has consistently played a strong and enduring role in the evolution of the innovation network, while the effects of multidimensional proximity indicators have fluctuated over time. Institutional proximity has a significantly positive impact, as cities with similar institutional environments are more conducive to collaborative innovation. Economic proximity generally has a positive effect, facilitating inter-city cooperation. However, the close cooperation between economic-center cities and resource-based cities in the northeastern

provinces exhibited a negative effect; greater economic disparities stimulated stronger cooperative relationships between cities. Geographical proximity only showed a positive impact in the later stages, indicating that it gradually became an important factor in promoting inter-city collaborative innovation. The impacts of industrial, technological, and social proximity varied, but overall, they exhibited a "proximity connection" network growth trend. The external innovation network supported the development of the internal network through knowledge spillovers and resource sharing. This was particularly evident in the northeastern provinces, where it promoted the evolution of the urban collaborative innovation network.

This study significantly deepens our understanding of the major city-level collaborative innovation networks within the strategic emerging industries in the three northeastern provinces of China. Through a comprehensive analysis of the spatio-temporal evolution patterns, network structural dynamics, and influencing factors of both external and internal innovation networks, it offers profound insights into how major cities can collaborate more effectively in the innovation process. The collaborative innovation networks among major cities play a crucial role in driving the sustainable development of these cities. This kind of innovation-led cooperation can help cities reduce resource waste and optimize industrial layouts, both of which are essential for long-term sustainable development. Building on these findings, future research could further refine our understanding of the innovation network by considering the following: (1) Individual patent applicants can broaden the scope of their research investigations by employing other methods, such as citation and inventor mobility tracking, to extract potential patterns of collaboration from previously excluded large-scale datasets. (2) Tripartite interactions within industry–university–research consortia should be systematically examined to explore their collaborative dynamics. (3) Nine strategic emerging industries (including next-generation information technology and high-end equipment manufacturing) require discrete network analysis through multi-network modeling approaches to identify domain-specific knowledge transfer pathways.

Supplementary Materials: The following supporting information can be downloaded at: https://www.mdpi.com/article/10.3390/land14040691/s1.

Author Contributions: X.Z. (Xiaodong Zhou) is responsible for topic selection, data processing, and manuscript writing. T.L. handles data proofreading (both authors contributed equally to this work and share the first authorship). P.Z. is responsible for manuscript revision. X.Z. (Xujia Zhang) is responsible for data analysis and verification. N.C. is responsible for manuscript polishing. All authors have read and agreed to the published version of the manuscript.

Funding: This study was supported by the National Natural Science Foundation of China (41601553); Heilongjiang Provincial Higher Education Institutions Basic Research Operating Expenses Program (No. 2022-KYYWF-0162).

Data Availability Statement: The data supporting the findings of this study are available from the PatSnap Global Patent Search and Analysis Database. All data generated or analyzed during this study are included in this published article.

Conflicts of Interest: The authors declare that there are no known competing financial interests or personal relationships that could have appeared to influence the work reported in this paper.

References

1. Xu, D.; Wang, J. Strategic emerging industries in China: Literature review and research prospect. *Am. J. Ind. Bus. Manag.* **2015**, *5*, 486. [CrossRef]
2. Porter, M. Conditions of the formation of high-tech industries clusters. In Proceedings of the Third International Conference on Management of Innovation and Technology (ICMT2002), Hangzhou, China, 25 October 2002; Volume 3.

3. Freeman, C. Networks of innovators: A synthesis of research issues. *Res. Policy* **1991**, *20*, 499–514. [CrossRef]
4. Zhou, C.; Zeng, G.; Shang, Y. Research progress and prospect of in novation network from the perspective of evolutionary economic geography. *Econ. Geogr.* **2019**, *39*, 27–36. (In Chinese) [CrossRef]
5. Taylor, P.J.; Hoyler, M.; Verbruggen, R. External Urban Relational Process: Introducing Central Flow Theory to Complement Central Place Theory. *Urban Stud.* **2010**, *47*, 2803–2818. [CrossRef]
6. Castells, M. Grassrooting the space of flows. *Urban Geogr.* **1999**, *20*, 294–302. [CrossRef]
7. Gertler, M.S.; Levitte, Y.M. Local nodes in global networks: The geography of knowledge flows in biotechnology innovation. *Ind. Innov.* **2005**, *12*, 487–507. [CrossRef]
8. Zhou, C.; Zeng, G.; Cao, X. Chinese inter-city innovation networks structure and city innovation capability. *Geogr. Res.* **2017**, *36*, 1297–1308. [CrossRef]
9. Cassi, L.; Plunket, A. Proximity, network formation and inventive performance: In search of the proximity paradox. *Ann. Reg. Sci.* **2014**, *53*, 395–422. [CrossRef]
10. Zhou, C.; Cao, X.; Zeng, G. Cluster networks and evolution path of China's electronic information industry innovation. *Geogr. Res.* **2019**, *38*, 2212–2225. (In Chinese)
11. Ma, J.; Zeng, G.; Hu, S.; Sun, K. Innovation Network Structure of Biomedical Industry and Its Influencing Factors in Yangtze River Delta. *Resour. Environ. Yangtze Basin* **2022**, *31*, 960–971. (In Chinese)
12. Luo, T.; Qu, J.; Cheng, S. Knowledge Network Embeddedness and Innovation Resilience. *IEEE Trans. Eng. Manag.* **2024**, *71*, 11454–11462. [CrossRef]
13. Li, D.; Wang, H.; Wang, J. Artificial Intelligence and Technological Innovation: Evidence from China's Strategic Emerging Industries. *Sustainability* **2024**, *16*, 7226. [CrossRef]
14. Calignano, G. Italian organisations within the European nanotechnology network: Presence, dynamics and effects. *Erde* **2014**, *145*, 241–259.
15. Broekel, T.; Hartog, M. Explaining the structure of inter-organizational networks using exponential random graph models. *Ind. Innov.* **2013**, *20*, 277–295. [CrossRef]
16. Glueckler, J.; Doreian, P. social network analysis and economic geography-positional, evolutionary and multi-level approaches. *J. Econ. Geogr.* **2016**, *16*, 1123–1134. [CrossRef]
17. Boschma, R.A. Proximity and innovation: A critical assessment. *Reg. Stud.* **2005**, *39*, 61–74. [CrossRef]
18. Boschma, R.; Frenken, K. The emerging empirics of evolutionary economic geography. *J. Econ. Geogr.* **2011**, *11*, 295–307. [CrossRef]
19. Santos, E.G.; Garcia, R.; Araujo, V.; Mascarini, S.; Costa, A. Spatial and non-spatial proximity in university-industry collaboration: Mutual reinforcement and decreasing effects. *Reg. Sci. Policy Pract.* **2021**, *13*, 1249–1262. [CrossRef]
20. Xi, Q.; Li, G.; Sun, Y.; LYU, S. Evolutionary characteristics of science and technology cooperation network of Beijing-Tianjin-Hebei region and its influencing factors. *Acta Geogr. Sin.* **2022**, *77*, 1359–1373. [CrossRef]
21. Broekel, T.; Boschma, R. Knowledge networks in the Dutch aviation industry: The proximity paradox. *J. Econ. Geogr.* **2012**, *12*, 409–433. [CrossRef]
22. Gui, Q.C.; Liu, C.; Du, D.B. Globalization of science and international scientific collaboration: A network perspective. *Geoforum* **2019**, *105*, 1–12. [CrossRef]
23. Li, J.; Zhang, P.; Sun, J.; Li, Q. Spatial Agglomeration, Industrial Networks, and Their Effects on Strategic Emerging Industries in China. *Trop. Geogr.* **2023**, *43*, 646–656.
24. Li, Z.; Yin, X. Spatial Disequilibrium and Spatio-temporal Evolution of China's Strategic Emerging Industries. *Stat. Decis.* **2022**, *38*, 102–107. (In Chinese) [CrossRef]
25. Li, Z.; Yang, S. Potential Dividend Development and Systematic Revitalization Strategies in Northeast China. *Soc. Sci. J.* **2018**, 67–77. (In Chinese)
26. Zhang, Y.; Tang, X.; Yang, J. Synergies of Technological and Institutional Innovation Driving Manufacturing Transformation: Insights from Northeast China. *J. Knowl. Econ.* **2024**, 1–35. [CrossRef]
27. Gao, X.; Chen, K. The complex network analysis of the structure evolution of collaborative innovation networks: An example of the industry-university-research collaboration of Chinese ICT industry. *Sci. Res. Manag.* **2015**, *36*, 28–36. [CrossRef]
28. Boschma, R.; Balland, P.A.; Kogler, D.F. Relatedness and technological change in cities: The rise and fall of technological knowledge in US metropolitan areas from 1981 to 2010. *Ind. Corp. Change* **2015**, *24*, 223–250. [CrossRef]
29. Scherngell, T.; Hu, Y.J. Collaborative Knowledge Production in China: Regional Evidence from a Gravity Model Approach. *Reg. Stud.* **2011**, *45*, 755–772. [CrossRef]
30. Tubadji, A.; Nijkamp, P. Cultural Gravity Effects among Migrants: A Comparative Analysis of the EU15. *Econ. Geogr.* **2015**, *91*, 343–380. [CrossRef]

31. Jiang, K.; Liang, S. Evolution of Technology Innovation Network and Multi-proximity Mechanism in Yangtze River Economic Belt. *Resour. Environ. Yangtze Basin* **2023**, *32*, 1796–1805. (In Chinese)
32. Jaffe, A. Technological Opportunity and Spillovers of R&D: Evidence from Firms' Patents, Profits and Market Value. *Am. Econ. Rev.* **1986**, *76*, 984–1001.
33. Watts, D.J.; Strogatz, S.H. Collective dynamics of 'small-world'networks. *Nature* **1998**, *393*, 440–442. [CrossRef] [PubMed]
34. Valverde, S.; Cancho, R.F.; Solé, R.V. Scale-free networks from optimal design. *Europhys. Lett.* **2002**, *60*, 512–517. [CrossRef]
35. Huanyong, H. The Distribution of Population in China, with Statistics and Maps. *Acta Geogr. Sin.* **1935**, *2*, 33–74. (In Chinese) [CrossRef]
36. Zhang, J.; Yan, Y.; Guan, J. Recombinant distance, network governance and recombinant innovation. *Technol. Forecast. Soc. Change* **2019**, *143*, 260–272. [CrossRef]
37. Maskell, P.; Malmberg, A. Localised learning and industrial competitiveness. *Cambr. J. Econ.* **1999**, *23*, 167–185. [CrossRef]
38. Anwar, S.T. The world is flat: A brief history of the twenty-first century. *J. Int. Manag.* **2007**, *13*, 78–89. [CrossRef]
39. Feng, P.; Yasar, M.; Rejesus, R.M. Innovation and regional economic convergence: Evidence from China. *Ann. Reg. Sci.* **2024**, *72*, 535–559. [CrossRef]
40. Liu, Z.; Zhang, W. The process and dynamics of high-tech industrial branching in China's metropolitan areas: A case study of biotechnology industry in Wuhan. *Geogr. Res.* **2018**, *37*, 1349–1363.
41. Shi, W.; Wang, Q. The framework for analyzing the mechanism of the evolution of inter-city relationship networks on regional economic resilience. *PLoS ONE* **2024**, *19*, e0308280. [CrossRef]
42. Frenken, K.; van Oort, F.; Verburg, T. Relate variety, unrelated variety and regional economic growth. *Reg. Stud.* **2007**, *41*, 685–697. [CrossRef]
43. Asheim, B.T. Smart specialisation, innovation policy and regional innovation systems: What about new path development in less innovative regions? *Innov. Eur. J. Soc. Sci. Res.* **2019**, *32*, 8–25. [CrossRef]
44. Cooke, P. Regional innovation systems, clusters, and the knowledge economy. *Ind. Corp. Change* **2001**, *10*, 945–974.
45. Breschi, S.; Lissoni, F. Mobility of skilled workers and co-invention networks: An anatomy of localized knowledge flows. *J. Econ. Geogr.* **2009**, *9*, 439–468. [CrossRef]
46. Fritsch, M.; Slavtchev, V. Determinants of the Efficiency of Regional Innovation Systems. *Reg. Stud.* **2011**, *45*, 905–918. [CrossRef]
47. Glaeser, E.L.; Kallal, H.D.; Scheinkman, J.A.; Shleifer, A. Growth in cities. *J. Political Econ.* **1992**, *100*, 1126–1152.

Disclaimer/Publisher's Note: The statements, opinions and data contained in all publications are solely those of the individual author(s) and contributor(s) and not of MDPI and/or the editor(s). MDPI and/or the editor(s) disclaim responsibility for any injury to people or property resulting from any ideas, methods, instructions or products referred to in the content.

MDPI AG
Grosspeteranlage 5
4052 Basel
Switzerland
Tel.: +41 61 683 77 34

Land Editorial Office
E-mail: land@mdpi.com
www.mdpi.com/journal/land

Disclaimer/Publisher's Note: The title and front matter of this reprint are at the discretion of the Guest Editors. The publisher is not responsible for their content or any associated concerns. The statements, opinions and data contained in all individual articles are solely those of the individual Editors and contributors and not of MDPI. MDPI disclaims responsibility for any injury to people or property resulting from any ideas, methods, instructions or products referred to in the content.

www.ingramcontent.com/pod-product-compliance
Lightning Source LLC
Chambersburg PA
CBHW040758250926
51850CB00007BA/1030
9783725849635